DYNAMICS *of* RIGID BODIES

S. A. Mollah

CELEBRATING 60 YEARS 1960-2020

Since BOOKS AND ALLIED 1960

[For Mathematics (Hons./Major) (CBCS) students of different Indian Universities. Also useful of NET, SLET, GATE, IAS & IPS examinees.]

DYNAMICS *of* RIGID BODIES

Dr. S. A. Mollah

Reader & Head (Retd.)
Department of Mathematics
St. Paul's C. M. College, Kolkata.

BOOKS & ALLIED (P) Ltd.

No. 1-E(1), 'Shubham Plaza' (1st Floor)
83/1 Beliaghata Main Road, Kolkata 700010 (India)
Tel : (033) 23740062 • 8274085530
e-mail : booksandallied1960@gmail.com • booksmktng@gmail.com

DYNAMICS *of* RIGID BODIES

First Published : July 2007
Second Edition : 2010
Reprinted : 2011. 2012, 2013
Third Edition : 2015
Reprinted : 2016, 2018, 2019
Revised Reprint : 2020

Publisher : Ruba Sen
BOOKS & ALLIED (P) LTD.
8/1 Chintamoni Das Lane, Kolkata 700009

Typesetter : BOOKS & ALLIED (P) LTD.
8/1 Chintamoni Das Lane, Kolkata 700009

Printer : EUREKA
91/A, Baithakkhana Road, Kolkata 700009

ISBN 978-81-87134-98-5

Price : **₹330.00**

In the Memory of

BABAJEE and MAA

Preface to the Third Edition

I feel great pleasure in bringing out the third edition of my book. The current edition of the book has been revised and updated to cover the latest syllabi of various Indian Universities for the Degree Honours and Engineering Examination.

In the present edition, I have added last 5–10 yrs University Question Papers (Hons.) of C.U., B.U. and K.U., with Hints to solutions.

Creative criticism and suggestion for further development of the book are cordially invited from colleagues, friends and students of the subject.

Lastly, I want to express my sincere thanks to my publisher Mr. Arunabha Sen, Director of Books & Allied (P) Ltd., for his co-operation in publishing this edition.

October, 2015 **Shahjahan Ali Mollah**

Preface to the Second Edition

I feel great pleasure in bringing out the second edition of my book. The current edition of the book has been thoroughly revised and updated to cover the latest syllabi of various Indian Universities for the Degree Honours and Engineering Examination.

In the present edition, I have inserted few topics in vectorial approach which were lacking in previous edition. I also inserted more problems which were lacking in the previous edition and the corrections were made which were misprinted in the previous edition.

I am grateful to those respected colleagues, friends and students, who drew my attention to the misprints of the previous edition. Creative criticism and suggestion for further development of the book are cordially invited from colleagues, friends and students of the subject.

Lastly, I want to express my sincere thanks to my publisher Mr. Arunabha Sen, Director of Books & Allied (P) Ltd., for his co-operation in publishing this edition.

January, 2010 **Shahjahan Ali Mollah**

Preface to the First Edition

DYNAMICS OF RIGID BODIES has been written primarily to meet the requirements and needs of the students who are preparing for the Honours degree in Mathematics in the Indian Universities, specially, Calcutta University, Jadavpur University, Burdwan University, Kalyani University and Vidyasagar University. This book will also be very helpful to those students who are going to appear for different competitive examinations namely IAS, IPS, etc.

I am thankful to the students who have constantly inspired me to complete this work by expressing the difficulties and problems they face in this subject. I have consulted a number of books in this subject including the classical works of Loney and Ramsey. I am grateful to my respected teachers Dr J.G. Chakravorty and Dr S.C. Bose who insisted me to complete this work in a very short time by providing their valuable suggestions and guidance.

I want to express my sincere thanks to my colleagues and friends Mr S. Kanjilal, Dr A. Sarkar, Dr A. Banerjee and Mr Anindya De of St. Paul's C.M. College, Mr M. Maity and Mr A. Debnath of Gurudas College, Dr U. Chatterjee and Dr S. Bhattacharya of Bangabasi College who adviced and inspired me to use my experience to fulfill the desires and demands of the students in the subject.

I also want to express my sincere thanks to my wife Mrs. H. Mollah and my son Mr T.I. Mollah whose constant inspiration forced me to complete the work in time.

Lastly I want to convey my sincere thanks to Mr Arunabha Sen, Director, Books & Allied (P) Ltd. for his active effort in publication of the book in a very short time.

I am looking forward to receiving suggestions and criticisms from respected colleagues, friends and students, for further improvement of the book.

July, 2007 **Shahjahan Ali Mollah**

Contents

Chapter 1

Moment of Inertia

1.1 Rigid Body

A body is a part of matter limited in every direction. The bodies exist in our surrounding are more or less elastic. They are capable of compression, extension or distortion under the action of external forces. The problems in respect of their motions are complicated. So we consider another kind of body called *rigid body* which is supposed to be incapable of compression, extension or distortion under any circumstances.

A rigid body is a conglomeration of particles called *elements bound* together by forces of cohesion and mutual attraction which, we know from Newton's third law of motion, are equal and opposite. We further consider that the elements of the rigid body do not alter their position, relative to each other, that is the distance between any two elements remains unaltered. So the total work done by the mutual action and reaction between the elements is zero.

The study of the dynamics of rigid bodies requires a knowledge of moment and product of inertia of rigid bodies.

1.2 Moment and Product of Inertia

(a) Moment of inertia

Suppose m is the mass of a particle and d be its distance from a fixed straight line AB, then md^2 is called the *moment of inertia* (m.i.) of the particle about AB.

If the body consists of a system of particles of masses $m_1, m_2, m_3, \ldots$ etc. at distances $d_1, d_2, d_3, \ldots$ etc. from the fixed straight line AB, the $\Sigma m_1 d_1^2$ is called the *moment of inertia* of the body about the straight line AB.

If however the body consists of a continuous mass then m is replaced by elementary mass dm at a distance d of a point (x, y, z) within the body and the summation is replaced by integration over the whole body, i. e.,

$$\int\limits_{\substack{\text{over the}\\\text{whole body}}} dm \cdot d^2.$$

If we consider

$$\sum m_1 d_1^2 = MK^2,$$

where M is the total mass of the body, then K is called the *radius of gyration* or *swing-radius* of the body about AB.

(b) Product of inertia

If (x, y) be the coordinates of an element of mass m with respect to mutually perpendicular straight line Ox and Oy, the Σmxy is defined as *product of inertia* of the body with respect to the lines Ox and Oy.

If (x, y, z) be the coordinates of an element of mass m with respect to three mutually perpendicular straight lines Ox, Oy, Oz as axes, then Σmyz, Σmzx, Σmxy are called the *product of inertia* of the body with respect to the pair of straight lines Oy&Oz, Oz&Ox and Ox&Oy, respectively.

In general, the *moment of inertia* (m.i.) of a body about Ox, Oy, Oz (as reference axes) are denoted by A, B and C respectively and the *product of inertia* of the body with respect to the axes y&z, z&x and x&y are denoted respectively by D, E and F, i. e., $D = \Sigma myz$, $E = \Sigma mzx$, $F = \Sigma mxy$.

1.3 Moment and Product of Inertia with respect to Three Mutually Perpendicular Axes

Let $P_1(x_1, y_1, z_1)$, $P_2(x_2, y_2, z_2)$, $P_3(x_3, y_3, z_3), \ldots$ etc. be the coordinates of the particles of mass $m_1, m_2, m_3, \ldots$ etc. with respect to a set of rectangular axes $Oxyz$.

$$\begin{aligned}\text{Then}\quad A &= \text{moment of inertia of the body about } Ox\\ &= \sum m_1\, d_1^2,\ \text{where } d_1 \text{ is the distance of } m_1 \text{ from } Ox.\end{aligned}$$

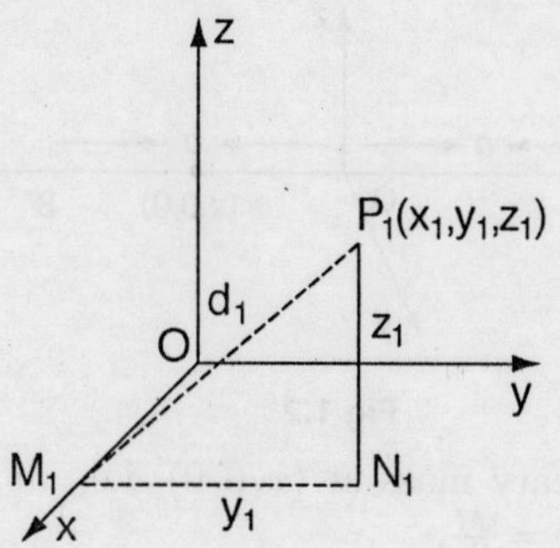

Fig 1.1

Now

$$d_1^2 = M_1N_1^2 + P_1N_1^2 = y_1^2 + z_1^2.$$

$$\therefore\ A = \sum m_1(y_1^2 + z_1^2). \tag{1.3.1}$$

Similarly,

$$B = \sum m_1(z_1^2 + x_1^2). \tag{1.3.2}$$

$$C = \sum m_1(x_1^2 + y_1^2). \tag{1.3.3}$$

The sums represented by

$$D = \sum m_1 y_1 z_1$$
$$E = \sum m_1 z_1 x_1$$
$$F = \sum m_1 x_1 y_1$$

are called the *product of inertia* of the body with respect to the plane yz, zx and xy, respectively.

1.4 Moments and Products of Inertia in some Simple Cases

(a) Uniform rod of mass M and length $2a$

Let M be the mass of the rod and $2a$ be the length of the uniform rod and O be the middle point (c.g.) of the rod. Let x-axis is taken along the rod. The coordinates of a point on the rod at distance x is $(x, 0, 0)$. Let m be the elementary mass at $(x, 0, 0)$.

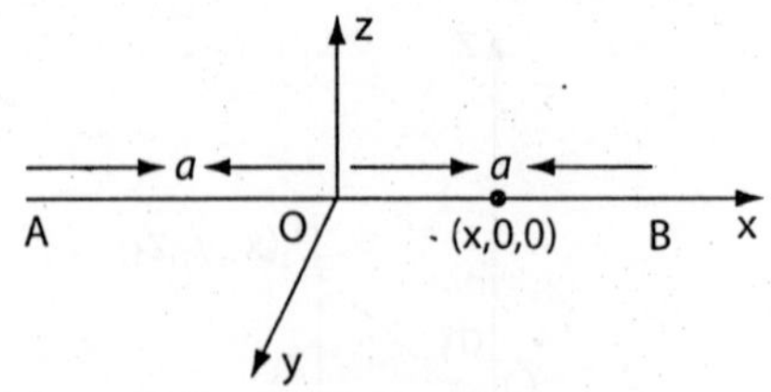

Fig 1.2

Let m be the elementary mass at $(x, 0, 0)$, i.e., m is the mass per unit length of the rod. $\therefore\ m = \frac{M}{2a}$.

$$\therefore\ A = \sum m(y^2 + z^2) = 0 \quad [\text{as } y = z = 0] \tag{1.4.1}$$

$$B = \sum m(z^2 + x^2) = \sum mx^2 \quad [\text{as } z = 0]$$

$$= \int_{-a}^{a} \frac{M}{2a} x^2\, dx = \frac{Ma^2}{3} \tag{1.4.2}$$

$$C = \sum m(x^2 + y^2) = \sum mx^2 = \frac{Ma^2}{3} \tag{1.4.3}$$

$$\text{and } D = \sum myz = 0, \quad E = \sum mzx = 0,$$

$$F = \sum mxy = 0 \quad [\text{as } y = 0, z = 0].$$

(b) Rectangular plate of mass M and of lengths $2a$, $2b$ at its centre

Let x-axis and y-axis be taken on the plane of the plate and parallel to the sides $2a$ and $2b$ respectively through O (c.g.) and z-axis perpendicular to the plate. The plate may be supposed to consists of thin parallel strips each is parallel to the length $2b$. The moment of inertia of the strip about x-axis is

$$\frac{b^2}{3} \times \text{mass of the strip} \quad [\text{by } (1.4.3)]$$

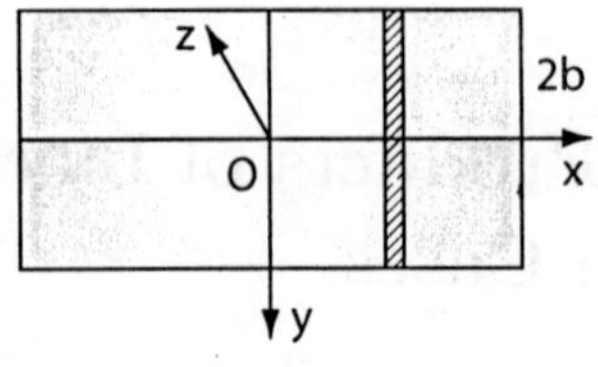

Fig 1.3

$$\therefore A = \text{m.i. of the plate about } x\text{-axis} = \sum \frac{b^2}{3} \times \text{mass of the strip}$$
$$= \frac{b^2}{3} \sum \text{mass of strip} = \frac{Mb^2}{3}. \qquad (1.4.4)$$

Similarly, if the plate be supposed to consists of thin parallel strips each is parallel to the length $2a$; i.e., x-axis, the m.i. of the strip about y-axis is $\frac{a^2}{3} \times$ mass of the strip.

$\therefore$ The moment of inertia of the plate about y-axis is

$$\begin{aligned} B &= \sum \frac{a^2}{3} \times \text{mass of the strip} \\ &= \frac{a^2}{3} \sum \text{mass of the strip} \\ &= \frac{Ma^2}{3}. \end{aligned} \qquad (1.4.5)$$

Let m be the elementary mass at $P(x, y, 0)$.

$$\begin{aligned} \therefore A &= \sum m(y^2 + z^2) = \sum my^2 \quad [\text{for the plate } z = 0] \\ B &= \sum m(z^2 + x^2) = \sum mx^2 \quad [\text{for the plate } z = 0] \\ \therefore C &= \sum m(x^2 + y^2) = \sum mx^2 + \sum my^2 \\ &= A + B = \frac{Ma^2}{3} + \frac{Mb^2}{3} = \frac{M}{3}(a^2 + b^2) \qquad (1.4.6) \\ \text{and } D &= \sum myz = 0, \ E = \sum mzx = 0 \text{ and } F = \sum mxy. \end{aligned}$$

Corresponding to each particle P of the mass m and of coordinates $(x, y, 0)$, we can find another particle P' of same mass m and of coordinates $(x, -y, 0)$. The sums of the product of inertia of these particles $= mxy + mx(-y) = 0$.

$\therefore$ The product of inertia of the plate with respect to the plane xy, $F = 0$.

(c) Rectangular parallelopiped of mass M and dimensions $2a$, $2b$ and $2c$ at its centre

Let O be the centre of gravity and the axes Ox, Oy and Oz be taken along parallel to the sides $2a$, $2b$ and $2c$, respectively. The rectangular parallelopiped is broken in parallel strips and each is perpendicular to x-axis.

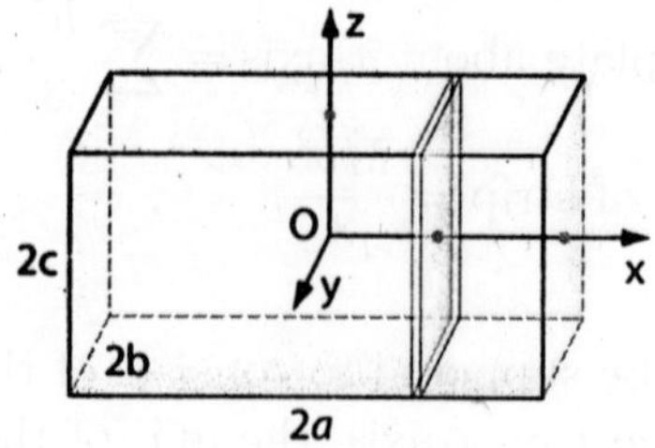

Fig 1.4

The m.i. of each strip about x-axis

$$= \frac{b^2 + c^2}{3} \times \text{mass of the strip} \quad [\text{by } (1.4.6)]$$

$\therefore$ A = m.i. of the rectangular parallelopiped about x-axis

$$= \sum \frac{b^2 + c^2}{3} \times \text{mass of the strip}$$

$$= \frac{b^2 + c^2}{3} \times \sum \text{mass of the strips}$$

$$= \frac{M(b^2 + c^2)}{3}. \tag{1.4.7}$$

Similarly,

$$B = \frac{M(c^2 + a^2)}{3} \quad \text{and} \quad C = \frac{M(a^2 + b^2)}{3}. \tag{1.4.8}$$

Now

$$D = \sum myz,$$

where m is the elementary mass at $P(x, y, z)$. Corresponding to a mass m at $P(x, y, z)$, we can find an equal mass m at $P'(x, y, -z)$. The sum of the product of inertia of these two masses with respect to yz-plane $= myz + my(-z) = 0$.

Hence,

$$D = \sum myz = 0.$$

Similarly,

$$E = \sum mzx = 0, \quad E = \sum mxy = 0.$$

(d) Circular plate of mass M and radius a at its centre

Let Ox and Oy-axis are taken in the plane of the plate and z-axis perpendicular to the plate.

Let us consider an elementary area $rd\theta\, dr$ at $P(r, \theta)$ or $P(x, y, 0)$.

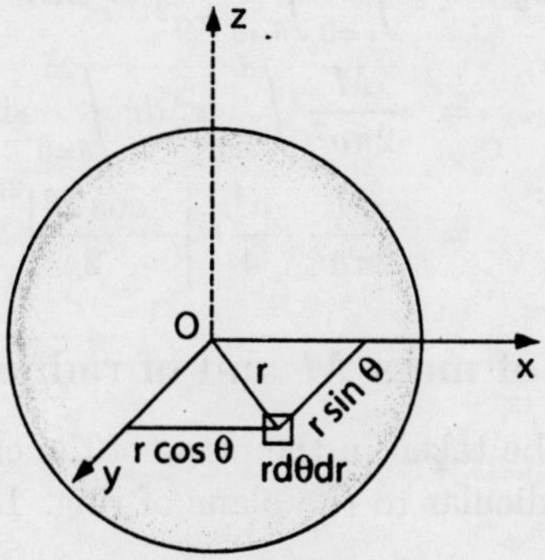

Fig 1.5

Let

$$m = \frac{\text{mass}}{\text{area}} = \frac{M}{\pi a^2}.$$

$\therefore$ The mass of the elementary area $rd\theta\, dr$

$$= m \cdot rd\theta\, dr = \frac{M}{\pi a^2} rd\theta\, dr.$$

$\therefore A =$ m.i. of the circular plate about x-axis

$$= \int_{r=0}^{a} \int_{\theta=0}^{2\pi} \frac{M}{\pi a^2} \cdot rd\theta\, dr (r \sin\theta)^2$$

$$= \frac{M}{\pi a^2} \int_{r=0}^{a} r^3\, dr \int_{\theta=0}^{2\pi} \sin^2\theta\, d\theta = \frac{Ma^2}{4}. \tag{1.4.9}$$

$B =$ m.i. of the plate about y-axis

$$= \int_{r=0}^{a} \int_{\theta=0}^{2\pi} \frac{M}{\pi a^2} \cdot rd\theta\, dr (r \cos\theta)^2$$

$$= \frac{M}{\pi a^2} \int_{r=0}^{a} r^3\, dr \int_{\theta=0}^{2\pi} \cos^2\theta\, d\theta = \frac{Ma^2}{4}. \tag{1.4.10}$$

$C =$ m.i. of the plate about z-axis

$$= \int_{r=0}^{a} \int_{\theta=0}^{2\pi} \frac{M}{\pi a^2} \cdot rd\theta\, dr (r^2)$$

$$= \frac{M}{\pi a^2} \int_{r=0}^{a} r^3\, dr \int_{0}^{2\pi} d\theta = \frac{Ma^2}{2} \tag{1.4.11}$$

$$D = \sum myz = 0, \quad E = \sum mzx = 0 \quad \text{as } z = 0.$$

$$
\begin{aligned}
F=\sum mxy &= \int_{r=0}^{a}\int_{\theta=0}^{2\pi}\frac{M}{\pi a^2}\cdot r\sin\theta\cdot r\cos\theta\cdot r d\theta\, dr\\
&= \frac{M}{2\pi a^2}\int_{r=0}^{a} r^3\,dr\int_{\theta=0}^{2\pi}\sin 2\theta\, d\theta\\
&= \frac{M}{2\pi a^2}\cdot\frac{a^4}{4}\cdot\left|-\frac{\cos 2\theta}{2}\right|_0^{2\pi}=0.
\end{aligned}
$$

(e) Circular ring of mass M and of radius a at its centre

The axes of x and y be taken in the plane of the circular ring at the centre O and z-axis perpendicular to the plane of ring. Let m be the mass of the unit arc of the ring.

$$\therefore \quad m=\frac{M}{2\pi a}.$$

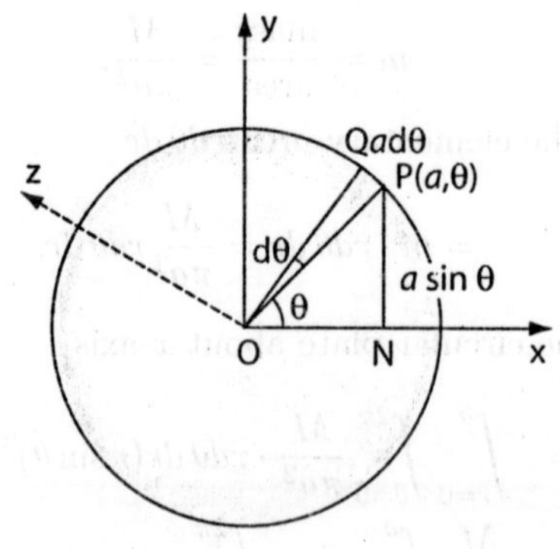

Fig 1.6

Let us consider an element PQ at $P(a,\theta)$.
$\therefore$ $A=$ m.i. of the ring about x-axis

$$
\begin{aligned}
&= \int_{\theta=0}^{2\pi} ma\,d\theta\cdot a^2\sin^2\theta=\frac{M}{2\pi a}\cdot a^3\int_{\theta=0}^{2\pi}\frac{1-\cos 2\theta}{2}d\theta\\
&= \frac{Ma^2}{4\pi}\left|\theta-\frac{\sin 2\theta}{2}\right|_{\theta}^{2\pi}=\frac{Ma^2}{4\pi}\cdot 2\pi=\frac{Ma^2}{2}. \qquad (1.4.12)
\end{aligned}
$$

Similarly,
$B=$ m.i. of the ring about y-axis

$$=\int_{\theta=0}^{2\pi} ma\,d\theta\cdot a^2\cos^2\theta=\frac{Ma^2}{2}. \qquad (1.4.13)$$

$C=$ m.i. of the ring about z-axis

$$=\int_{\theta=0}^{2\pi} ma\,d\theta\cdot a^2=\int_0^{2\pi}\frac{M}{2\pi a}a\,d\theta\cdot a^2=\frac{Ma^2}{2\pi}\cdot 2\pi=Ma^2. \qquad (1.4.14)$$

(f) Right circular cylinder $x^2 + y^2 = a^2$ of height h, mass M at its centre

The axes are taken at the centre O (c.g.) as shown (Fig 1.7). Let us consider a strip of thickness dz parallel to xy-plane at a distance z from the origin O.

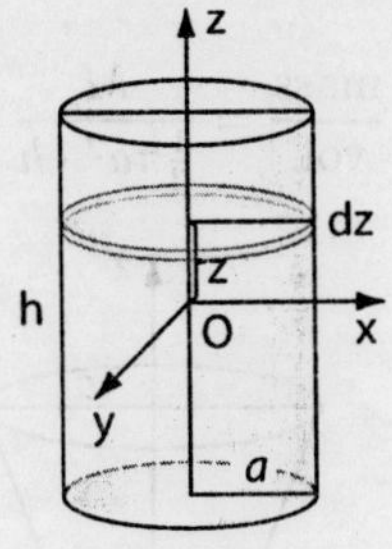

Fig 1.7

From symmetry $A = B$

$$\therefore \quad \sum m(y^2 + z^2) = \sum m(z^2 + x^2),$$

$$\text{or,} \quad \sum my^2 = \sum mx^2 = \frac{\sum m(x^2 + y^2)}{2} = \frac{C}{2}, \tag{1.4.15}$$

where m is mass of the cylinder per unit volume, i. e., $m = \frac{M}{\pi a^2 h}$.

Now

$$C = \sum \frac{a^2}{2} \times \text{mass of the strip [by (1.4.11)]}$$

$$= \frac{a^2}{2} \times \sum \text{mass of the strip} = \frac{Ma^2}{2}. \tag{1.4.16}$$

$$\therefore \quad \sum my^2 = \sum mz^2 = \frac{Ma^2}{4}.$$

Now

$$\sum mz^2 = \int_{-h/2}^{h/2} \frac{M}{\pi a^2 h} \cdot \pi a^2 \, dz \cdot z^2$$

$$\left[\text{as the mass of the strip} = m \cdot \pi a^2 \, dz = \frac{M}{\pi a^2 h} \cdot \pi a^2 \, dz\right]$$

$$= \frac{M}{h} \int_{-h/2}^{h/2} z^2 \, dz = \frac{Mh^2}{12}.$$

$$\therefore \quad A = \sum m(y^2 + z^2) = \sum my^2 + \sum mz^2$$

$$= \frac{Ma^2}{4} + \frac{Mh^2}{12} = \frac{M}{4}\left(a^2 + \frac{h^2}{3}\right) = B. \tag{1.4.17}$$

(g) Right circular cone: $x^2 + y^2 = z^2 \tan^2 \alpha$ of mass M, height h and radius of the base a

We take the axes at the vertex O as shown in (Fig 1.8). We consider a circular strip of thickness dz and of radius r at a distance z from the vertex.

Let

$$m = \frac{\text{mass}}{\text{vol.}} = \frac{M}{\frac{1}{3}\pi a^2 \cdot h} = \frac{3M}{\pi a^2 h}. \tag{1.4.18}$$

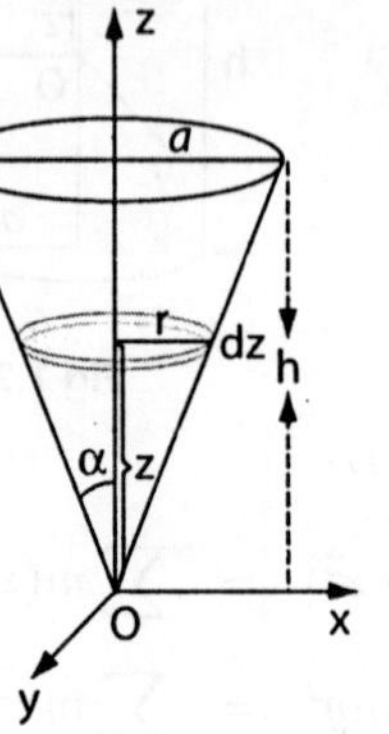

Fig 1.8

From symmetry, $A = B$,

$$\begin{aligned} \text{i.e.,}\quad \sum m(y^2+z^2) &= \sum m(z^2+x^2), \\ \text{or,}\quad \sum my^2 = \sum mx^2 &= \frac{1}{2}\left[\sum mx^2 + \sum my^2\right] \\ &= \frac{1}{2}\sum m(x^2+y^2) = \frac{C}{2}. \end{aligned} \tag{1.4.19}$$

m.i. of the strip about z-axis

$$\begin{aligned} &= \frac{r^2}{2} \times (\text{mass of the strip}) = \frac{r^2}{2} \cdot \pi r^2\, dz \cdot m \quad [\text{by (1.4.11)}] \\ &= \frac{\pi}{2} \cdot r^4 \cdot \frac{3M}{\pi a^2 h} \cdot dz = \frac{3M}{2a^2 h} \cdot r^4\, dz \\ &= \frac{3M}{2a^2 h} \cdot (z \tan\alpha)^4\, dz \quad [\because\ r = h\tan\alpha]. \end{aligned}$$

Hence C = m.i. of the cone about z-axis

$$= \int_0^h \frac{3M}{2a^2 h} \tan^4\alpha \cdot z^4\, dz = \frac{3M}{2a^2 h} \tan^4\alpha \cdot \frac{h^5}{5}$$

$$= \frac{3M}{10a^2}(h\tan\alpha)^4 = \frac{3M}{10}a^2 \quad [\because\quad a = h\tan\alpha] \tag{1.4.20}$$

$$= \frac{3M}{10}h^2\tan^2\alpha = \frac{3M}{10}a^2. \tag{1.4.21}$$

Now

$$
\begin{aligned}
\sum mz^2 &= \int_0^h \frac{3M}{\pi a^2 h} \cdot \pi r^2 \, dz z^2 = \frac{3M}{a^2 h} \int_0^h (z \tan\alpha)^2 z^2 \, dz \\
&= \frac{3M}{a^2 h} \tan^2\alpha \int_0^h z^4 \, dz = \frac{3M}{5a^2} h^4 \tan^2\alpha \\
&= \frac{3M}{5a^2} (h \tan\alpha)^2 \cdot h^2 = \frac{3M}{5} h^2 \quad [\text{as } a = h \tan\alpha]. \qquad (1.4.22)
\end{aligned}
$$

$$
\begin{aligned}
\therefore \quad A &= \sum m(y^2 + z^2) = \sum my^2 + \sum mz^2 \\
&= \frac{C}{2} + \frac{3M}{5} h^2 = \frac{3M}{20} a^2 + \frac{3M}{5} h^2 = B. \qquad (1.4.23)
\end{aligned}
$$

From symmetry $D = E = F = 0$.

(h) Sphere $x^2 + y^2 + z^2 = a^2$, of mass M, at its centre

We take the axes at the centre O as shown in (Fig 1.9).

Let

$$
m = \frac{\text{mass}}{\text{vol.}} = \frac{M}{\frac{4}{3} \cdot \pi a^3} = \frac{3M}{4\pi a^3}.
$$

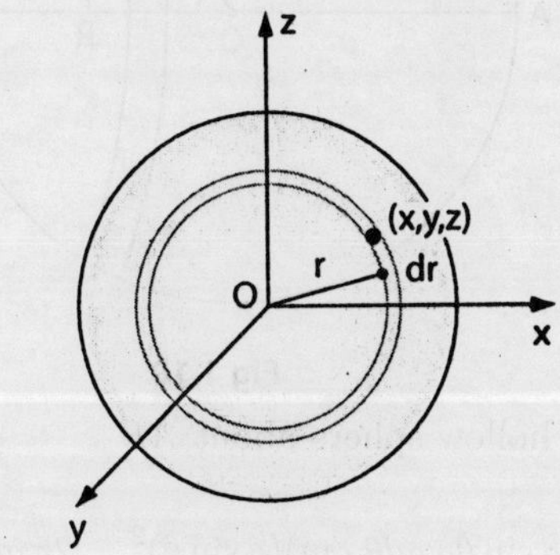

Fig 1.9

Let us consider a spherical shell of radius r and thickness dr. From symmetry,

$$
\begin{aligned}
A = B = C &= \tfrac{1}{3}(A + B + C) \\
&= \tfrac{1}{3} \sum m(y^2 + z^2 + z^2 + x^2 + x^2 + y^2) \\
&= \tfrac{2}{3} \sum m(x^2 + y^2 + z^2) \\
&= \tfrac{2}{3} \sum mr^2,
\end{aligned}
$$

where r is the radius of the spherical shell considered.

$$\therefore\ A = B = C = \frac{2}{3}\int_0^a \frac{3M}{4\pi a^3}\cdot(4\pi r^2\,dr)\cdot r^2 = \frac{2M}{a^3}\cdot\frac{a^5}{5} = \frac{2}{5}Ma^2. \tag{1.4.24}$$

And from symmetry $D = E = F = 0$.

(i) Hollow sphere of mass M and radius a about a diameter

Let us consider an elementary arc $PQ = ad\theta$ at a point P. When this arc revolved about the diameter AB, generates a circular ring of radius $a\sin\theta$ and of width $ad\theta$.

Let

$$m = \frac{\text{mass}}{\text{area}} = \frac{M}{4\pi a^2}.$$

$\therefore$ mass of the elementary ring $= m\cdot 2\pi a\sin\theta\cdot ad\theta$.

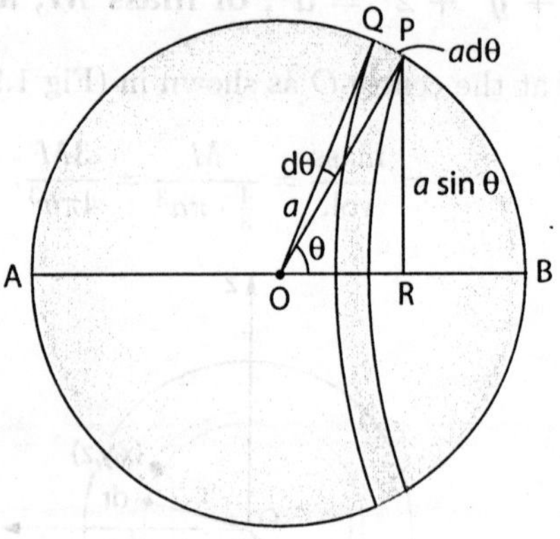

Fig 1.10

Hence m.i. of the hollow sphere about AB

$$= \int_0^\pi (2\pi a\sin\theta\cdot ad\theta\cdot m)(a\sin\theta)^2 = 2\pi m a^4\int_0^\pi \sin^3\theta\,d\theta$$

$$= 2\pi\cdot\frac{M}{4\pi a^2}\cdot a^4\int_0^\pi \frac{3\sin\theta - \sin 3\theta}{4} = \frac{2}{3}Ma^2 \tag{1.4.25}$$

(j) Hemisphere $x^2 + y^2 + z^2 = a^2$, $z > 0$, mass M, radius a

We take the axes at O, centre of the base of the hemisphere, as shown in (Fig 1.11). We consider a circular strip of radius r, perpendicular to z-axes and of thickness dz at a hight z from O.

Let

$$m = \frac{\text{mass}}{\text{vol.}} = \frac{M}{\frac{2}{3}\pi a^3} = \frac{3M}{2\pi a^3}. \tag{1.4.26}$$

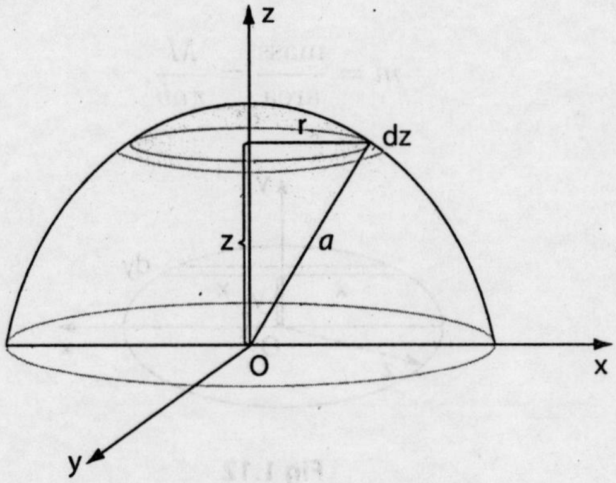

Fig 1.11

From symmetry, $A = B$, i. e.,

$$\sum m(y^2 + z^2) = \sum m(z^2 + x^2),$$

$$\text{or,} \quad \sum my^2 = \sum mx^2 = \frac{1}{2}\sum m(x^2 + y^2) = \frac{C}{2}. \quad (1.4.27)$$

Now

$$\begin{aligned} C &= \sum \frac{r^2}{2} \times \text{mass of the strip} = \int_0^a \frac{3M}{2\pi a^3} \cdot \pi r^2 \, dz \cdot \frac{r^2}{2} \\ &= \frac{3M}{4a^3} \int_0^a r^4 \, dz = \frac{3M}{4a^3} \int_0^a \left(a^2 - z^2\right)^2 dz \quad [\because\ a^2 = r^2 + z^2] \\ &= \frac{2}{5} Ma^2. \end{aligned} \quad (1.4.28)$$

$$\begin{aligned} \sum mz^2 &= \int_0^a \frac{3M}{2\pi a^3} \cdot \pi r^2 \, dz \cdot z^2 = \frac{3M}{2a^3} \int_0^a (a^2 - z^2) z^2 \, dz \\ &= \frac{3M}{2a^3} \int_0^a (a^2 z^2 - z^4) \, dz = \frac{1}{5} Ma^2. \end{aligned}$$

$$\begin{aligned} \therefore \quad A &= \sum m(y^2 + z^2) = \sum my^2 + \sum mz^2 = \frac{C}{2} + \frac{1}{5}Ma^2 \\ &= \frac{2}{10}Ma^2 + \frac{1}{5}Ma^2 = \frac{2}{5}Ma^2 = B. \end{aligned} \quad (1.4.29)$$

From symmetry $D = E = F = 0$.

(k) Elliptic plate: $\frac{x^2}{a^2} + \frac{y^2}{b^2} = 1$, of mass M, at its centre

We consider the x and y-axis in the plane of the elliptic plate at O, the centre of the plate and z-axis perpendicular to the plate.

Let

$$m = \frac{\text{mass}}{\text{area}} = \frac{M}{\pi ab}. \tag{1.4.30}$$

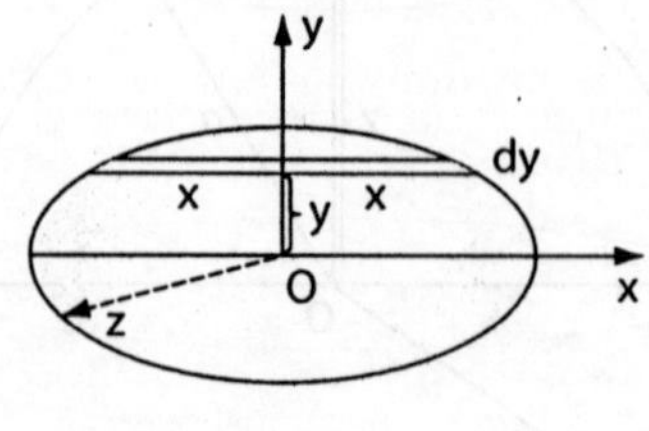

Fig 1.12

Let us consider a strip of thickness dy parallel to x-axis at a distance y from origin O. As on the plate $z = 0$, we have

$$\begin{aligned} A &= \sum m(y^2 + z^2) = \sum my^2 \\ B &= \sum m(z^2 + x^2) = \sum mx^2 \\ \text{and } C &= \sum m(x^2 + y^2) = \sum mx^2 + \sum my^2 \\ &= A + B. \end{aligned} \tag{1.4.31}$$

Now,

$$\begin{aligned} A &= \sum my^2 = \int_{-b}^{b} \frac{M}{\pi ab} \cdot 2x dy \cdot y^2 = \frac{2M}{\pi ab} \int_{-b}^{b} a\sqrt{1 - \frac{y^2}{b^2}} \cdot y^2\, dy \\ &= \frac{4M}{\pi b^2} \int_0^b \sqrt{b^2 - y^2} y^2\, dy \quad \begin{array}{l}(\text{putting } y = b\sin\theta, dy = b\cos\theta\, d\theta, \\ \theta \to 0 \text{ as } y \to 0; \theta \to \pi/2 \text{ as } y \to b)\end{array} \\ &= \frac{4M}{\pi b^2} \int_0^{\pi/2} b\cos\theta \cdot b^2 \sin^2\theta \cdot b\cos\theta\, d\theta \\ &= \frac{4M}{\pi b^2} \cdot b^4 \int_0^{\pi/2} \sin^2\theta \cos^2\theta\, d\theta \\ &= \frac{4M}{\pi} \cdot b^2 \cdot \frac{1}{4} \cdot \frac{1}{2} \cdot \frac{\pi}{2} = \frac{Mb^2}{4}. \end{aligned} \tag{1.4.32}$$

Similarly,

$$B = \sum mx^2 = \frac{Ma^2}{4} \tag{1.4.33}$$

$$\text{and } C = A + B = \frac{M}{4}(b^2 + c^2). \tag{1.4.34}$$

(l) Ellipsoid: $\frac{x^2}{a^2} + \frac{y^2}{b^2} + \frac{z^2}{c^2} = 1$, mass M at its centre

We consider the reference axes at the centre of the ellipsoid O as shown in (Fig 1.13).

Let

$$m = \frac{\text{mass}}{\text{vol.}} = \frac{M}{\frac{4}{3}\pi abc} = \frac{3M}{4\pi abc}. \tag{1.4.35}$$

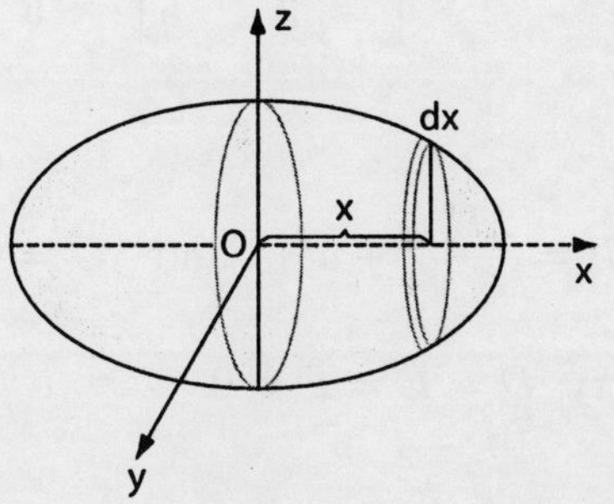

Fig 1.13

Let us consider an elliptical strip of thickness dx at a distance x from the centre and perpendicular to x-axis.

The equation of the elliptical strip is

$$\frac{y^2}{b^2} + \frac{z^2}{c^2} = 1 - \frac{x^2}{a^2} = \lambda^2 \quad \text{(say)},$$

$$\text{or,} \quad \frac{y^2}{(\lambda b)^2} + \frac{z^2}{(\lambda c)^2} = 1,$$

where λb, λc are the length of the semi-axes of the elliptical strip.

$\therefore$ The m.i. of the strip about x-axis

$$= \frac{m_1}{4}(\lambda^2 b^2 + \lambda^2 c^2) = \frac{m_1}{4}\lambda^2(b^2 + c^2) \quad \text{[by (1.4.34)]}$$

where m_1 = mass of the strip $= m \cdot \pi \cdot \lambda b \cdot \lambda c \cdot dx$

$$= \frac{3M}{4\pi abc} \cdot \pi\lambda^2 \cdot bc\, dx$$

$\therefore$ The m.i. of the strip about x-axis

$$= \frac{3M}{4\pi abc} \cdot \pi\lambda^2 \cdot bc \cdot dx \cdot \frac{\lambda^2(b^2 + c^2)}{4} = \frac{3M}{16a}(b^2 + c^2)\lambda^4 \cdot dx.$$

Hence the m.i. of the ellipsoid about x-axis

$$\begin{aligned} A &= \int_{-a}^{a} \frac{3M}{16a}(b^2+c^2)\lambda^4\,dx = \frac{6M}{16a}(b^2+c^2)\int_0^a \left(1-\frac{x^2}{a^2}\right)^2 dx \\ &= \frac{3M}{8a}(b^2+c^2)\int_0^a \left(1-\frac{2x^2}{a^2}+\frac{x^4}{a^4}\right) dx \\ &= \frac{3M}{8a}(b^2+c^2)\left[a-\frac{2}{3}a+\frac{a}{5}\right] = \frac{M}{5}(b^2+c^2). \end{aligned} \tag{1.4.36}$$

Similarly,

$$B = \frac{M}{5}(c^2+a^2) \quad \text{and} \quad C = \frac{M}{5}(a^2+b^2). \tag{1.4.37}$$

From symmetry, $D = E = F = 0$.

1.5 Routh's Rule

Dr. Routh suggested a rule for remembering the moment of inertia of certain symmetrical solid bodies.

Rule: Moment of Inertia (m.i.) about an axis of symmetry.

$$= \text{Mass} \times \frac{\text{sum of the squares of perpendicular semi-axes}}{3, 4, 5}$$

the denominator being 3 or 4 or 5 according as the body is rectangular (including a rod and square), or ellipsoid (including circular) or elliptic (including spherical).

1.6 Moment of Inertia of Simple Geometrical Bodies

If M demotes the mass of the body and density of the body supposed to be uniform throughout.

The moment of inertia of a

1	Thin rod of length $2a$	About an axis through the centre of mass and perpendicular to the rod is	$\frac{Ma^2}{3}$
		About an axis through one end and perpendicular to the rod is	$\frac{4Ma^2}{3}$

		About an axis through the centre mass and parallel to the side $2a$ is	$\frac{Mb^2}{3}$
2	Rectangular lamina of sides $2a$ and $2b$	About an axis through the centre of mass and parallel to the side $2b$ is	$\frac{M}{3}a^2$
		About an axis through the centre of mass and perpendicular to lamina is	$\frac{M}{3}(a^2+b^2)$
		About an axis through centre of mass and parallel to edge $2a$ is	$\frac{M}{3}(b^2+c^2)$
3	Rectangular parallelopiped with edges $2a$, $2b$, $2c$.	About an axis through centre of mass and parallel to the edge $2b$ is	$\frac{M}{3}(c^2+a^2)$
		About an axis through centre of mass and parallel to the edge $2c$ is	$\frac{M}{3}(a^2+b^2)$
		About a diameter is	$M\frac{a^2}{2}$
4	Thin circular ring of radius a	About an axis through centre and perpendicular to the plane of the circle is	Ma^2
		About a diameter is	$\frac{Ma^2}{4}$
5	Circular disc of radius a	About an axis through the centre and perpendicular to the plane of the disc is	$\frac{Ma^2}{2}$
		About major axis of length $2a$ is	$M\frac{b^2}{4}$
6	Elliptic disc of axes $2a$, $2b$.	About minor axis of length $2b$ is	$M\frac{a^2}{4}$
		About an axis through centre and perpendicular to the plane of the disc is	$M\frac{a^2+b^2}{4}$
7	Hollow sphere of radius a.	About a diameter is	$\frac{2Ma^2}{3}$
8	Solid sphere of radius a	About a diameter is	$\frac{2Ma^2}{5}$

9	Solid ellipsoid with axes $2a$, $2b$, $2c$	About the axis $2a$ is	$\frac{M}{5}(b^2+c^2)$
		About the axis $2b$ is	$\frac{M}{5}(c^2+a^2)$
		About the axis $2c$ is	$\frac{M}{5}(a^2+b^2)$

1.7 Dirichlet's Integral

If $\xi+\eta+\zeta<1$, then

$$\int_{\xi=0}^{1}\int_{\eta=0}^{1-\xi}\int_{\zeta=0}^{1-\xi-\eta}\xi^{p-1}\eta^{q-1}\zeta^{r-1}d\xi\,d\eta\,d\zeta=\frac{\Gamma(p)\Gamma(q)\Gamma(r)}{\Gamma(p+q+r+1)}.$$

1.8 Important Solved Problems: Application of Dirichlet's Integral

▶ **Problem 1.8.1** *Use Dirichlet's integral to prove that the moments and products of inertia of the octant of the ellipsoid*

$$\frac{x^2}{a^2}+\frac{y^2}{b^2}+\frac{z^2}{c^2}=1$$

bounded by the positive sides of the axial plane are given by

$$A=\frac{M}{5}(b^2+c^2),\quad B=\frac{M}{5}(c^2+a^2),\quad C=\frac{M}{5}(a^2+b^2),$$

$$D=\frac{2M}{5\pi}bc,\qquad E=\frac{2M}{5\pi}ca,\qquad F=\frac{2M}{5\pi}ab;$$

where M is the mass of the octant.

Solution: Let

$$m=\frac{\text{mass}}{\text{vol.}}=\frac{M}{\frac{1}{8}\cdot\frac{4}{3}\pi abc}=\frac{6M}{\pi abc}. \tag{1.8.1}$$

We consider the reference axes as shown in the Fig 1.14. Let $P(x,y,z)$ be any point.

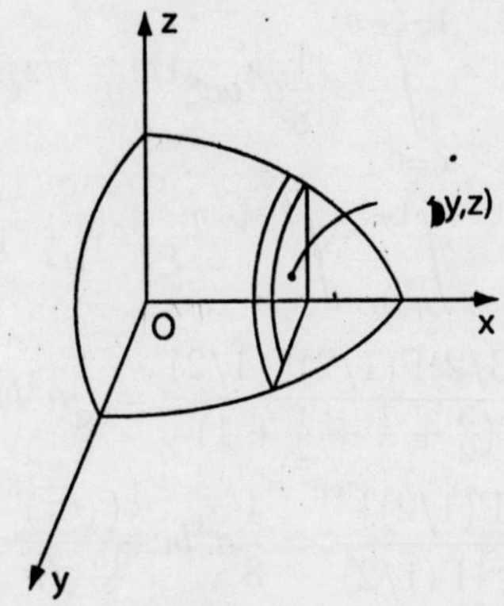

Fig 1.14

Let us consider the integral

$$\underset{\substack{\text{over the}\\ \text{octant}}}{\iiint} x^2\,dx\,dy\,dz = \int_{x=0}^{a}\int_{y=0}^{b\sqrt{1-x^2/a^2}}\int_{z=0}^{c\sqrt{1-\frac{x^2}{a^2}-\frac{y^2}{b^2}}} x^2\,dx\,dy\,dz.$$

Putting

$$\zeta = \frac{x^2}{a^2},\quad \eta = \frac{y^2}{b^2},\quad \xi = \frac{z^2}{c^2},$$

we get when

$$\left.\begin{matrix} x=0, & \xi = 0 \\ x=a, & \xi=1 \end{matrix}\right\};\quad \left.\begin{matrix} y=0, & \eta=0 \\ y=b\sqrt{1-\frac{x^2}{a^2}}, & \eta = 1-\frac{x^2}{a^2} = 1-\xi \end{matrix}\right\}$$

and

$$\left.\begin{matrix} z=0, & \zeta=0 \\ z=c\sqrt{1-\frac{x^2}{a^2}-\frac{y^2}{b^2}}, & \zeta = 1-\frac{x^2}{a^2}-\frac{y^2}{b^2} = 1-\xi-\eta \end{matrix}\right\}.$$

$\therefore$ The above integral

$$\underset{\substack{\text{over the}\\ \text{octant}}}{\iiint} x^2\,dx\,dy\,dz = \int_{\xi=0}^{1}\int_{\eta=0}^{1-\xi}\int_{\zeta=0}^{1-\xi-\eta} \xi a^2\cdot abc\cdot\frac{1}{8}\xi^{-1/2}\eta^{-1/2}\zeta^{-1/2}\,d\xi\,d\eta\,d\zeta$$

[Since $2x\,dx = a^2\,d\xi$.

$$\therefore\ dx = \frac{a^2\,d\xi}{2x} = \frac{a^2}{2a\xi^{1/2}}d\xi = \frac{a}{2}\cdot\xi^{-1/2}d\xi.$$

Similarly,

$$dy = \frac{b}{2}\eta^{-1/2}d\eta,\quad dz = \frac{c}{2}\zeta^{-1/2}d\zeta]$$

$$= \int_{\xi=0}^{1}\int_{\eta=0}^{1-\xi}\int_{\zeta=0}^{1-\xi-\eta} \frac{1}{8}a^3bc\xi^{1/2}\eta^{-1/2}\zeta^{-1/2}\,d\xi\,d\eta\,d\zeta$$

$$= \frac{1}{8}a^3bc\int_{\xi=0}^{1}\int_{\eta=0}^{1-\xi}\int_{\zeta=0}^{1-\xi-\eta} \xi^{\frac{3}{2}-1}\eta^{\frac{1}{2}-1}\zeta^{\frac{1}{2}-1}\,d\xi\,d\eta\,d\zeta$$

$$= \frac{1}{8}a^3bc\frac{\Gamma(3/2)\Gamma(1/2)\Gamma(1/2)}{\Gamma(\frac{3}{2}+\frac{1}{2}+\frac{1}{2}+1)} = \frac{1}{8}a^3bc\cdot\frac{\frac{1}{2}\Gamma(1/2)\Gamma(1/2)\Gamma(1/2)}{\Gamma(7/2)}$$

$$= \frac{1}{8}a^3bc\frac{\frac{1}{2}\{\Gamma(1/2)\}^3}{\frac{5}{2}\frac{3}{2}\frac{1}{2}\Gamma(1/2)} = \frac{1}{8}a^3bc\frac{\frac{1}{2}(\sqrt{\pi})^3}{\frac{15}{8}\sqrt{\pi}} = \frac{1}{30}a^3bc\pi.$$

Similarly,

$$\iiint_{\substack{\text{over the}\\ \text{octant}}} y^2\,dx\,dy\,dz = \frac{\pi}{30}ab^3c;\quad \iiint_{\substack{\text{over the}\\ \text{octant}}} z^2\,dx\,dy\,dz = \frac{\pi}{30}abc^3.$$

$\therefore$ The moment of inertia of the octant about x-axis

$$A = \iiint_{\substack{\text{over the}\\ \text{octant}}} m(y^2+z^2)\,dx\,dy\,dz = \frac{6M}{\pi abc}\cdot\frac{\pi}{30}(ab^3c+abc^3)$$

$$= \frac{M}{5}(b^2+c^2). \tag{1.8.2}$$

Similarly,

$$B = \frac{M}{5}(c^2+a^2) \quad\text{and}\quad C = \frac{M}{5}(a^2+b^2). \tag{1.8.3}$$

$$D = \iiint_{\substack{\text{over the}\\ \text{octant}}} myz\,dx\,dy\,dz$$

$$= \int_{\xi=0}^{1}\int_{\eta=0}^{1-\xi}\int_{\zeta=0}^{1-\xi-\eta} \frac{6M}{\pi abc}\cdot b\eta^{1/2}\cdot c\zeta^{1/2}\cdot\frac{1}{8}abc\xi^{-1/2}\eta^{-1/2}\zeta^{-1/2}d\xi\,d\eta\,d\zeta$$

$$= \frac{3M}{4\pi}bc\int_{\xi=0}^{1}\int_{\eta=0}^{1-\xi}\int_{\zeta=0}^{1-\xi-\eta} \xi^{\frac{1}{2}-1}\cdot\eta^{1-1}\cdot\zeta^{1-1}\,d\xi\,d\eta\,d\zeta$$

$$= \frac{3M}{4\pi}bc\cdot\frac{\Gamma(1/2)\Gamma(1)\Gamma(1)}{\Gamma(\frac{1}{2}+1+1+1)} = \frac{3M}{4\pi}bc\frac{\Gamma(1/2)}{\Gamma(7/2)}$$

$$= \frac{3M}{4\pi}bc\cdot\frac{\Gamma(1/2)}{\frac{5}{2}\cdot\frac{3}{2}\cdot\frac{1}{2}\Gamma(1/2)} = \frac{2M}{5\pi}bc. \tag{1.8.4}$$

Similarly,

$$E = \frac{2M}{5\pi} ca, \quad F = \frac{2M}{5\pi} ab. \tag{1.8.5}$$

► **Problem 1.8.2** *Prove that the moments and products of inertia of the quadrant of the ellipse*

$$\frac{x^2}{a^2} + \frac{y^2}{b^2} = 1$$

bounded by positive sides of the axes are given by

$$A = \frac{M}{4} b^2, \quad B = \frac{M}{4} a^2, \qquad C = \frac{M}{4}(a^2 + b^2),$$

$$D = E = 0 \quad and \quad F = \frac{M}{2} ab.$$

Solution: We take the axes of x and y in the plane of the ellipse and z-axes perpendicular to the plane of the ellipse at the centre O as shown in the Fig 1.15.

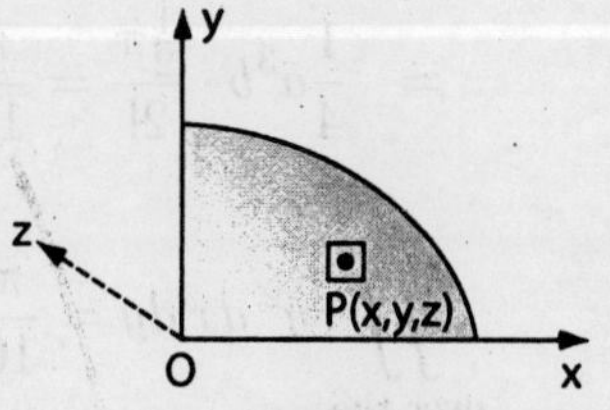

Fig 1.15

Let

$$m = \frac{\text{mass}}{\text{area}} = \frac{M}{\frac{\pi ab}{4}} = \frac{4M}{\pi ab}. \tag{1.8.6}$$

Let $P(x, y, 0)$ be any point on the quadrant.
Now

$$\iint\limits_{\substack{\text{over the} \\ \text{quadrant}}} x^2 \, dx \, dy = \int\limits_{x=0}^{a} \int\limits_{y=0}^{b\sqrt{1-x^2/a^2}} x^2 \, dx \, dy.$$

Putting

$$\xi = \frac{x^2}{a^2}, \quad \text{or,} \quad 2x \, dx = a^2 \, d\xi,$$

$$\text{or,} \quad dx = \frac{a^2}{2x} d\xi = \frac{a^2}{2 \cdot a\sqrt{\xi}} d\xi = \frac{a}{2} \xi^{-1/2} \, d\xi$$

$$x = 0, \xi = 0; \quad x = a, \xi = 1$$

and

$$\eta = \frac{y^2}{b^2}, \quad \text{or,} \quad 2y\,dy = b^2\,d\eta, \quad \text{or,} \quad dy = \frac{b^2\eta\,d\eta}{2b\sqrt{\eta}} = \frac{b}{2}\eta^{-1/2}d\eta$$

when $y = 0$, $\eta = 0$;

and $y = b\sqrt{1 - \frac{x^2}{a^2}}$; $\quad \eta = \left(1 - \frac{x^2}{a^2}\right) = 1 - \xi.$

$$\therefore \int_{x=0}^{1} \int_{y=0}^{b\sqrt{1-\frac{x^2}{a^2}}} x^2\,dx\,dy = \int_{\xi=0}^{1} \int_{\eta=0}^{1-\xi} a^2\xi \cdot a \cdot \frac{1}{2}\xi^{-1/2} \cdot \frac{1}{2} \cdot b\eta^{-1/2} d\xi\,d\eta$$

$$= \frac{1}{4}a^3b \int_{\xi=0}^{1} \int_{\eta=0}^{1-\xi} \xi^{\frac{3}{2}-1} \cdot \eta^{\frac{1}{2}-1}\,d\xi\,d\eta$$

$$= \frac{1}{4}a^3b\frac{\Gamma(\frac{3}{2})\Gamma(\frac{1}{2})}{\Gamma(\frac{3}{2}+\frac{1}{2}+1)} = \frac{1}{4} \cdot a^3b\frac{\frac{1}{2}\Gamma(\frac{1}{2})\Gamma(\frac{1}{2})}{\Gamma(3)}$$

$$= \frac{1}{4}a^3b \cdot \frac{\frac{1}{2}\pi}{2!} = \frac{\pi}{16}a^3b \qquad \left[\because \Gamma\left(\frac{1}{2}\right) = \sqrt{\pi}\right].$$

Similarly,

$$\iint_{\substack{\text{over the}\\ \text{quadrant}}} y^2\,dx\,dy = \frac{\pi}{16}ab^3.$$

Now moment of inertia of the quadrant about x-axis

$$A = \iint_{\substack{\text{over the}\\ \text{quadrant}}} m(y^2 + z^2)\,dx\,dy = \iint_{\substack{\text{over the}\\ \text{quadrant}}} my^2\,dx\,dy \quad [\text{As } z = 0]$$

$$= \frac{4M}{\pi ab} \cdot \frac{\pi}{16}ab^3 = \frac{M}{4}b^2. \tag{1.8.7}$$

m.i. of the quadrant about y-axis and z-axis

$$B = \iint_{\substack{\text{over the}\\ \text{quadrant}}} m(z^2 + x^2)\,dx\,dy = \iint_{\substack{\text{over the}\\ \text{quadrant}}} mx^2\,dx\,dy \quad [\text{as } z = 0]$$

$$= \frac{M}{4}a^2. \tag{1.8.8}$$

$$C = \iint_{\substack{\text{over the}\\ \text{quadrant}}} m(x^2 + y^2)\,dx\,dy = \frac{4M}{\pi ab}\left[\frac{\pi}{16}ab^3 + \frac{\pi}{16}a^3b\right]$$

$$= \frac{M}{4}(a^2 + b^2). \tag{1.8.9}$$

Product of inertia

$$D = \iint\limits_{\substack{\text{over the}\\ \text{quadrant}}} myz\,dx\,dy = 0, \quad E = \iint\limits_{\substack{\text{over the}\\ \text{quadrant}}} mzx\,dx\,dy = 0 \quad [\text{as } z = 0].$$

$$E = \iint\limits_{\substack{\text{over the}\\ \text{quadrant}}} mxy\,dx\,dy = \frac{4M}{\pi ab}\int_{x=0}^{a}\int_{y=0}^{b\sqrt{1-\frac{x^2}{a^2}}} xy\,dx\,dy$$

$$= \frac{4M}{\pi ab}\int_{\xi=0}^{1}\int_{\eta=0}^{1-\xi} b\cdot a\cdot \xi^{1/2}\cdot\eta^{1/2}\cdot\frac{1}{2}a\xi^{-1/2}\cdot\frac{1}{2}b\eta^{-1/2}\,d\xi\,d\eta$$

$$= \frac{M}{\pi}ab\int_{\xi=0}^{1}\int_{\eta=0}^{1-\xi}\xi^{1-1}\eta^{1-1}d\xi\,d\eta = \frac{M}{\pi}ab\frac{\Gamma(1)\Gamma(1)}{\Gamma(1+1+1)}$$

$$= \frac{M}{\pi}ab\cdot\frac{1}{\Gamma(3)} = \frac{M}{\pi}\frac{ab}{2} = \frac{M}{2\pi}ab. \qquad (1.8.10)$$

1.9 Moments and Products of Inertia of a Body about Line or Lines

(a) *If A, B, C and D, E, F are the moments and products of inertia respectively of a body with respect to reference axes Ox, Oy, Oz then*

$$Al^2 + Bm^2 + Cn^2 - 2Dmn - 2Enl - 2Flm,$$

is the moment of inertia of the body about the line

$$\frac{x}{l} = \frac{y}{m} = \frac{z}{n},$$

where l, m, n be the d.c's of the line. [CH 1996,'04: VH '01. '06; KH '04]

Let $P(x, y, z)$ be the position of an elementary mass m_0 of the body with respect to the reference axes Ox, Oy, Oz. PN be the perpendicular to the line OA having d.c's l, m, n.

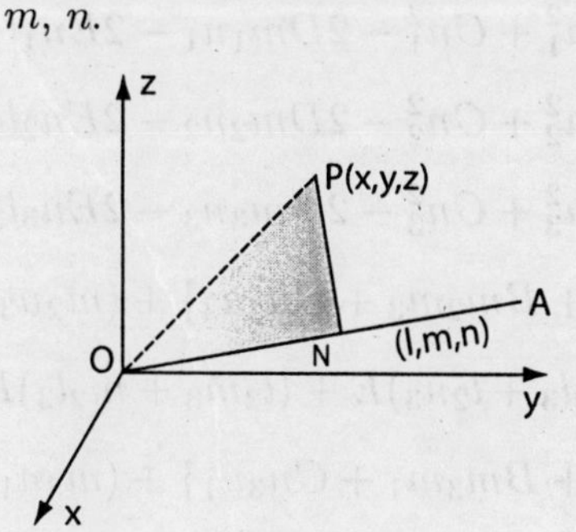

Fig 1.16

The m.i. of the body about OA

$$
\begin{aligned}
&= \sum m_0 PN^2 = \sum m_0 \left(OP^2 - ON^2\right) \\
&= \sum m_0[x^2 + y^2 + z^2 - (lx + my + nz)^2] \\
&\qquad [\text{As } ON = \text{projection of } OP \text{ on } OA] \\
&= \sum m_0 \left\{\left(x^2 + y^2 + z^2\right)\left(l^2 + m^2 + n^2\right) - (lx + my + nz)^2\right\} \\
&= l^2 \sum m_0 \left(y^2 + z^2\right) + m^2 \sum m_0 \left(z^2 + x^2\right) + n^2 \sum m_0 \left(x^2 + y^2\right) \\
&\qquad -2mn \sum m_0 yz - 2nl \sum m_0 zx - 2lm \sum m_0 xy \\
&= Al^2 + Bm^2 + Cn^2 - 2Dmn - 2Enl - 2Flm.
\end{aligned}
$$

Remark 1.9.1 *If $D = E = F = 0$, then $\frac{x}{l} = \frac{y}{m} = \frac{z}{n}$ or the line OA is the principal axis and the moment of inertia about the principal axis = $Al^2 + Bm^2 + Cn^2$.*

(b) *If A, B, C and D, E, F are the moments and products of inertia of the body with respect to the axes Ox, Oy and Oz, then A_1, B_1, C_1 and D_1, E_1, F_1 the moments and products of inertia of the body with respect to the three mutually perpendicular lines Ox_1, Oy_1, and Oz_1 having d.c's (l_1, m_1, n_1), (l_2, m_2, n_2) and (l_3, m_3, n_3) respectively are given by.*

$$
\begin{aligned}
A_1 &= Al_1^2 + Bm_1^2 + Cn_1^2 - 2Dm_1n_1 - 2En_1l_1 - 2Fl_1m_1 \\
B_1 &= Al_2^2 + Bm_2^2 + Cn_2^2 - 2Dm_2n_2 - 2En_2l_2 - 2Fl_2m_2 \\
C_1 &= Al_3^2 + Bm_3^2 + Cn_3^2 - 2Dm_3n_3 - 2En_3l_3 - 2Fl_3m_3. \\
D_1 &= -\{Al_2l_3 + Bm_2m_3 + Cn_2n_3\} + (m_2n_3 + n_2m_3)D \\
&\qquad +(n_2l_3 + l_2n_3)E + (l_2m_3 + m_2l_3)F \\
E_1 &= -\{Al_3l_1 + Bm_3m_1 + Cn_3n_1\} + (m_3n_1 + n_3m_1)D \\
&\qquad +(n_3l_1 + l_3n_1)E + (l_1m_3 + m_1l_3)F \\
F_1 &= -\{Al_1l_2 + Bm_1m_2 + Cn_1n_2\} + (m_2n_1 + n_2m_1)D \\
&\qquad +(n_1l_2 + l_1n_2)E + (l_1m_2 + m_1l_2)F.
\end{aligned}
$$

[CUH 1998]

Let P be the position of an elementary mass m_0. Let (x, y, z) and (x_1, y_1, z_1) be the coordinates of P with respect to $Oxyz$ and $Ox_1y_1z_1$ as axes respectively.

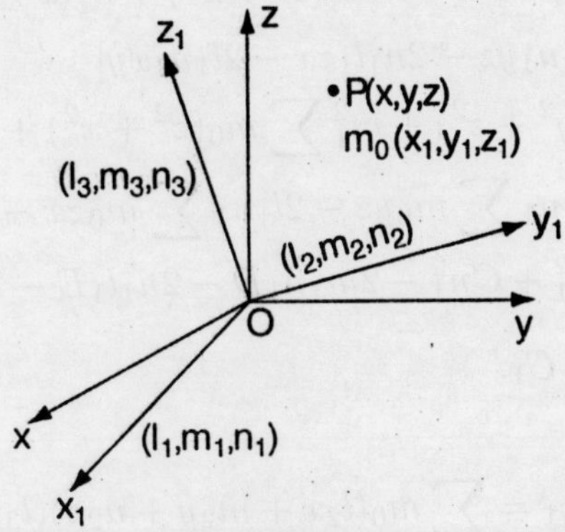

Fig 1.17

Let (l_1, m_1, n_1), (l_2, m_2, n_2) and (l_3, m_3, n_3) are the d.c's of Ox_1, Oy_1 and Oz_1 respectively.

The x_1, y_1, z_1 are given by the scheme

	x	y	z
x_1	l_1	m_1	n_1
y_1	l_2	m_2	n_2
z_1	l_3	m_3	n_3

$$\text{Thus} \quad \left.\begin{aligned} x_1 &= l_1x + m_1y + n_1z \\ y_1 &= l_2x + m_3y + n_2z \\ z_1 &= l_3x + m_3y + n_3z \end{aligned}\right\}, \qquad (1.9.1)$$

$$\left.\begin{aligned} &\text{where} \quad l_1^2 + m_1^2 + n_1^2 = l_2^2 + m_2^2 + n_2^2 = l_3^2 + m_3^2 + n_3^2 = 1 \\ &\qquad\quad l_1^2 + l_2^2 + l_3^2 = m_1^2 + m_2^2 + m_3^2 = n_1^2 + n_2^2 + n_3^2 = 1 \\ &\text{and} \quad l_1l_2 + m_1m_2 + n_1n_2 = l_2l_3 + m_2m_3 + n_2n_3 \\ &\qquad = l_3l_1 + m_3m_1 + n_3n_1 = l_1m_1 + l_2m_2 + l_3m_3 \\ &\qquad = m_1n_1 + m_2n_2 + m_3n_3 = l_1n_1 + l_2n_2 + l_3n_3 = 0 \end{aligned}\right\}, \qquad (1.9.2)$$

Thus

$$\begin{aligned} A_1 &= \sum m_0(y_1^2 + z_1^2) \\ &= \sum m_0[(l_2x + m_2y + n_2z)^2 + (l_3x + m_3y + n_3z)^2] \\ &= \sum m_0[(l_2^2 + l_3^2)x^2 + (m_2^2 + m_3^2)y^2 + (n_2^2 + n_3^2)z^2 \\ &\qquad +2(m_2n_2 + m_3n_3)yz + 2(l_2n_2 + l_3n_3)zx + 2(l_2m_2 + l_3m_3)xy] \\ &= \sum m_0[(1 - l_1^2)x^2 + (1 - m_1^2)y^2 + (1 - n_1^2)z^2 \\ &\qquad -2m_1n_1yz - 2n_1l_1zx - 2l_1m_1xy] \quad \text{[by (1.9.1) and (1.9.2)]} \end{aligned}$$

$$\begin{aligned}
&= \sum m_0[(m_1^2+n_1^2)x^2+(n_1^2+l_1^2)y^2+(l_1^2+m_1^2)z^2 \\
&\qquad -2m_1n_1yz-2n_1l_1zx-2l_1n_1xy] \quad [\text{by}(1.9.2)] \\
&= \sum m_0[l_1^2(y^2+z^2)+m_1^2(z^2+x^2)+n_1^2(x^2+y^2) \\
&\qquad -2m_1n_1yz-2n_1l_1zx-2l_1n_1xy] \\
&= l_1^2\sum m_0(y^2+z^2)+m_1^2\sum m_0(z^2+x^2)+n_1^2\sum m_0(x^2+y^2) \\
&\qquad -2m_1n_1\sum m_0yz-2l_1n_1\sum m_0zx-2l_1n_1\sum m_0xy \\
&= Al_1^2+Bm_1^2+Cn_1^2-2m_1n_1D-2n_1l_1E-2l_1m_1F.
\end{aligned}$$

Similarly for B_1 and C_1.

Now,

$$\begin{aligned}
D_1 &= \sum m_0y_1z_1=\sum m_0(l_2x+m_2y+n_2z)(l_3x+m_3y+n_3z) \\
&= \sum m_0[(l_2l_3x^2+m_2m_3y^2+n_2n_3z^2) \\
&\qquad +(m_2n_3+m_3n_2)yz+(n_3l_2+n_2l_3)zx+(l_2m_3+l_3m_2)xy] \\
&= l_2l_3\sum m_0x^2+m_2m_3\sum m_0y^2+n_2n_3\sum m_0z^2 \\
&\qquad +(m_2n_3+n_2m_3)\sum m_0yz+(n_3l_2+n_2l_3)\sum m_0zx \\
&\qquad +(l_2m_3+l_3m_2)\sum m_0xy. \qquad (1.9.3)
\end{aligned}$$

Now,

$$\begin{aligned}
\sum m_0x^2 &= \sum m_0[(x^2+y^2+z^2)-(y^2+z^2)] \\
&= \sum m_0(x^2+y^2+z^2)-\sum m_0(y^2+z^2) \\
&= \frac{A+B+C}{2}-A.
\end{aligned}$$

Similarly,

$$\sum m_0y^2=\frac{A+B+C}{2}-B \quad \text{and} \quad \sum m_0z^2=\frac{A+B+C}{2}-C.$$

Substituting these values in (1.9.3) we get,

$$\begin{aligned}
D_1 &= \frac{A+B+C}{2}[l_2l_3+m_2m_3+n_2n_3]-(Al_2l_3+Bm_2m_3+Cn_2n_3) \\
&\qquad +(m_2n_3+n_2m_3)D+(l_2n_3+n_2l_3)E+(l_2m_3+m_2l_3)F \\
&= -(Al_2l_3+Bm_2m_3+Cn_2n_3) \\
&\qquad +(m_2n_3+n_2m_3)D+(l_2n_3+n_2l_3)E+(l_2m_3+m_2l_3)F \\
&\qquad [\text{by } (1.9.2)].
\end{aligned}$$

Similarly we get, E_1 and F_1.

(c) ***Moments and products of inertia of a plane lamina about two perpendicular lines through the origin when the moments and products of inertia of the lamina with respect to a given rectangular axes at the origin are known.***

Let A and B the moments of inertia and D the product of inertia of a plane lamina with respect to the axes Ox and Oy are known. Let Ox_1 and Oy_1 be two perpendicular lines at the origin O, where Ox_1 makes an angle θ with Ox.

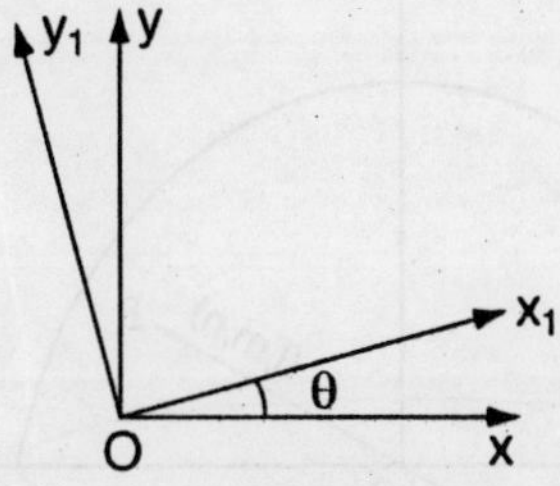

Fig 1.18

Thus d.c's of Ox_1 are $\quad l_1 = \cos\theta, \quad m_1 = \sin\theta$

and d.c's of Oy_1 are $\quad l_2 = -\sin\theta, \quad m_2 = \cos\theta.$

Let A_1 and B_1 be the moments of inertia of the lamina about Ox_1, and Oy_1 and F_1 be the product of inertia with respect to Ox_1, and Oy_1.

Thus

$$\begin{aligned}
A_1 &= Al_1^2 + Bm_1^2 - 2Fl_1m_1 \\
&= A\cos^2\theta + B\sin^2\theta - 2F\cos\theta\sin\theta \\
&= A\frac{1+\cos 2\theta}{2} + B\frac{1-\cos 2\theta}{2} - F\sin 2\theta \\
&= \frac{A+B}{2} + \frac{1}{2}(A-B)\cos 2\theta - F\sin 2\theta.
\end{aligned}$$

$$\begin{aligned}
B_1 &= Al_2^2 + Bm_2^2 - 2Fl_2m_2 \\
&= A\sin^2\theta + B\cos^2\theta + 2F\sin\theta\cos\theta \\
&= A\frac{1-\cos 2\theta}{2} + B\frac{1+\cos 2\theta}{2} + F\sin 2\theta \\
&= \frac{A+B}{2} - \frac{A-B}{2}\cos 2\theta + F\sin 2\theta
\end{aligned}$$

$$\begin{aligned}
\text{and} \quad F_1 &= -(Al_1l_2 + Bm_1m_2) + (l_1m_2 + m_1l_2)F \\
&= -[-A\sin\theta\cos\theta + B\sin\theta\cos\theta] + [\cos^2\theta - \sin^2\theta]F \\
&= \frac{A-B}{2}\sin 2\theta + F\cos 2\theta.
\end{aligned}$$

1.10 Few Problems

(a) *Find the* m.i. *of a hemisphere of mass M about any line through its centre.*

We take the x and y axes through the centre O in the base of the hemisphere and z-axis perpendicular to the xy-plane in that part of it in which the hemisphere lines. Let the d.c's of the given line AB be (l, m, n).

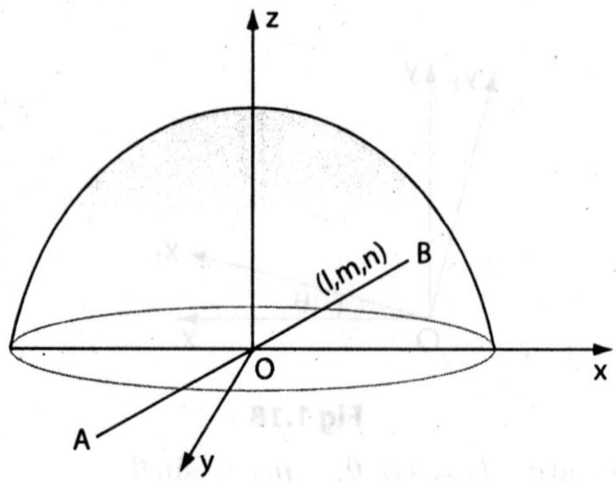

Fig 1.19

We know that

$$A = B = C = \frac{2}{5}Ma^2 \quad \text{[cf. (1.4.28) and (1.4.29)]},$$

where M is the mass and a be the radius of the hemisphere and

$$D = E = F = 0.$$

$\therefore$ Required m.i. of the hemisphere about AB

$$\begin{aligned} &= Al^2 + Bm^2 + Cn^2 \\ &= A(l^2 + m^2 + n^2) \quad [\text{As } A = B = C] \\ &= A = \frac{2}{5}Ma^2. \end{aligned}$$

(b) *Find the* m.i. *of a cube of mass M about a line through its centre.*

We take the axis Ox, Oy and Oz at the centre O and parallel to the sides of the cube. Let AB be the line having d.c's (l, m, n) about which the m.i. to be calculated. The cube is divided into thin strips parallel to yz-plane.

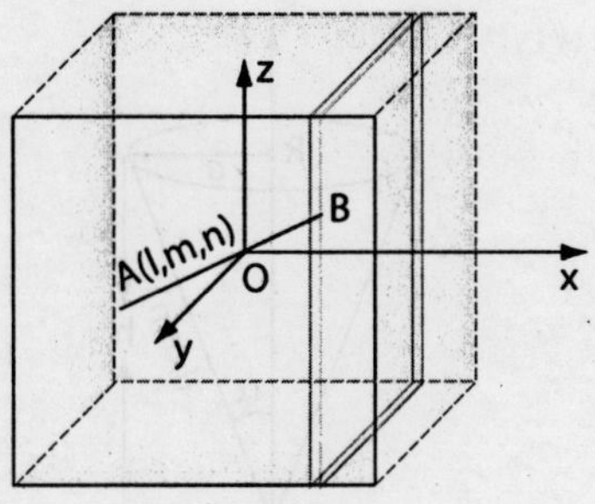

Fig 1.20

$\therefore$ m.i. of the cube about x-axis

$$\begin{aligned} A &= \sum \frac{2a^2}{3} \times \text{mass of the strip} \\ &= \frac{2a^2}{3} \sum \text{(mass of the strips)} = \frac{2M}{3} a^2. \end{aligned}$$

From symmetry

$$A = B = C = \frac{2M}{3} a^2 \quad \text{and} \quad D = E = F = 0.$$

$\therefore$ Required m.i. about AB

$$= Al^2 + Bm^2 + Cn^2 = A(l^2 + m^2 + n^2) = \frac{2}{3} Ma^2.$$

(c) ***Find the m.i. of a Right Circular Cone of mass M about its generator.*** **[C.H. 1966; KUH '01]**

The equation of the cone is

$$x^2 + y^2 = z^2 \tan^2 \alpha,$$

where h is the height of the cone, a is the radius of the base and M be the mass of the cone.

Here

$$A = B = 3M \left(\frac{h^2}{5} + \frac{a^2}{20} \right) \text{ and } C = \frac{3}{10} Ma^2 \quad \text{[cf. (1.4.20) and (1.4.23)]}$$

and $D = E = F = 0$.

Let us calculate the m.i. of the cone about the generator OP. If l, m, n be the d.c's of OP, then

$$n = \cos \alpha = \frac{OR}{OP} = \frac{h}{\sqrt{h^2 + a^2}}.$$

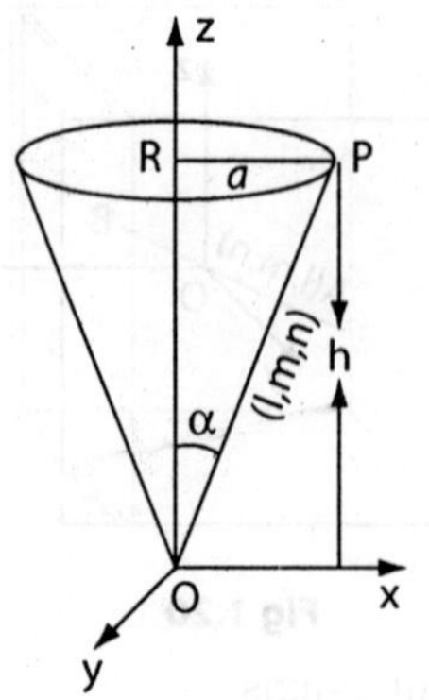

Fig 1.21

$\therefore$ Required m.i. about OP

$$\begin{aligned}
&= Al^2 + Bm^2 + Cn^2 = A(l^2+m^2) + Cn^2 \qquad [\because A = B] \\
&= A(1-n^2) + Cn^2 = A - (A-C)n^2 \\
&= 3M\left(\frac{h^2}{5}+\frac{a^2}{20}\right) - \left[3M\left(\frac{h^2}{5}+\frac{a^2}{20}\right) - \frac{3}{10}Ma^2\right]\cdot\frac{h^2}{h^2+a^2} \\
&= 3M\left(\frac{h^2}{5}+\frac{a^2}{20}\right)\left[1-\frac{h^2}{a^2+h^2}\right] + \frac{3Ma^2}{10}\cdot\frac{h^2}{h^2+a^2} \\
&= 3M\left(\frac{h^2}{5}+\frac{a^2}{20}\right)\left[\frac{a^2}{a^2+h^2}\right] + \frac{3Ma^2}{10}\cdot\frac{h^2}{h^2+a^2} \\
&= \frac{3Ma^2}{a^2+h^2}\left[\frac{h^2}{5}+\frac{a^2}{20}+\frac{h^2}{10}\right] \\
&= \frac{3Ma^2}{20}\cdot\frac{6h^2+a^2}{h^2+a^2}.
\end{aligned}$$

(*d*) ***Show that the m.i. of elliptic area of mass M and semi-axes a and b about a diameter of length 2r is*** $\frac{M}{4}\frac{a^2b^2}{r^2}$.

Let the equation of the ellipse is $\frac{x^2}{a^2}+\frac{y^2}{b^2}=1$.

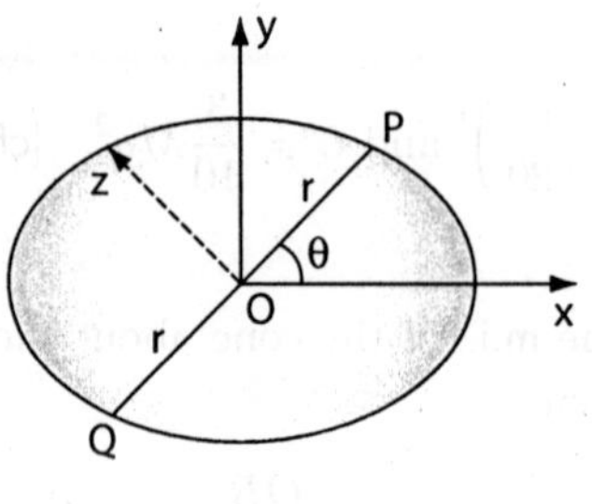

Fig 1.22

We know that m.i. of the elliptic area of semi-axes a and b about Ox and Oy axes are

$$A = \frac{M}{4}b^2, \quad B = \frac{M}{4}a^2$$

and the product of inertia

$$F = 0. \tag{1.10.1}$$

Let PQ be the diameter of length $2r$ which makes an angle θ with Ox axis.

$\therefore$ d.c's of PQ are $(\cos\theta, \sin\theta, 0)$.

$\therefore$ The required m.i. of the elliptic area about PQ is

$$\begin{aligned} I &= Al^2 + Bm^2 - 2Flm \\ &= A\cos^2\theta + B\sin^2\theta - 2F\sin\theta\cos\theta \\ &= \frac{M}{4}[b^2\cos^2\theta + a^2\sin^2\theta]. \end{aligned}$$

As r is the length of the semi-diameter, the coordinates of P are

$$(r\cos\theta, r\sin\theta).$$

As P lies on the ellipse

$$\frac{x^2}{a^2} + \frac{y^2}{b^2} = 1$$

we get

$$\frac{r^2\cos^2\theta}{a^2} + \frac{r^2\sin^2\theta}{b^2} = 1, \quad \text{or,} \quad b^2\cos^2\theta + a^2\sin^2\theta = \frac{a^2b^2}{r^2}.$$

$$\therefore \quad I = \frac{M}{4}\,\frac{a^2b^2}{r^2}.$$

(*e*) ***Show that the* m.i. *of an ellipse of mass M and semi-axes a and b about a tangent is $\frac{5}{4}Mp^2$, where p is the perpendicular from the centre to the tangent. Show further that the sum of* m.i. *of an elliptic area about any two tangents at right angle is always constant.*** **[VUH 2001]**

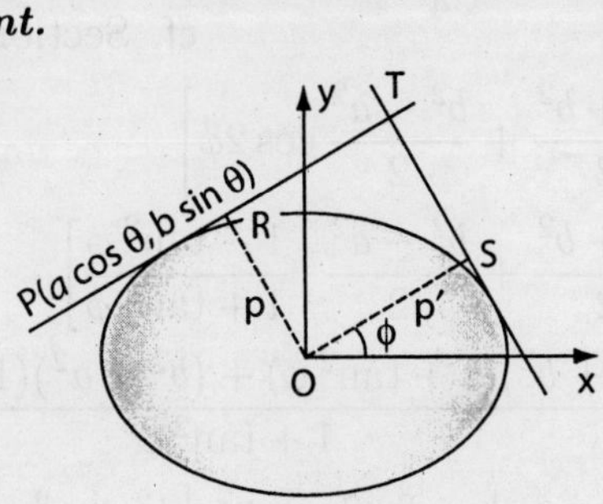

Fig 1.23

Let the equation of the ellipse is

$$\frac{x^2}{a^2} + \frac{y^2}{b^2} = 1. \tag{1.10.2}$$

We consider the tangent at $P(a\cos\theta, b\sin\theta)$.
The equation of the tangent at $P(a\cos\theta, b\sin\theta)$ is

$$\frac{x\cos\theta}{a} + \frac{y\sin\theta}{b} = 1. \tag{1.10.3}$$

$\therefore$ Perpendicular distance from centre to the tangent

$$p = \frac{1}{\sqrt{\frac{\cos^2\theta}{a^2} + \frac{\sin^2\theta}{b^2}}} = \frac{ab}{\sqrt{b^2\cos^2\theta + a^2\sin^2\theta}}. \tag{1.10.4}$$

The equation of the diameter parallel to the tangent at $P(a\cos\theta, b\sin\theta)$ is

$$\frac{x\cos\theta}{a} + \frac{y\sin\theta}{b} = 0.$$

$\therefore$ The slope of the diameter is

$$\tan\phi = -\frac{\frac{\cos\theta}{a}}{\frac{\sin\theta}{b}} = -\frac{b}{a}\cot\theta, \tag{1.10.5}$$

where ϕ is the angle made by the diameter parallel to the tangent at P with the x-axis.

$\therefore$ d.c's of the tangent at P are $(\cos\phi, \sin\phi, 0)$.
$\therefore$ I_0 = m.i. of the ellipse about the diameter

$$\begin{aligned}
&= A\cos^2\phi + B\sin^2\phi - 2F\sin\phi\cos\phi \\
&= \frac{Mb^2}{4}\cos^2\phi + \frac{Ma^2}{4}\sin^2\phi \quad [\because A = \tfrac{Mb^2}{4}, B = \tfrac{Ma^2}{4} \text{ and } F = 0 \\
&\qquad\qquad\qquad\qquad\qquad\qquad \text{cf. Section 1.4, problem (k)}] \\
&= \frac{M}{4}\left[\frac{a^2+b^2}{2} + \frac{b^2-a^2}{2}\cos 2\phi\right] \\
&= \frac{M}{4}\left[\frac{a^2+b^2}{2} + \frac{b^2-a^2}{2}\cdot\frac{1-\tan^2\phi}{1+\tan^2\phi}\right] \\
&= \frac{M}{8}\left[\frac{(a^2+b^2)(1+\tan^2\phi) + (b^2-a^2)(1-\tan^2\phi)}{1+\tan^2\phi}\right] \\
&= \frac{M}{8}\left[\frac{2b^2 + 2a^2\tan^2\phi}{1+\tan^2\phi}\right] = \frac{M}{4}\left[\frac{b^2 + a^2\cdot\frac{b^2}{a^2}\cot^2\theta}{1+\frac{b^2}{a^2}\cot^2\theta}\right] \quad \text{[by (1.10.5)]} \\
&= \frac{M}{4}\cdot\frac{a^2b^2(1+\cot^2\theta)}{a^2+b^2\cot^2\theta} = \frac{M}{4}\cdot\frac{a^2b^2}{b^2\cos^2\theta + a^2\sin^2\theta} \\
&= \frac{M}{4}p^2 \quad \text{[by (1.10.4)]}.
\end{aligned}$$

$$\therefore\ I = \text{m.i. about the tangent}$$
$$= I_0 + Mp^2 = \frac{M}{4}p^2 + Mp^2 = \frac{5}{4}Mp^2.$$

Let I_1 = m.i. of the ellipse about a tangent perpendicular to the first tangent and let two tangents meet at T, then T lies on the diameter of the director circle $x^2 + y^2 = a^2 + b^2$, so $OT^2 = a^2 + b^2$.

$$\therefore\quad I_1 = \frac{5}{4}Mp_1^2,$$

where p_1 is the perpendicular on the second tangent.

$$\begin{aligned}\therefore\quad I + I_1 &= \frac{5}{4}M(p^2 + p_1^2)\\ &= \frac{5}{4}M \cdot OT^2, \quad \text{(since the quadrilateral is rectangle)}\\ &= \frac{5}{4}M(a^2 + b^2) = \text{constant.}\end{aligned}$$

1.11 Theorem of Parallel and Perpendicular Axes

(*a*) *Theorem of Parallel Axes.* If A_0, B_0, C_0 and D_0, E_0, F_0 are the moments and products of inertia of a body with respect to axes Gx_0, Gy_0, Gz_0 at the centre of mass $G(a, b, c)$ then the moments and products of inertia of the body with respect to the parallel axes Ox, Oy, Oz at the point $O(0, 0, 0)$; A, B, C and D, E, F respectively are given by [KUH '08]

$$A = A_0 + M(b^2 + c^2) \quad B = B_0 + M(c^2 + a^2) \quad C = C_0 + M(a^2 + b^2)$$
$$D = D_0 + Mbc \qquad E = E_0 + Mca \qquad F = F_0 + Mab,$$

where M is the mass of the body.

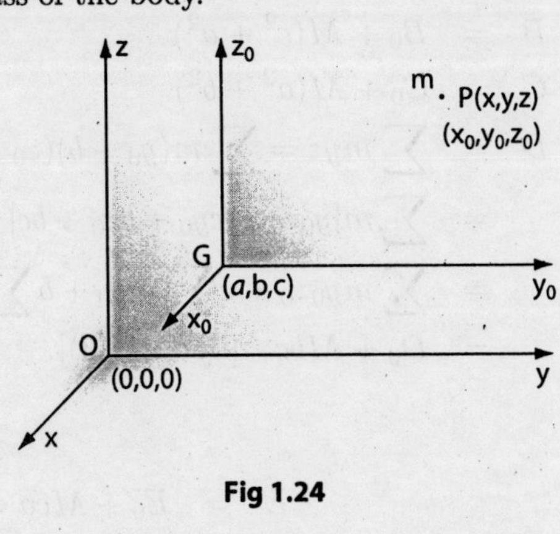

Fig 1.24

Let (x_0, y_0, z_0) be the coordinates of an elementary mass m at P with respect to the rectangular frame Gx_0, Gy_0, Gz_0 at the centre of mass G and (x, y, z) be the coordinates of the same mass with respect to the rectangular frame Ox, Oy, Oz at O, parallel to Gx_0, Gy_0, Gz_0. Let (a, b, c) be the coordinates of G with respect to Ox, Oy and Oz.

$$\therefore \left.\begin{aligned} x &= x_0 + a \\ y &= y_0 + b \\ z &= z_0 + c \end{aligned}\right\} \tag{1.11.1}$$

and

$$\sum mx_0 = \sum my_0 = \sum mz_0 = 0. \tag{1.11.2}$$

$$\therefore \left.\begin{aligned} \sum mx &= \sum m(x_0 + a) = \sum mx_0 + \sum ma = Ma \\ \sum my &= \sum m(y_0 + b) = \sum my_0 + \sum mb = Mb \\ \sum mz &= \sum m(z_0 + c) = \sum mz_0 + \sum mc = Mc. \end{aligned}\right\}. \tag{1.11.3}$$

Now

$$\begin{aligned} A &= \sum m(y^2 + z^2) = \sum m[(y_0 + b)^2 + (z_0 + c)^2] \\ &= \sum m(y_0^2 + z_0^2) + 2b\sum my_0 + 2c\sum m_0 z_0 + \sum m(b^2 + c^2) \\ &= \sum m(y_0^2 + z_0^2) + (b^2 + c^2)\sum m \quad \text{[by (1.11.3)]} \\ &= A_0 + M(b^2 + c^2). \end{aligned}$$

Similarly

$$\begin{aligned} B &= B_0 + M(c^2 + a^2) \\ C &= C_0 + M(a^2 + b^2) \\ \text{and} \quad D &= \sum myz = \sum m(y_0 + b)(z_0 + c) \\ &= \sum m[y_0 z_0 + cy_0 + bz_0 + bc] \\ &= \sum my_0 z_0 + c\sum my_0 + b\sum mz_0 + bc\sum m \\ &= D_0 + Mbc \quad \text{[by (1.11.3)]}. \end{aligned}$$

Similarly

$$\begin{aligned} E &= E_0 + Mca \\ \text{and} \quad F &= F_0 + Mab. \end{aligned}$$

Thus the moments and products of inertia about Ox, Oy, Oz are equal to the moments and products of inertia about parallel axes Gx_0, Gy_0, Gz_0

through centre of mass G + the moments and products of inertia of the whole mass M placed at centre of mass G about $\boldsymbol{Ox}$, $\boldsymbol{Oy}$ and $\boldsymbol{Oz}$.

(*b*) ***Theorem of perpendicular axes.*** If A and B be the moments of inertia of a distribution of matter in one plane about any two perpendicular lines in the plane and C be the moment of inertia of the same distribution of matter about a line perpendicular to the plane, then $C = A + B$.

Let the plane of the distribution be taken as the plane $z = 0$. Let m be an elementary mass on the plane, for which $z = 0$. Let any two perpendicular lines be taken in the plane as x and y-axes and any line perpendicular to the plane and through the interesting of x and y-axes as z-axis.

Now

$$\begin{aligned} A &= \sum m(y^2+z^2) = \sum my^2 \quad \text{as } z=0 \\ B &= \sum m(z^2+x^2) = \sum mx^2 \quad \text{as } z=0. \\ \therefore \quad C &= \sum m(y^2+z^2) = \sum my^2 + \sum mz^2 = A+B. \end{aligned}$$

Hence the theorem.

1.12 Solved Problems

▶ **Problem 1.12.1** *Moment of inertia of a thin uniform rod of mass M about one of its extremities.*

Solution: Let M be the mass of the rod AB of length $2a$.

I_0 = m.i. of the rod about an axis through c.g. G or the middle point of the rod perpendicular to it

$= \frac{Ma^2}{3}$,

i.e., $I_0 = \frac{Ma^2}{3}$.

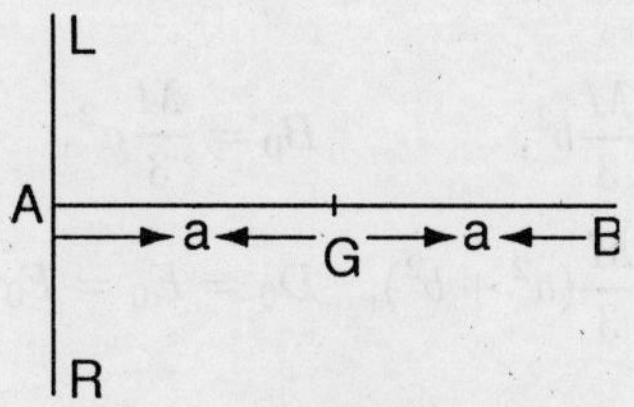

Fig 1.25

$\therefore$ I = m.i. of the rod about a parallel line LR through the extremity A.
$= I_0 + Ma^2 = \frac{1}{3}Ma^2 + Ma^2 = \frac{4}{3}Ma^2$.

► **Problem 1.12.2** *Moment of inertia of a circular plate of mass M and radius a about a tangent.*

Solution: Let M be the mass of the plate and a be its radius.

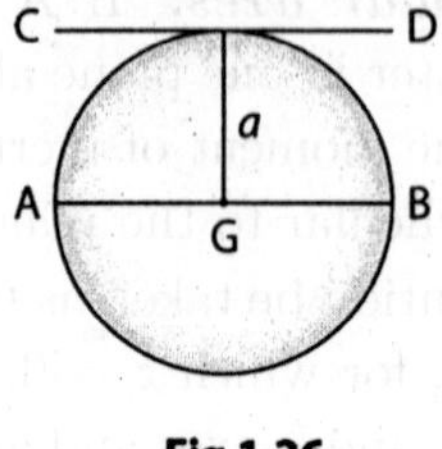

Fig 1.26

∴ m.i. of the plate about the diameter AB through c.g. G and parallel to the tangent CD

$$I_0 = \frac{M}{4}a^2.$$

∴ m.i. of the plate about CD

$$= I_0 + Ma^2 = \frac{M}{4}a^2 + Ma^2 = \frac{5}{4}Ma^2.$$

► **Problem 1.12.3** *Moments of inertia of a rectangular plate of mass M and sides 2a and 2b about a line through a corner.*

Solution: Let $ABCD$ be a rectangular plate of mass M and of sides $2a$ and $2b$. Let G_0x_0 and G_0y_0 taken as axes at centre G_0 of the plate and parallel to the sides AB and AD.

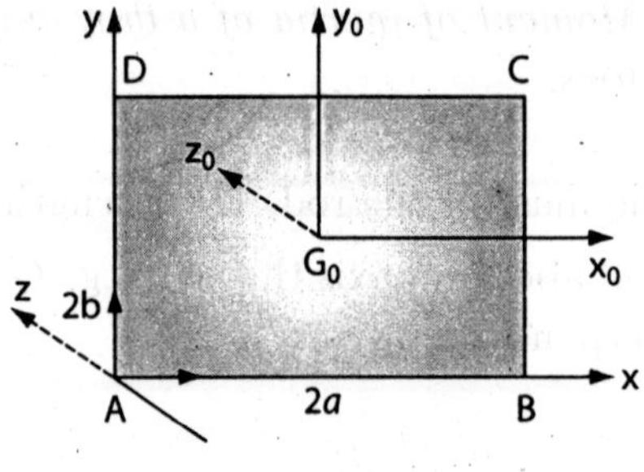

Fig 1.27

Here

$$A_0 = \frac{M}{3}b^2, \qquad B_0 = \frac{M}{3}a^2,$$

$$C_0 = \frac{M}{3}(a^2 + b^2), \quad D_0 = E_0 = F_0 = 0.$$

$$\therefore \quad A = A_0 + Mb^2 = \frac{M}{3}b^2 + Mb^2 = \frac{4M}{3}b^2$$

$$B = B_0 + Ma^2 = \frac{M}{3}a^2 + Ma^2 = \frac{4M}{3}a^2$$

$$C = C_0 + M(a^2+b^2) = \frac{M}{3}(a^2+b^2) + M(a^2+b^2)$$

$$= \frac{4}{3}M(a^2+b^2)$$

$$D = E = 0$$

$$\text{and} \quad F = F_0 + Mab = Mab,$$

where A, B, C and D, E, F are the moments and products of inertia about the parallel axes at the corner A.

► **Problem 1.12.4** *Moments and products of inertia of a rectangular parallelopiped of mass M and sides 2a, 2b, 2c about rectangular axes at a corner.*

Solution: Let $ABCDEFGH$ be a rectangular parallelopiped of mass M and of sides $2a$, $2b$, $2c$. Let A_0, B_0, C_0 and D_0, E_0, F_0 are moments and products of inertia of the parallelopiped about rectangular axes G_0x_0, G_0y_0, G_0z_0 at the centre G_0. Let A, B, C and D, E, F are the moments and products of inertia of the parallelopiped about axes Ax, Ay, Az at a corner A parallel to G_0x_0, G_0y_0, G_0z_0.

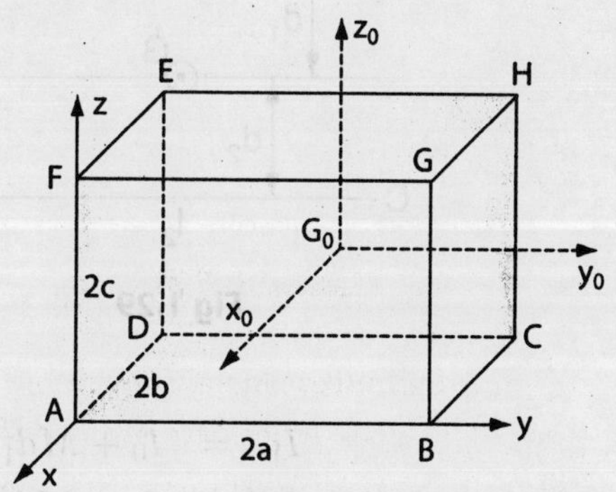

Fig 1.28

We know

$$A_0 = \frac{M}{3}(b^2+c^2), \quad B_0 = \frac{M}{3}(c^2+a^2),$$

$$C_0 = \frac{M}{3}(a^2+b^2), \quad D_0 = E_0 = F_0 = 0.$$

$$\therefore \quad A = A_0 + M(b^2+c^2) = \frac{M}{3}(b^2+c^2) + M(b^2+c^2) = \frac{4M}{3}(b^2+c^2)$$
$$B = B_0 + M(c^2+a^2) = \frac{M}{3}(c^2+a^2) + M(c^2+a^2) = \frac{4M}{3}(c^2+a^2)$$
$$C = C_0 + M(a^2+b^2) = \frac{M}{3}(a^2+b^2) + M(a^2+b^2) = \frac{4M}{3}(a^2+b^2)$$
$$D = D_0 + Mbc = Mbc$$
$$E = E_0 + Mca = Mca$$
$$F = F_0 + Mab = Mab.$$

As $D_0 = E_0 = F_0 = 0$.

► **Problem 1.12.5** *If I_1 and I_2 be the m.i. of a body of mass M about two parallel lines then $I_1 - Md_1^2 = I_2 - Md_2^2$, where d_1 and d_2 are the distances of c.g. from two straight lines.*

Solution: Let I_1 and I_2 be the m.i. of the body about two parallel straight lines AB and CD. Let I_0 be the m.i. of the body about a line through its c.g. and parallel to the straight lines AB and CD.

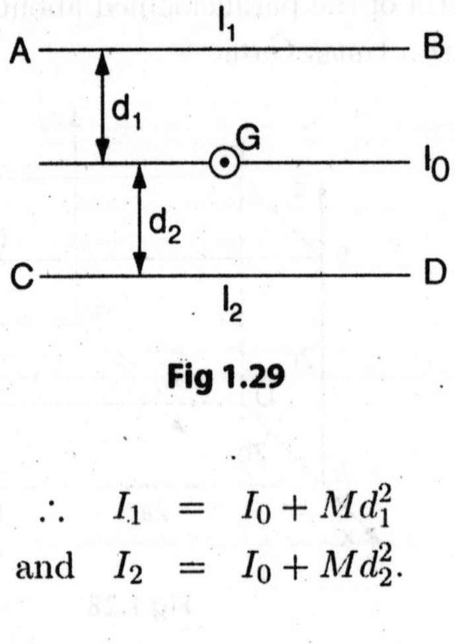

Fig 1.29

$$\therefore \quad I_1 = I_0 + Md_1^2$$
$$\text{and} \quad I_2 = I_0 + Md_2^2.$$

$$\therefore \quad I_1 - Md_1^2 = I_2 - Md_2^2 = I_0.$$
$$\therefore \quad I_1 - Md_1^2 = I_2 - Md_2^2.$$

1.13 Principal Axis

If there exist three rectangular coordinate axis at a point be such that all the product of inertias Σmyz, Σmzx and Σmxy vanish, then these rectangular axes are said to be the *principal axes* of the body at the given point.

The moments of inertia about the principal axes are called the *Principal moments of Inertia.*

The three planes each containing two principal axes are called the *principal planes* at the point.

Thus, if A, B, C denote the principal m.i. at the point O, then m.i. I about a line OP, having d.c's l, m, n is given by

$$I = Al^2 + Bm^2 + Cn^2.$$

Remark 1.13.1 *Planes of symmetry are the principal planes.*

Let a homogeneous body be symmetrical about the xy-plane. Thus for every element of mass m on one side of the plane, whose coordinates are (x, y, z) we can find another element of same mass m on the other side of the plane whose coordinates are $(x, y, -z)$. Hence for this body

$$\begin{aligned} \sum mxz &= 0 \\ \text{and} \quad \sum myz &= 0. \end{aligned}$$

Similarly for the other planes. Thus it is clear that the line of symmetry of a body are principal axes.

In case of lamina in the xy-plane, the z-coordinate of every element is zero; then

$$\begin{aligned} \sum mxz &= 0 \\ \text{and} \quad \sum myz &= 0. \end{aligned}$$

Hence the plane of the lamina is the principal plane.

In case of ellipsoid, the three principal section being the planes of symmetry, are the principal planes and the three principal diameters are the principal axes of inertia.

For the elliptic lamina, the major and minor axes are the principal axes of inertia.

1.14 Momental Ellipsoid and Principal Axes of Momental Ellipsoid

[KUH '02, '05, '06]

Definitions: ***Momental ellipsoid*** and ***momental cylinder.***

The *momental ellipsoid* or the ellipsoid of inertia at any point O of a given distribution of mass is defined as the *locus of a point P* such that OP

is inversely proportional to the square root of the moment of inertia of the distribution about the line OP, provided that there is no line through O about which the moment of inertia vanishes. If there is a line through O about which the moment of inertia vanishes, the locus of P on lines through O other than the line is called the *momental cylinder.* The point O may or may not be a point of the material distribution.

On a straight line through O a point P is taken such that the square root of the moment of inertia of the body about the line I is inversely proportional to OP

$$\text{i.e.,}\quad \sqrt{I} \propto \frac{1}{OP}, \quad \text{or,} \quad I = \frac{M\epsilon^4}{OP^2}, \tag{1.14.1}$$

where ϵ is a constant and M is the mass of the body. Then the locus of P is called the *momental ellipsoid.*

To prove that the locus of P is an ellipsoid. **[CUH 1996]**

Let A, B, C and D, E, F are the moments and products of inertia of a body with respect to Ox, Oy, Oz as axes and (l, m, n) be the d.c's of OP, then the m.i. of the body about OP is given by

$$\begin{aligned} I &= Al^2 + Bm^2 + Cn^2 - 2mnD - 2nlE - 2lmF \\ &= \frac{M\epsilon^4}{OP^2} \quad [\text{by (1.14.1)}]. \end{aligned} \tag{1.14.2}$$

If (x, y, z) be the coordinates of P, then $x = OP \cdot l$, $y = OP \cdot m$ and $z = OP \cdot n$.

$\therefore$ The equation (1.14.2) becomes to

$$Ax^2 + By^2 + Cz^2 - 2Dyz - 2Ezx - 2Fxy = M\epsilon^4, \tag{1.14.3}$$

which is the locus of P. As A, B, C are m.i. and are always positive, then the equation (1.14.3) essentially represents an ellipsoid. This ellipsoid is called the *momental ellipsoid* at O.

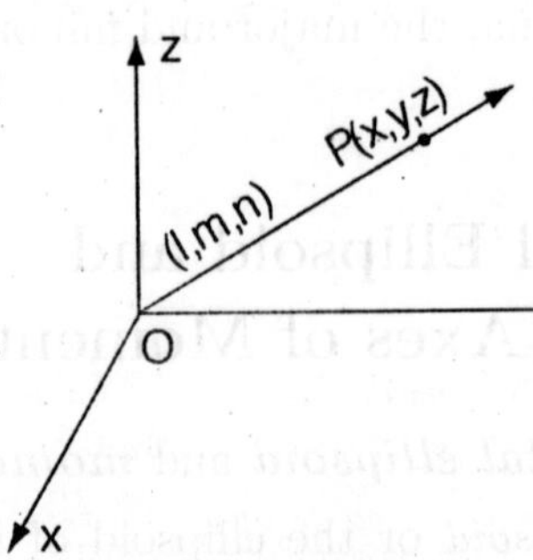

Fig 1.30

Now we know that for every ellipsoid there exists three mutually perpendicular diameters such that referred to these diameters as axes the equation of the ellipsoid will be free from the terms yz, zx and xy. These axes are the principal axes of the ellipsoid. Thus referred to the principal axes at O, the equation of the momental ellipsoid (1.14.3) takes the form

$$Ax^2 + By^2 + Cz^2 = M\epsilon^4. \tag{1.14.4}$$

Thus for every body there exists a set of three mutually perpendicular axes, at every point O, which are the three principal diameters of the momental ellipsoid at O such that the products of inertia of the body about these axes, taken in pair, vanish.

Remark 1.14.1 *When the three principal moments of inertia at any point O are equal, then the ellipsoid (1.14.4) becomes to*

$$A(x^2 + y^2 + z^2) = M\epsilon^4, \tag{1.14.5}$$

which is a sphere.

Remark 1.14.2 *If A and B the moments of inertia of a plane lamina with respect to rectangular axes Ox and Oy on the plane and F be the product of inertia of the lamina with respect to these two axes, then the equation of the momental ellipse is*

$$Ax^2 + By^2 - 2Fxy = M\epsilon^4. \tag{1.14.6}$$

Remark 1.14.3 *In case of a thin straight rod or a line distribution of mass, we take the line as x-axes. Then for every elementary mass m on the line $y = z = 0$.*

$$\begin{aligned}
\therefore\ A &= \sum m(y^2 + z^2) & B &= \sum m(z^2 + x^2) & C &= \sum m(x^2 + y^2)\\
&= 0 & &= \sum mx^2 & &= \sum mx^2 = B\\
D &= \sum myz & E &= \sum mzx & F &= \sum mxy\\
&= 0 & &= 0 & &= 0.
\end{aligned}$$

Thus the equation (1.14.3), i. e., the momental quadratic of the line distribution at O is

$$B(y^2 + z^2) = M\epsilon^4, \tag{1.14.7}$$

which is a right circular cylinder and called momental cylinder with its axis along the thin rod. This is a surface of revolution about x-axis. Thus every straight line perpendicular to the rod at any point on it is a principal axis at that point.

Determination of Principal Axes

Let

$$2f(l,m,n) = Al^2 + Bm^2 + Cn^2 - 2Dmn - 2Enl - 2Flm = I.$$

The d.c.'s of the principal axes (l, m, n) are given by

$$\frac{\frac{\partial f}{\partial l}}{l} = \frac{\frac{\partial f}{\partial m}}{m} = \frac{\frac{\partial f}{\partial n}}{n},$$

or, $$\frac{Al - Fm - En}{l} = \frac{-Fl + Bm - Dn}{m} = \frac{-El - Dm + Cn}{n}$$

$$= \frac{l(Al - Fm - En) + m(-Fl + Bm - Dn) + n(-El - Dm + Cn)}{l^2 + m^2 + n^2}$$

$$= Al^2 + Bm^2 + Cn^2 - 2mnD - 2nlE - 2lmF = I.$$

$$\begin{aligned} \therefore \quad (A - I)l - Fm - En &= 0 \\ -Fl + (B - I)m - Dn &= 0 \\ -El - Dm + (C - I)n &= 0. \end{aligned}$$

Eliminating l, m, n we get

$$\begin{vmatrix} A - I & -F & -E \\ -F & B - I & -D \\ -E & -D & C - I \end{vmatrix} = 0. \tag{1.14.8}$$

The equation (1.14.8) is a cubic equation in I, which gives three values of I. To each of which there corresponds a principal axis, i.e., its gives moments of inertia about principal axes of the momental ellipsoid (1.14.3). These moments of inertia are called the principal *moments of inertia.*

Remark 1.14.4 *The equation* (1.14.8) *can be written as*

$$I^3 - (A + B + C)I^2 + (AB + BC + CA - D^2 - E^2 - F^2)I$$
$$-(ABC - 2DEF - AD^2 - BE^2 - CF^2) = 0.$$

Thus the following quantities are invariants:

(i) $A + B + C$;

(ii) $AB + BC + CA - D^2 - E^2 - F^2$ *and*

(iii) $ABC - 2DEF - AD^2 - BE^2 - CF^2$.

1.15 Solved Problems on Momental Ellipsoid

▶ **Problem 1.15.1** *Prove that the momental ellipsoid at the centre of the elliptic plate is*

[KUH '05]

$$\frac{x^2}{a^2}+\frac{y^2}{b^2}+\left(\frac{1}{a^2}+\frac{1}{b^2}\right)z^2 = \textit{constant}.$$

Solution: We take two principal axes Ox and Oy along the major and minor axes at O and z-axis perpendicular to the plate.

$$\therefore \quad A=\frac{Mb^2}{5}, \quad B=\frac{Ma^2}{5}, \quad C=\frac{M}{5}(a^2+b^2)$$

and $D=E=F=0$.

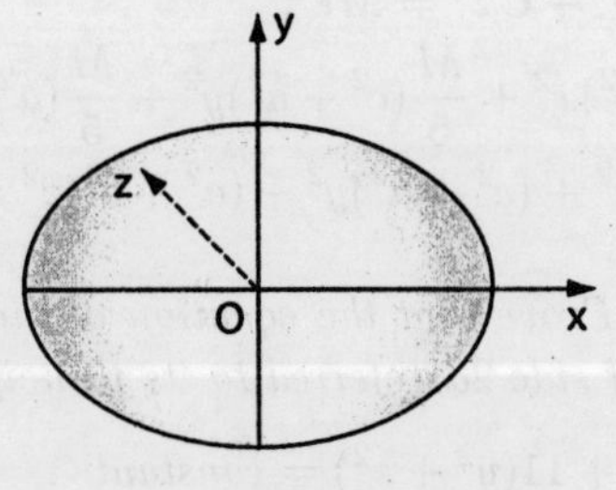

Fig 1.31

$\therefore$ The momental ellipsoid is

$$Ax^2+By^2+Cz^2=M\epsilon^4,$$

$$\text{or,} \quad \frac{M}{5}b^2\cdot x^2+\frac{Ma^2}{5}y^2+\frac{M}{5}(a^2+b^2)\cdot z^2=M\epsilon^4,$$

$$\text{or,} \quad b^2x^2+a^2y^2+(a^2+b^2)z^2=\text{constant},$$

$$\text{or,} \quad \frac{x^2}{a^2}+\frac{y^2}{b^2}+\left(\frac{1}{a^2}+\frac{1}{b^2}\right)z^2=\text{constant}.$$

▶ **Problem 1.15.2** *Prove that the momental ellipsoid at the centre of the solid ellipsoid is*

$$(b^2+c^2)x^2+(c^2+a^2)y^2+(a^2+b^2)z^2=\textit{constant}.$$

Solution: Let the equation of the ellipsoid be

$$\frac{x^2}{a^2}+\frac{y^2}{b^2}+\frac{z^2}{c^2}=1.$$

We take the axes along the principal axes of the ellipsoid.

$$\therefore \quad A=\frac{M}{5}(b^2+c^2), \quad B=\frac{M}{5}(c^2+a^2), \quad C=\frac{M}{5}(a^2+b^2)$$

and $D=E=F=0$.

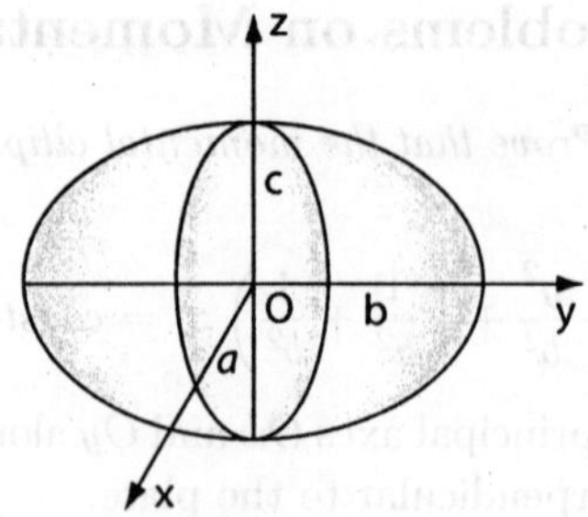

Fig 1.32

$\therefore$ The momental ellipsoid is

$$Ax^2 + By^2 + Cz^2 = M\epsilon^4,$$
$$\text{or,}\quad \frac{M}{5}(b^2+c^2)x^2 + \frac{M}{5}(c^2+a^2)y^2 + \frac{M}{5}(a^2+b^2)z^2 = M\epsilon^4,$$
$$\text{or,}\quad (b^2+c^2)x^2 + (c^2+a^2)y^2 + (a^2+b^2)z^2 = \text{constant}.$$

► **Problem 1.15.3** *Prove that the equation of the momental ellipsoid at the corner of a cube, of side 2a referred to its principal axes is*

$$2x^2 + 11(y^2+z^2) = \textit{constant}. \qquad \textbf{[BUH 1997]}$$

Solution: Let G be the c.g. of the cube of sides $2a$ and M be the mass of the cube. Let O be a corner of the cube.

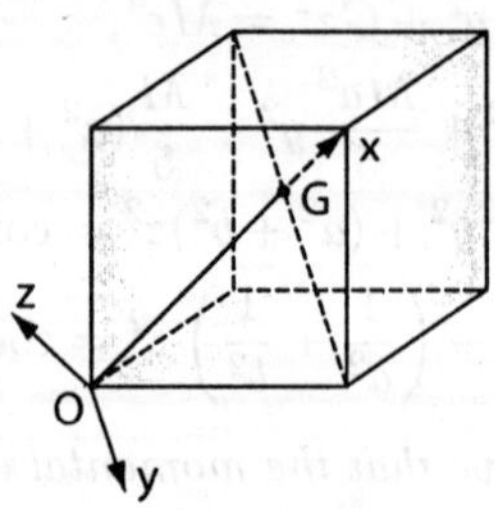

Fig 1.33

We know that moment of inertia of a cube of sides $2a$ about any axes through c.g. G is $\frac{2}{3}Ma^2$,

$$\text{i.e.,}\quad A_0 = B_0 = C_0 = \frac{2}{3}Ma^2.$$

We take OG as x-axes and two mutually perpendicular lines Oy and Oz through O as the axes of y and z.

$$\therefore\quad OG = \sqrt{a^2+a^2+a^2} = \sqrt{3}a.$$

Thus
the m.i. of the cube about Ox,

$$A = A_0 + 0 = \frac{2}{3}Ma^2;$$

the m.i. of the cube about Oy,

$$B = B_0 + M \cdot OG^2 = \frac{2}{3}Ma^2 + M \cdot 3a^2 = \frac{11}{3}Ma^2;$$

the m.i. of the cube about Oz,

$$C = C_0 + M \cdot OG^2 = \frac{2}{3}Ma^2 + M \cdot 3a^2 = \frac{11}{3}Ma^2.$$

Thus momental ellipsoid is

$$Ax^2 + By^2 + Cz^2 = \text{constant},$$
$$\text{or,}\quad \frac{2}{3}Ma^2x^2 + \frac{11}{3}Ma^2(y^2 + z^2) = \text{constant},$$
$$\text{or,}\quad 2x^2 + 11(y^2 + z^2) = \text{constant}.$$

► **Problem 1.15.4** *Show that the momental ellipsoid at a point on the rim of a hemisphere*

$$2x^2 + 7(y^2 + z^2) - \frac{15}{4}xz = constant. \qquad \text{[VUH '06; KUH '07]}$$

Solution: We take the diameter of the base of the hemisphere OO' as x-axis and y-axis is the tangent at O on the rim and z-axis perpendicular to xy-plane, where O' is the centre of the base of the hemisphere.

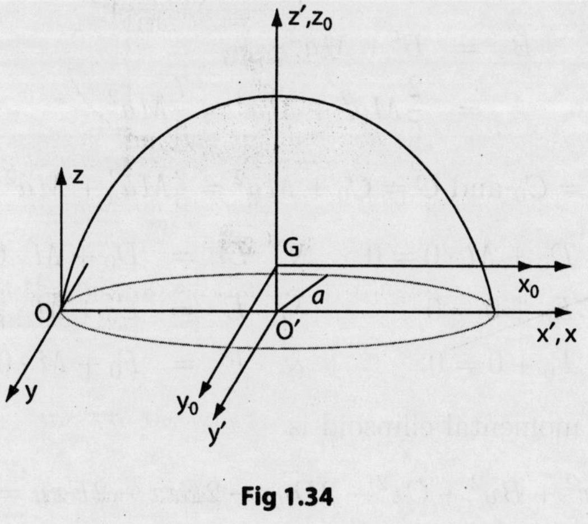

Fig 1.34

Let a be the radius of the base of the hemisphere.

$$\therefore \quad O'G = \frac{3}{8}a, \tag{1.15.1}$$

where G is the c.g. of the hemisphere.

Let A, B, C and D, E, F are the moments and products of inertia of the hemisphere with respect to Ox, Oy and Oz as axes and A', B', C' and D', E', F' are the moments and products of inertia with respect to axes $O'x'$, $O'y'$ and $O'z'$ at centre of the base O' and A_0, B_0, C_0 and D_0, E_0, F_0 are the moments and products of inertia with respect to axes Gx_0, Gy_0 and Gz_0 at c.g. G.

Again we know that

$$A' = B' = C' = \frac{2}{5}Ma^2 \text{ and } D' = E' = F' = 0. \text{ [cf. Art. 1.4(j)]} \tag{1.15.2}$$

Now

$$\begin{aligned} A' &= A_0 + M \cdot O'G^2 = A_0 + M \cdot \left(\frac{3}{8}a\right)^2 \\ \text{and} \quad A &= A_0 + M\left(\frac{3a}{8}\right)^2. \quad \therefore \quad A = A' = \frac{2}{3}Ma^2. \\ B' &= B_0 + M\left(\frac{3}{8}a\right)^2 \qquad (1.15.3) \\ \text{and} \quad B &= B_0 + M\left[a^2 + \left(\frac{3}{8}a\right)^2\right]. \\ \therefore \quad B &= B' + Ma^2 \qquad (1.15.4) \\ &= \frac{2}{5}Ma^2 + Ma^2 = \frac{7}{5}Ma^2. \end{aligned}$$

Again $C' = C_0$ and $C = C_0 + Ma^2 = \frac{2}{5}Ma^2 + Ma^2 = \frac{7}{5}Ma^2$.

$$\begin{aligned} D' &= D_0 + M \cdot 0 = 0 \quad & \& \quad D &= D_0 + M \cdot 0 \cdot \tfrac{3}{8}a = D_0 = 0 \\ E' &= E_0 + 0 = 0 \quad & \& \quad E &= E_0 + M \cdot \tfrac{3}{8}a \cdot a = \tfrac{3}{8}Ma^2 \\ F' &= F_0 + 0 = 0. \quad & \& \quad F &= F_0 + M \cdot 0 = 0. \end{aligned}$$

Thus the momental ellipsoid is

$$Ax^2 + By^2 + Cz^2 - 2Dyz - 2Ezx - 2Fxy = \text{constant}.$$

$$\frac{2}{5}Ma^2 \cdot x^2 + \frac{7}{5}Ma^2 \cdot y^2 + \frac{7}{5}Ma^2z^2 - 0 - 2 \cdot \frac{3}{8}Ma^2zx - 2 \cdot 0 = \text{constant}.$$

$$2x^2 + 7(y^2 + z^2) - \frac{15}{4}zx = \text{constant}.$$

► **Problem 1.15.5** *Prove that the momental ellipsoid at a point on the circular edge of a solid cone is*

$$(3a^2 + 2h^2)x^2 + (23a^2 + 2h^2)y^2 + 26a^2z^2 - 10ahxz = \textit{constant}.$$

Solution: Let O be a point on the circular edge of the solid cone of height h having semi-vertical angle α and a be the radius of the circular edge.

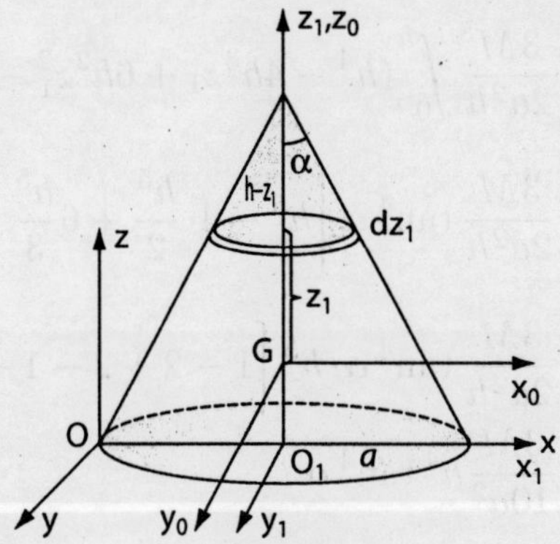

Fig 1.35

Let A, B, C and D, E, F are the moments and products of inertia of the cone with respect to the axes Ox, Oy and Oz. A_1, B_1, C_1 and D_1, E_1, F_1 be the moments and products of inertia of the cone with respect to O_1x_1, O_1y_1 and O_1z_1 as axes at O_1, the centre of the circular edge. A_0, B_0, C_0 and D_0, E_0, F_0 are the moments and products of inertia of the cone with respect to axes Gx_0, Gy_0, Gz_0 at c.g. G of the cone.

$$\therefore \quad O_1G = .\frac{h}{4}.$$

Let M be the mass of the cone and m mass per unit value of the cone.

$$\therefore \quad m = \frac{M}{\frac{1}{3}\pi a^2 \cdot h} = \frac{3M}{\pi a^2 h}.$$

Now from symmetry $A_1 = B_1$

$$\therefore \quad \sum m(y_1^2 + z_1^2) = \sum m(z_1^2 + x_1^2).$$

$$\therefore \quad \sum my_1^2 = \sum mx_1^2 = \frac{1}{2}\left[\sum m(x_1^2 + y_1^2)\right] = \frac{C_1}{2}.$$

Let us consider a circular strip parallel to the circular edge of thickness dz_1 at a height z_1 from the centre of the circular edge.

$$\therefore\ C_1 = \int_0^h \frac{3M}{\pi a^2 h}\cdot \pi (h-z_1)^2 \tan^2\alpha\cdot dz_1\cdot \frac{(h-z_1)^2}{2}\tan^2\alpha$$

$$= \frac{3M}{2a^2h}\tan^4\alpha \int_0^h (h-z_1)^4 dz_1$$

$$= \frac{3M}{2a^2h}\int_0^h (h^4 - 4h^3 z_1 + 6h^2 z_1^2 - 4hz_1^3 + z_1^4)dz_1$$

$$= \frac{3M}{2a^2h}\tan^4\alpha \left[h^5 - 4\cdot\frac{h^5}{2} + 6\frac{h^5}{3} - 4\frac{h^5}{4} + \frac{h^5}{5}\right]$$

$$= \frac{3M}{2a^2h}\tan^4\alpha\cdot h^5\left[1 - 2 + 2 - 1 + \frac{1}{5}\right]$$

$$= \frac{3M}{10a^2}h^4\tan^4\alpha$$

$$= \frac{3M}{10a^2}\cdot a^4$$

$$= \frac{3M}{10}a^2 \quad [\because a = h\tan\alpha].$$

$$\sum m z_1^2 = \frac{3M}{\pi a^2 h}\cdot \int_0^h \pi (h-z_1)^2 \tan^2\alpha\cdot z_1^2 dz_1$$

$$= \frac{3M}{a^2h}\int_0^h (h^2 - 2hz_1 + z_1^2)z_1^2 dz_1$$

$$= \frac{3M}{a^2h}\tan^2\alpha \int_0^h [h^2 z_1^2 - 2hz_1^3 + z_1^4]dz_1$$

$$= \frac{3M}{a^2h}\left[\frac{h^5}{3} - \frac{2h^5}{4} + \frac{h^5}{5}\right]\tan^2\alpha$$

$$= \frac{3M}{a^2h}\tan^2\alpha\cdot h^5\left[\frac{1}{3} - \frac{1}{2} + \frac{1}{5}\right]$$

$$= \frac{3M}{a^2h}\cdot h^5\cdot\tan^2\alpha\left[\frac{10-15+6}{30}\right]$$

$$= \frac{Mh^4}{10a^2}\cdot\tan^2\alpha = \frac{Mh^4\tan^2\alpha}{10\cdot h^2\tan^2\alpha}$$

$$= \frac{Mh^2}{10} \quad [\because a = h\tan\alpha].$$

$$
\begin{aligned}
\therefore \quad A_1 &= \sum m(y_1^2 + z_1^2) \\
&= \sum m y_1^2 + \sum m z_1^2 \\
&= \frac{C_1}{2} + \frac{Mh^2}{10} \\
&= \frac{3M}{20} a^2 + \frac{Mh^2}{10} \\
&= \frac{M(3a^2 + 2h^2)}{20} = B_1 \\
\text{and} \quad C_1 &= \frac{3Ma^2}{10}.
\end{aligned}
$$

Now

$$
A_1 = A_0 + M\left(\frac{h}{4}\right)^2,
$$

$$
A = A_0 + M\left(\frac{h}{4}\right)^2.
$$

$$
\therefore \quad A = A_1 = \frac{M}{20}(3a^2 + 2h^2) \quad \text{and} \quad B_1 = B_0 + M\left(\frac{h}{4}\right)^2
$$

$$
\begin{aligned}
B &= B_0 + M\left[\left(\frac{h}{4}\right)^2 + a^2\right] \\
&= B_0 + M\cdot\left(\frac{h}{4}\right)^2 + Ma^2 \\
&= B_1 + Ma^2 = \frac{M}{20}[3a^2 + 2h^2] + Ma^2 \\
&= \frac{M}{20}[23a^2 + 2h^2]. \\
C_1 &= C_0 + 0; \\
C &= C_0 + Ma^2 = \frac{3M}{10}a^2 + Ma^2 = \frac{13Ma^2}{10}.
\end{aligned}
$$

Again,

$$
D_1 = E_1 = F_1 = 0, \quad \therefore \ D_1 = D_0 = 0, \quad E_1 = E_0 + 0, \quad F_1 = F_0 + 0.
$$

$$
\therefore \quad D = 0, \quad E = E_0 + M\cdot\frac{h}{4}\cdot a = \frac{Mah}{4}, \quad F = F_0 + 0 = 0.
$$

Thus the momental ellipsoid is

$$Ax^2 + By^2 + Cz^2 - 2Dyz - 2Ezx - 2Fxy = \text{constant}.$$

$$\therefore \frac{M}{20}[3a^2 + 2h^2]x^2 + \frac{M}{20}[23a^2 + 2h^2]y^2 + \frac{13Ma^2}{10}z^2 - 2\cdot\frac{Mah}{4}zx$$
$$= \text{constant},$$

$$\text{or, } (3a^2 + 2h^2)x^2 + (23a^2 + 2h^2)y^2 + 26a^2z^2 - 10ahxz$$
$$= \text{constant}.$$

▶ **Problem 1.15.6** *Show that the momental ellipsoid at a point on the edge of the circular base of a thin hemispherical shell is*

$$2x^2 + 5(y^2 + z^2) - 3xz = \text{constant}.$$ **[KUH '06]**

Solution: Let M and a be the weight and radius of the hemispherical shell.

Let O' be the point on the circular edge. Let A_0, B_0, C_0, D_0, E_0 and F_0 are the moment of inertias and product of inertias about Ox_0, Oy_0 and Oz_0 and A, B, C and D, E, F are the moment of inertias and product of inertias

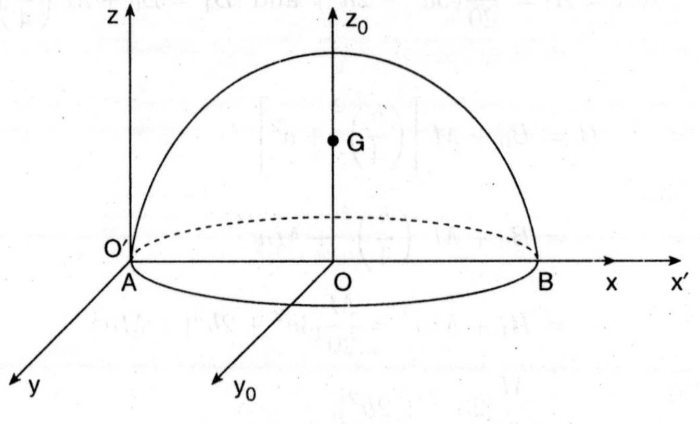

Fig 1.35(a)

about $O'x, O'y$ and $O'z$.

$$\therefore\ A_0 = B_0 = \frac{2}{5}Ma^2,$$

$$C_0 = \frac{2}{5}Ma^2$$

$$\text{and } D_0 = E_0 = F_0 = 0 \quad [\text{cf. §1.4(j)}]$$

$$\therefore\ A = A_0 = \frac{2}{3}Ma^2,$$

$$B = B_0 + Ma^2 = \frac{2}{3}Ma^2 + Ma^2$$

$$= \frac{5}{3}Ma^2$$

$$C = C_0 + Ma^2 = \frac{2}{3}Ma^2 + Ma^2$$

$$= \frac{5}{3}Ma^2$$

Again the coordinates of c.g. G is $\left(a, 0, \frac{1}{2}a\right)$

$$\therefore\ D = F = 0,$$

$$E = Ma\cdot\frac{1}{2}a = \frac{1}{2}Ma^2$$

Therefore the equation of the momental elliposoid is

$$\frac{2}{3}Ma^2x^2 + \frac{5}{3}Ma^2y^2 + \frac{5}{3}Ma^2z^2 - 2\cdot\frac{1}{2}Ma^2zx = \text{constant}$$

$$2x^2 + 5(y^2 + z^2) - 3zx = \text{constant.}$$

1.16 Condition that a given Straight Line may be a principal axis of the Material System at any point of its Length, and if so, find the other Two Principal Axes

OR

If A, B, C and D, E, F are the moments and products of inertia of material system with respect to axes Ox, Oy, Oz then z-axis is a principal axis at some point of its length if $D\bar{x} = E\bar{y}$, where $(\bar{x}, \bar{y}, \bar{z})$ are the coordinates of centre of gravity of the system.

If this condition is satisfied, then z-axis is a principal axis at the point $(0, 0, h)$ of it, where

$$h = \frac{D}{M\bar{x}} = \frac{E}{M\bar{y}}$$

and the other two principal axes at $(0, 0, h)$ are given by

$$\tan 2\theta = \frac{2F}{B - A}$$

where θ is the angle between one of the principal axis makes with Ox, where M is the mass of the body.

[CUH 1997, 1999, 2001, 2003, 2005, 2007; VUH 2002, 2005; KUH 2003]

The given straight is taken as z-axis and any point O on it is taken as origin. Let at some point $O'(0, 0, h)$ on z-axis Oz be a principal axis where $OO' = h$. Let other two principal axes be $O'x'$ and $O'y'$ and θ is the angle between Ox and $O'x'$. Let the coordinates of an elementary mass m at P be (x, y, z) with respect to axes Ox, Oy and Oz and (x', y', z') with respect to the axes $O'x', O'y'$ and $O'z'$. Let the coordinates of c.g. G of the material system be $(\bar{x}, \bar{y}, \bar{z})$.

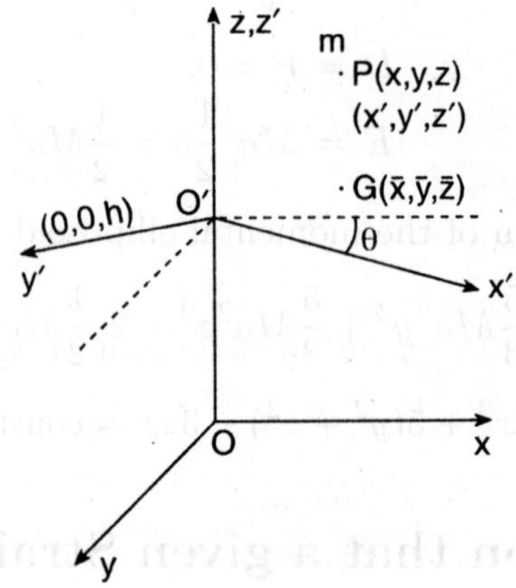

Fig 1.36

$$\therefore \sum mx = M\bar{x},$$

$$\sum my = M\bar{y},$$

$$\sum mz = M\bar{z}.$$

$$\left.\begin{aligned} \therefore\ x' &= x\cos\theta + y\sin\theta \\ y' &= y\cos\theta - x\sin\theta \\ z' &= z - h \end{aligned}\right\} \tag{1.16.1}$$

As $O'x', O'y'$ and $O'z'$ are principal axes, then

$$D' = \Sigma my'z' = 0 \tag{1.16.2}$$

$$E' = \Sigma mz'x' = 0 \tag{1.16.3}$$

$$F' = \Sigma mx'y' = 0. \tag{1.16.4}$$

From (1.16.2),

$$D' = \Sigma m(-x\sin\theta + y\cos\theta)(z-h) = 0,$$

$$\text{or,}\quad -\sin\theta\Sigma mxz + \cos\theta\Sigma myz + h\sin\theta\Sigma mx - h\cos\theta\Sigma my = 0,$$

$$\text{or,}\quad -E\sin\theta + D\cos\theta = -Mh\sin\theta\cdot\bar{x} + Mh\cos\theta\cdot\bar{y}$$

$$= Mh\cos\theta\cdot\bar{y} - Mh\sin\theta\cdot\bar{x}. \tag{1.16.5}$$

From (1.16.3),

$$E' = \Sigma m(z-h)(x\cos\theta + y\sin\theta) = 0,$$

$$\text{or,}\quad E\cos\theta + D\sin\theta = Mh\bar{x}\cos\theta + Mh\bar{y}\sin\theta. \tag{1.16.6}$$

Multiplying (1.16.5) by $\cos\theta$ and (1.16.6) by $\sin\theta$ and then adding we get,

$$\begin{aligned} D &= Mh(-\bar{x}\sin\theta + \bar{y}\cos\theta)\cos\theta + Mh(\bar{x}\cos\theta + \bar{y}\sin\theta)\sin\theta \\ &= Mh\bar{y}. \end{aligned} \tag{1.16.7}$$

Similarly multiplying (1.16.5) by $-\sin\theta$ and (1.16.6) by $\cos\theta$ and adding we get.

$$E = Mh\bar{x}. \tag{1.16.8}$$

$\therefore\ D\bar{x} = E\bar{y}.$

Again by (1.16.7) and (1.16.8), we have

$$h = \frac{D}{M\bar{y}} = \frac{E}{M\bar{x}} \quad (\text{if } \bar{x} = \bar{y} \neq 0).$$

From (1.16.4) we get

$$F' = \sum m(x\cos\theta + y\sin\theta)(-x\sin\theta + y\cos\theta) = 0,$$

$$\text{or,}\quad \sum m[-x^2\sin\theta\cos\theta + xy\cos^2\theta - xy\sin^2\theta + y^2\sin\theta\cos\theta] = 0,$$

$$\text{or,}\quad (\cos^2 - \sin^2\theta)\sum mxy + \sin\theta\cos\theta\sum m(y^2 - x^2) = 0,$$

$$\text{or,}\quad 2\cos 2\theta\sum mxy + \sin 2\theta\sum m[(y^2+z^2) - (z^2+x^2)] = 0,$$

$$\text{or,}\quad 2\cos 2\theta\cdot F + \sin 2\theta(A-B) = 0,$$

$$\text{or,}\quad \tan 2\theta = \frac{2F}{B-A}.$$

Remark 1.16.1 *If a straight line through the centre of gravity of a body be a principal axis at any point of its length, then it is a principal axis at all points of its length. If z-axis be the principal axis at O, then $D = E = 0$. Thus $h = 0$; hence there will be no O'. On the other hand, if $\bar{x} = \bar{y} = 0$, then h is indeterminate so that for every value of h, the z-axis is a principal axis through the c.g. G.*

1.17 Equimomental System [VUH 2003]

Two bodies or two material systems are said to be *equimomental* or *kinematically equivalent* if the moment of inertia of one body about any line is equal to the moment of inertia of the other body about the same line.

The conditions that two bodies or two material systems may be equimomental are:

(i) that they should have the same centre of gravity;

(ii) that they should have the same mass;

(iii) that they should have the same principal axes of the moment of inertia at their centre of gravity.

Proof. **(i)** Let us consider two bodies or material systems A and B. If possible, let G_A and G_B be centre of gravities of the two bodies A and B and be different. We draw two parallel straight line A_1G_A and B_1G_B through G_A and G_B and perpendicular to G_AG_B. We join G_AG_B. Let M_A and M_B be the mass of the bodies A and B. Let I_A and I_B be the moments of inertia of any of these bodies about A_1G_A and B_1G_B respectively. As the two bodies are equimomental, the moment of inertia of the bodies about A_1G_A and also about B_1G_B are equal.

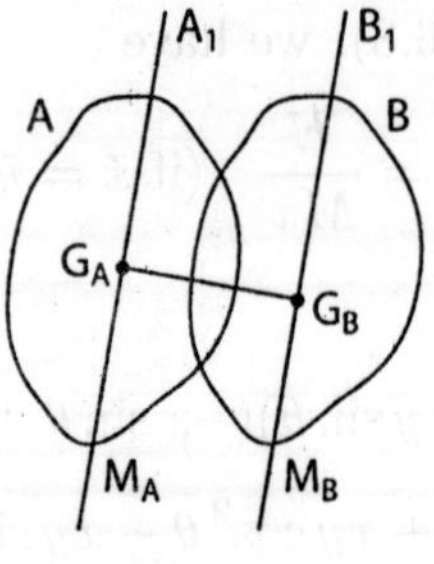

Fig 1.37

Now

$$\begin{aligned} I_A &= \text{moment of inertia of the body } B \text{ about } A_1G_A \\ &= I_B + M_B \cdot G_AG_B^2. \end{aligned} \tag{1.17.1}$$

Again

$$\begin{aligned} I_B &= \text{moment of inertia of the body } A \text{ about } B_1G_B \\ &= I_A + M_A \cdot G_AG_B^2. \end{aligned} \tag{1.17.2}$$

Adding (1.17.1) and (1.17.2) we get

$$
\begin{aligned}
& I_A + I_B = I_A + I_B + (M_A + M_B) G_A G_B^2, \\
\text{or,} \quad & (M_A + M_B) G_A G_B^2 = 0, \\
\text{or,} \quad & G_A G_B = 0 \quad \text{as } M_A + M_B \neq 0.
\end{aligned}
$$

Hence the two system have the same centre of gravity.

(ii) Let G be the centre of gravity of two systems of masses M_1 and M_2. Any line AB is drawn through G and another line CD is also drawn parallel to AB at distance d.

Let I and I_1 be the moments of inertia of the body about AB and CD respectively. As the bodies are equimomental we have

$$
\begin{aligned}
\therefore \quad I_1 &= \text{moment of inertia of the first body about } CD \\
&= \text{moment of inertia of the first body about} AB + M_1 d^2 \\
&= I + M_1 d^2. \qquad (1.17.3)
\end{aligned}
$$

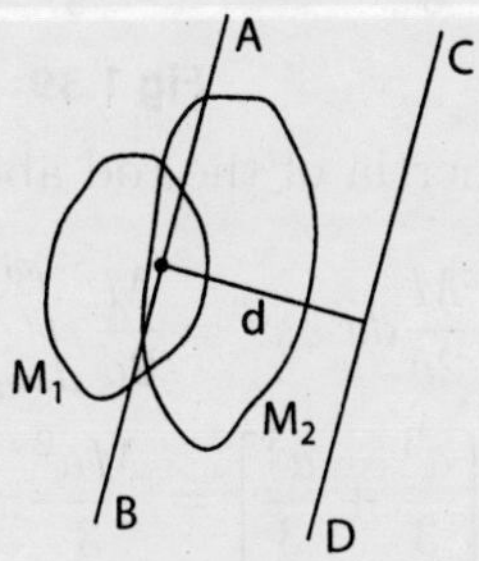

Fig 1.38

Again

$$
\begin{aligned}
I_1 &= \text{moment of inertia of the second body about } CD \\
&= \text{m.i. of the second body about } AB + M_2 d^2 \\
&= I + M_2 d^2. \qquad (1.17.4)
\end{aligned}
$$

Subtracting (1.17.3) and (1.17.4) we get

$$(M_1 - M_2)\, d^2 = 0. \quad \therefore\ M_1 = M_2 \quad \text{as } d \neq 0.$$

(iii) Since, by definition, the principal axes at centre of gravity are axes of greatest and least moment of inertia, then must be identical for two system. And consequently the principal moment of inertia are also be same.

1.18 Solved Problems on Equimomental System

► **Problem 1.18.1** *Prove that an uniform rod, of mass M, is kinematically equivalent to three masses, two masses $M/6$ each rigidly connected at the each end of the rod and third mass $2M/3$ places at the middle point of the rod.*

Solution: Let AB be the rod of mass M and $2a$ be the length of the rod. We take the middle point of the rod as origin O. We consider an elementary thickness dx at a distance x from the origin O and Oy axis perpendicular to the rod.

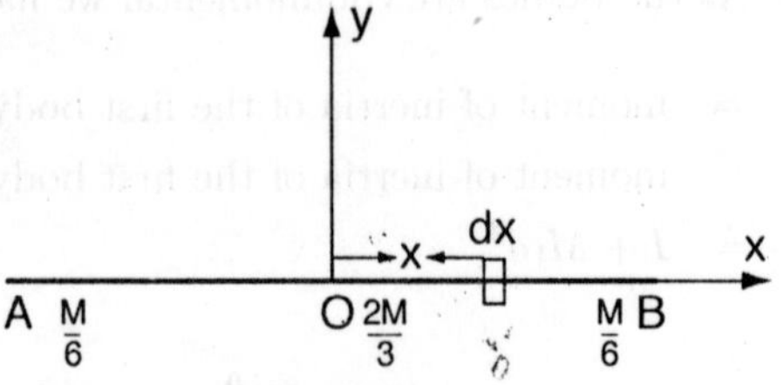

Fig 1.39

$\therefore$ The moment of inertia of the rod about Oy

$$= \int_{-a}^{a} \frac{M}{2a} dx \cdot x^2 = \frac{M}{2a} \int_{-a}^{a} x^2\, dx = \frac{M}{2a} \left[\frac{x^3}{3} \right]_{-a}^{a}$$

$$= \frac{M}{2a} \left[\frac{a^3}{3} + \frac{a^3}{3} \right] = \frac{Ma^2}{3}.$$

Now moment of inertia of the mass $M/6$ at A + moment of inertia of the mass $M/6$ at B + moment of inertia of the mass $2M/3$ at O

$$= \frac{M}{6} a^2 + \frac{M}{6} a^2 + \frac{2M}{3} \cdot 0 = \frac{Ma^2}{3}.$$

Hence the rod is equimomental of three masses, $M/6$ at $A, M/6$ at B and $2M/3$ at the middle point of the rod AB.

► **Problem 1.18.2** *Prove that an uniform triangular lamina of mass M is equimomental with three particles each of mass $M/3$ placed at middle points on the sides of the triangular lamina.* **[CUH 2002; VUH 2003, 2004]**

Solution: Let ABC is the given triangular lamina of mass M. We take Bx and By two perpendicular lines at B as x and y-axes. Let it be required to find the moment of inertia of the triangular lamina ABC about BD. We produce AC to meet Bx at D. We also assume that ABD be a lamina of same density as that of the given lamina ABC.

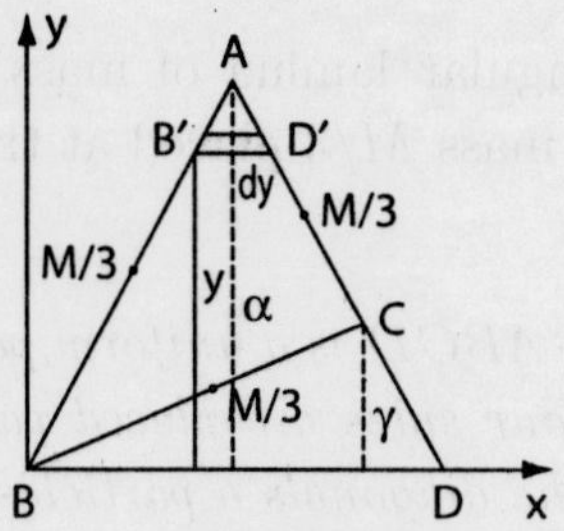

Fig 1.40

Thus moment of inertia of ABD about BD = moment of inertia of ABC about BD + moment of inertia of BCD about BD.

Thus moment of inertia of ABC about BD = moment of inertia of ABD about BD − moment of inertia of BCD about BD.

Let $BD = a$, and α and γ be the height of A and C from BD respectively.

We consider a strip $B'D'$ parallel to BD at a height y from BD and of thickness dy.

Let

$$\therefore \quad m = \frac{\text{mass}}{\text{area}} = \frac{M}{\frac{1}{2}a\alpha - \frac{1}{2}a\gamma} = \frac{2M}{a(\alpha-\gamma)}.$$

$\therefore$ The moment of inertia of $B'D'$ about BD

$$= m \cdot \frac{\alpha - y}{\alpha} a \cdot dy \cdot y^2 \left[\because \frac{B'D'}{\alpha - y} = \frac{BD}{y}. \quad \therefore B'D' = \frac{(\alpha-y)a}{\alpha}\right].$$

$\therefore$ The moment of inertia of ABD about BD

$$= \int_0^\alpha m \frac{\alpha - y}{\alpha} a \cdot y^2\, dy = \frac{ma}{\alpha}\left[\alpha \cdot \frac{\alpha^3}{3} - \frac{\alpha^4}{4}\right] = \frac{ma}{12}\alpha^3.$$

Similarly the moment of inertia BCD about $BD = \frac{ma}{12}\gamma^3$.

Thus the moment of inertia of ABC about BD

$$= \frac{ma}{12}\alpha^3 - \frac{ma}{12}\gamma^3 = \frac{ma}{12}(\alpha^3 - \gamma^3).$$

$\therefore$ The moment of inertia of ABC about BD

$$= \frac{2M \cdot a}{12a(\alpha - \gamma)}(\alpha^3 - \gamma^3) = \frac{M}{6}(\alpha^2 + \gamma^2 + \alpha\gamma). \tag{1.18.1}$$

Now the moment of inertia of three particle each of mass $M/3$ placed at the middle points of the sides of the triangular lamina ABC about BD

$$\begin{aligned} &= \frac{M}{3}\left(\frac{\alpha}{2}\right)^2 + \frac{M}{3}\left(\frac{\alpha+\gamma}{2}\right)^2 + \frac{M}{3}\left(\frac{\gamma}{2}\right)^2 \\ &= \frac{M}{12}[\alpha^2 + (\alpha+\gamma)^2 + \gamma^2] \\ &= \frac{M}{6}(\alpha^2 + \alpha\gamma + \gamma^2). \end{aligned} \tag{1.18.2}$$

Therefore the triangular lamina of mass M is equimomental with the three particles each of mass $M/3$ placed at the middle points of the sides of the lamina.

► **Problem 1.18.3** *ABCD is a uniform parallelogram, of mass M, at the middle points of the four sides are placed particles each of mass M/6 and at the intersection of the diagonals a particles of mass M/3; show that these five particles and the parallelogram are equimomental.*

Solution: Let $ABCD$ be the uniform parallelogram of mass M. And P, Q, R, S are the middle points of sides AB, BC, CD and DA respectively and the diagonals BD and AC intersect at O.

$$\therefore \quad \Delta ABD = \Delta BCD.$$

$$\therefore \quad \text{mass of triangle } ABD = \text{mass of triangle } BCD = \frac{M}{2} = M_1(\text{say}).$$

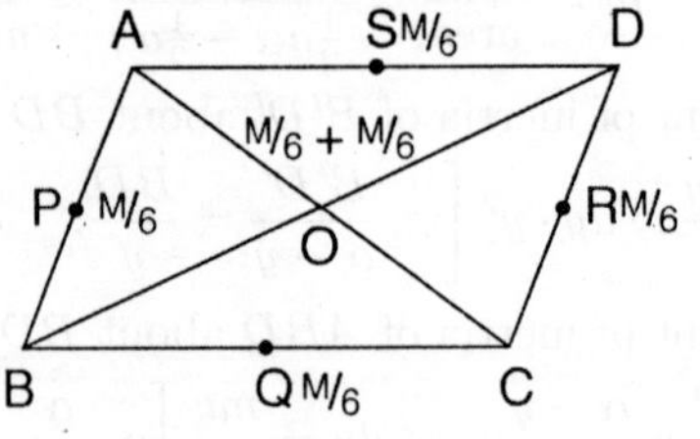

Fig 1.41

$\therefore$ The triangle ABD is equimomental to the masses

$$\frac{M_1}{3} \text{ at } P, \quad \frac{M_1}{3} \text{ at } S, \quad \frac{M_1}{3} \text{ at } O,$$

$$\text{i.e.,} \quad \frac{M}{6} \text{ at } P, \quad \frac{M}{6} \text{ at } S, \quad \frac{M}{6} \text{ at } O.$$

Similarly the triangle BCD is equimomental to the masses

$$\frac{M}{6} \text{ at } Q, \quad \frac{M}{6} \text{ at } R \quad \text{and} \quad \frac{M}{6} \text{ at } O.$$

Thus the uniform parallelogram is equimomental to the masses

$$\frac{M}{6} \text{ at } P, \quad \frac{M}{6} \text{ at } Q, \quad \frac{M}{6} \text{ at } R, \quad \frac{M}{6} \text{ at } S, \quad \frac{M}{6} + \frac{M}{6} = \frac{M}{3} \text{ at } O.$$

Thus the uniform parallelogram is equimomental to four masses $M/6$ placed at the middle points of the sides and the mass $M/3$ placed at the intersection of the diagonals of the parallelogram.

► **Problem 1.18.4** *Prove that a solid uniform tetrahedron of mass M, is equimomental with five particles namely four particles each of mass $M/20$ placed at vertices and a mass $(4/5)M$ placed at the centre of gravity of the tetrahedron.*

Solution: Let $OABC$ be the uniform tetrahedron. Through O, we take three mutually perpendicular lines Ox, Oy and Oz as the coordinate axes. Let the coordinates of A, B, C are $(x_1, y_1, z_1), (x_2, y_2, z_2)$ and (x_3, y_3, z_3) respectively and Δ_0 be the area of triangle ABC and p_0 be its perpendicular distance from the origin O.

We draw a thin strip $A'B'C'$, of thickness dp parallel to ABC and let Δ be the area of $A'B'C'$ and p be the perpendicular distance of strip $A'B'C'$ from origin O. If (x_1', y_1', z_1'), (x_2', y_2', z_2') and (x_3', y_3', z_3') are the coordinates of A', B', C' respectively, then

$$\frac{x_1'}{p} = \frac{x_1}{p_0}; \quad \frac{y_1'}{p} = \frac{y_1}{p_0}, \quad \frac{z_1'}{p} = \frac{z_1}{p_0},$$

$$\text{i.e.,} \quad x_1' = \frac{p}{p_0}x_1; \quad y_1' = \frac{p}{p_0}y_1, \quad z_1' = \frac{p}{p_0}z_1$$

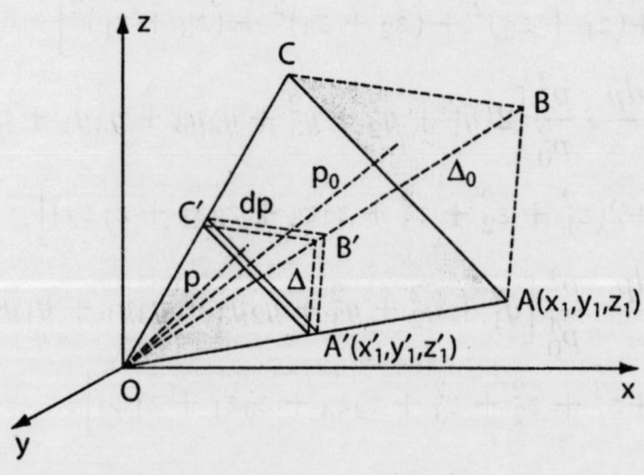

Fig 1.42

Thus the coordinates of A', B', C' are respectively

$$\left(\frac{p}{p_0}x_1, \frac{p}{p_0}y_1, \frac{p}{p_0}z_1\right), \quad \left(\frac{p}{p_0}x_2, \frac{p}{p_0}y_2, \frac{p}{p_0}z_2\right) \quad \text{and} \quad \left(\frac{p}{p_0}x_3, \frac{p}{p_0}y_3, \frac{p}{p_0}z_3\right).$$

Again from geometry

$$\frac{\text{Area of } A'B'C'}{\text{Area of } ABC} = \frac{\Delta}{\Delta_0} = \frac{C'A'^2}{CA^2} = \frac{p^2}{p_0^2}.$$

$$\therefore \quad \Delta = \frac{p^2}{p_0^2}\Delta_0. \tag{1.18.3}$$

Let

$$m = \frac{\text{mass}}{\text{Volume}} = \frac{M}{\frac{1}{3}\Delta_0 p_0} = \frac{3M}{\Delta_0 p_0}. \tag{1.18.4}$$

Then the mass of the strip $A'B'C' = m\Delta dp$.

Thus $A'B'C'$ is equimomental with three particles each of mass $\frac{m\Delta dp}{3}$ placed at the middle points of the sides of the triangle $A'B'C'$.

$\therefore$ The moment of inertia of $A'B'C'$ about x-axis is

$$= \frac{m\Delta \cdot dp}{3}\left[\left\{\frac{p}{p_0}\cdot\left(\frac{y_1+y_2}{2}\right)\right\}^2 + \left\{\frac{p}{p_0}\left(\frac{z_1+z_2}{2}\right)\right\}^2 + \left\{\frac{p}{p_0}\left(\frac{y_2+y_3}{2}\right)\right\}^2 \right.$$
$$\left. + \left\{\frac{p}{p_0}\left(\frac{z_2+z_3}{2}\right)\right\}^2 + \left\{\frac{p}{p_0}\left(\frac{y_3+y_1}{2}\right)\right\}^2 + \left\{\frac{p}{p_0}\left(\frac{z_3+z_1}{2}\right)\right\}^2\right]$$
$$= \frac{m\Delta \cdot dp}{12}\cdot\frac{p^2}{p_0^2}\left[(y_1+y_2)^2 + (y_2+y_3)^2 + (y_3+y_1)^2 \right.$$
$$\left. + (z_1+z_2)^2 + (z_2+z_3)^2 + (z_3+z_1)^2\right]$$
$$= \frac{m\cdot\Delta\cdot dp}{12}\cdot\frac{p^2}{p_0^2}\left[2(y_1^2+y_2^2+y_3^2+y_2y_3+y_3y_1+y_1y_2) \right.$$
$$\left. + 2(z_1^2+z_2^2+z_3^2+z_2z_3+z_3z_1+z_1z_2)\right]$$
$$= \frac{m\cdot\Delta_0 dp}{6}\cdot\frac{p^4}{p_0^4}\left[y_1^2+y_2^2+y_3^2+y_2y_3+y_3y_1+y_1y_2 \right.$$
$$\left. + z_1^2+z_2^2+z_3^2+z_2z_3+z_3z_1+z_1z_2\right].$$

Thus moment of inertia of the whole tetrahedron about x-axis

$$= \frac{m\Delta_0}{6p_0^4}\left[y_1^2+y_2^2+y_3^2+z_1^2+z_2^2+z_3^2+y_2y_3+y_3y_1+y_1y_2 \right.$$
$$\left. + z_2z_3+z_3z_1+z_1z_2\right]\int_0^{p_0} p^4\, dp$$
$$= \frac{m\Delta_0 p_0}{30}\left[y_1^2+y_2^2+y_3^2+z_1^2+z_2^2+z_3^2+y_2y_3+y_3y_1+y_1y_2 \right.$$
$$\left. + z_2z_3+z_3z_1+z_1z_2\right]$$
$$= \frac{3M}{\Delta_0 p_0}\,\frac{\Delta_0 p_0}{30}\left[y_1^2+y_2^2+y_3^2+z_1^2+z_2^2+z_3^2+y_2y_3+y_3y_1+y_1y_2 \right.$$
$$\left. + z_2z_3+z_3z_1+z_1z_2\right] \quad \text{[by (1.18.4)]}.$$

$$\therefore\ I = \frac{M}{10}\Big[y_1^2+y_2^2+y_3^2+z_1^2+z_2^2+z_3^2+y_2y_3+y_3y_1+y_1y_2+z_2z_3+z_3z_1+z_1z_2\Big]. \tag{1.18.5}$$

Now let us consider the moment of inertia of five particles of which four of mass $M/20$ placed at the vertices of the tetrahedron and fifth of mass $4M/5$ placed at the c.g. of the tetrahedron. The moment of inertia of the four particles of mass $M/20$ at the vertices about x-axis.

$$\begin{aligned} &= \frac{M}{20}[(0^2+0^2)+(y_1^2+z_1^2)+(y_2^2+z_2^2)+(y_3^2+z_3^2)] \\ &= \frac{M}{20}[y_1^2+y_2^2+y_3^2+z_1^2+z_2^2+z_3^2] \end{aligned} \tag{1.18.6}$$

moment of inertia of mass $4M/5$ place at c.g.

$$\left(\frac{x_1+x_2+x_3}{4}, \frac{y_1+y_2+y_3}{4}, \frac{z_1+z_2+z_3}{4}\right)$$

of the tetrahedron about x-axis

$$\begin{aligned} &= \frac{4M}{5}\left[\left(\frac{y_1+y_2+y_3}{4}\right)^2+\left(\frac{z_1+z_2+z_3}{4}\right)^2\right] \\ &= \frac{M}{20}\big[y_1^2+y_2^2+y_3^2+z_1^2+z_2^2+z_3^2 \\ &\qquad +2y_2y_3+2y_3y_1+2y_1y_2+2z_2z_3+2z_3z_1+2z_1z_2\big]. \end{aligned} \tag{1.18.7}$$

Adding (1.18.6) and (1.18.7) we get the m.i. of the five particles about x-axis

$$\begin{aligned} &= \frac{M}{20}\big[2y_1^2+2y_2^2+2y_3^2+2z_1^2+2z_2^2+2z_3^2 \\ &\qquad +2y_2y_3+2y_3y_1+2y_1y_2+2z_2z_3+2z_3z_1+2z_1z_2\big] \\ &= \frac{M}{10}[y_1^2+y_2^2+y_3^2+z_1^2+z_2^2+z_3^2+y_2y_3+y_3y_1+y_1y_2+z_2z_3+z_3z_1+z_1z_2] \\ &= \text{moment of inertia of tetrahedron about } x\text{-axis} \quad [\text{by } (1.18.5)]. \end{aligned}$$

▶ **Problem 1.18.5** *Find the moment of inertia of a regular tetrahedron of mass M about a side, given that the length of side of regular tetrahedron is a and is equimomental with four particles each of mass $M/20$ at vertices and mass $4M/5$ at c.g. of the tetrahedron.*

Solution: Let $OABC$ be the regular tetrahedron of mass M and sides of length a. Let G' be the c.g. of the base ABC and G be the c.g. of the tetrahedron. Let four particles each of mass $M/20$ placed at the vertices O,

A, B, C and the mass $4M/5$ placed at the c.g. G of the tetrahedron. Let D is the middle point of side BC.

$$\therefore \quad G'D = \frac{1}{3}AD = \frac{1}{3} \cdot a \sin 60^\circ = \frac{\sqrt{3}}{6}a = \frac{a}{2\sqrt{3}}.$$

$$OG'^2 = OD^2 - G'D^2 = \left(\frac{\sqrt{3}a}{2}\right)^2 - \frac{a^2}{12} = \frac{3a^2}{4} - \frac{a^2}{12} = \frac{2}{3}a^2.$$

$$\therefore \quad OG' = \sqrt{\frac{2}{3}}a. \quad \therefore \quad GG' = \frac{1}{4} \cdot OG' = \frac{1}{4} \cdot \sqrt{\frac{2}{3}}a.$$

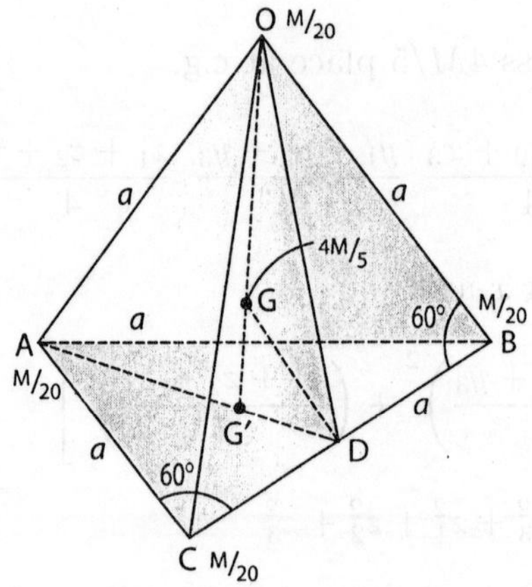

Fig 1.43

Now

$$GD^2 = GG'^2 + G'D^2 = \frac{1}{16} \cdot \frac{2}{3}a^2 + \frac{a^2}{12} = \frac{a^2}{8}.$$

$$\therefore \quad GD = \frac{a}{2\sqrt{2}} \quad \text{and} \quad AD = \frac{\sqrt{3}}{2}a.$$

$\therefore$ m.i. of the tetrahedron about BC = m.i. of four particles each of mass $M/20$ at the vertices about BC + m.i. of the particles $4M/5$ placed at

$$\begin{aligned}\text{c.g. } G &= \frac{M}{20} \cdot AD^2 + \frac{M}{20} \cdot OD^2 + \frac{4M}{5}GD^2 \\ &= \frac{M}{20}\left[\frac{3}{4}a^2 + \frac{3a^2}{4}\right] + \frac{4M}{5} \cdot \frac{a^2}{8} \\ &= \frac{3Ma^2}{40} + \frac{Ma^2}{10} = \frac{7Ma^2}{40}.\end{aligned}$$

► **Problem 1.18.6** *Each of three equal uniform rods AB, BC, CD of mass M and length $2a$ and each is perpendicular to the other two, find the m.i. at the centre of inertia of the system about the axes parallel to the rods.*

Solution: Let AB, BC and CD be the rods of mass M and length $2a$. We take Bx, By and Bz as reference axes. Let G be the c.g. of the system. We take Gx_0, Gy_0 and Gz_0 as the axes through c.g. of the system and parallel to the reference axes. Let $(\bar{x}, \bar{y}, \bar{z})$ be the coordinates of c.g. G. Let A, B, C; A_0, B_0, C_0 are the m.i. of the system about Bx, By, Bz and Gx_0, Gy_0 and Gz_0 respectively.

$$\therefore \bar{x} = \frac{M \cdot a + M \cdot 0 + M \cdot 0}{3M} = \frac{a}{3}$$

$$\bar{y} = \frac{M \cdot 0 + Ma + M \cdot 2a}{3M} = a$$

$$\bar{z} = \frac{M \cdot 0 + M \cdot 0 + Ma}{3M} = \frac{a}{3}.$$

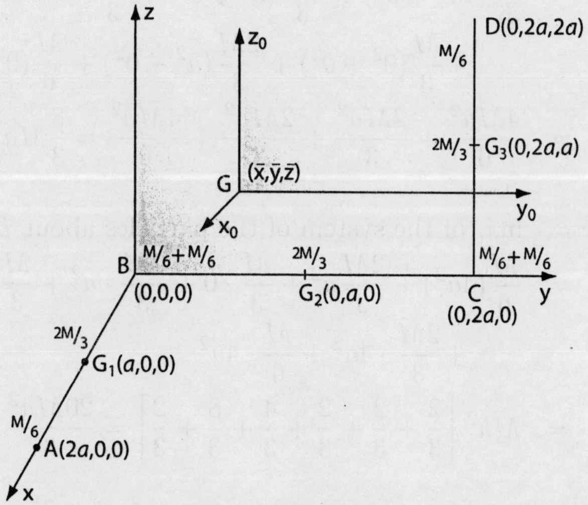

Fig 1.44

As a uniform rod of mass M is equimomental with two particles of mass $M/6$ placed at the extremities of the rod and a particle of mass $2M/3$ placed at c.g. of the rod, then the system is equimomental the particles of mass.

$$\frac{M}{6} \text{ at } A(2a,0,0), \quad \frac{2M}{3} \text{ at } G_1(a,0,0), \quad \frac{M}{6}+\frac{M}{6}=\frac{M}{3} \text{ at } B(0,0,0)$$

$$\frac{2M}{3} \text{ at } G_2(0,a,0), \quad \frac{M}{6}+\frac{M}{6}=\frac{M}{3} \text{ at } C(0,2a,0),$$

$$\frac{2M}{3} \text{ at } G_3(0,2a,a), \quad \frac{M}{6} \text{ at } D(0,2a,2a).$$

where G_1, G_2, G_3 are the c.g. of the rod, AB, BC and CD respectively.

$$\begin{aligned}
\therefore \quad A &= \text{m.i. of the system of the particles about } Bx \\
&= \frac{M}{6}(0^2+0^2)+\frac{2M}{3}(0^2+0^2)+\frac{M}{3}(0^2+0^2)+\frac{2M}{3}(a^2+0^2) \\
&\quad +\frac{M}{3}(4a^2+0^2)+\frac{2M}{3}(4a^2+a^2)+\frac{M}{6}(4a^2+4a^2) \\
&= \frac{2M}{3}a^2+\frac{4Ma^2}{3}+\frac{10Ma^2}{3}+\frac{4Ma^2}{3}=\frac{20Ma^2}{3};
\end{aligned}$$

$$\begin{aligned}
B &= \text{m.i. of the system of the particles about } By \\
&= \frac{M}{6}(4a^2+0^2)+\frac{2M}{3}(a^2+0^2)+\frac{M}{3}(0^2+0^2)+\frac{2M}{3}(0^2+0^2) \\
&\quad +\frac{M}{3}(0^2+0^2)+\frac{2M}{3}(a^2+0^2)+\frac{M}{6}(0^2+4a^2) \\
&= \frac{4Ma^2}{6}+\frac{2Ma^2}{3}+\frac{2Ma^2}{3}+\frac{4Ma^2}{6}=\frac{8}{3}Ma^2;
\end{aligned}$$

$$\begin{aligned}
C &= \text{m.i. of the system of the particles about } Bz \\
&= \frac{M}{6}(4a^2)+\frac{2M}{3}a^2+\frac{M}{3}\cdot 0+\frac{2M}{3}\cdot a^2+\frac{M}{3}\cdot 4a^2 \\
&\quad +\frac{2M}{3}\cdot 4a^2+\frac{M}{6}\cdot 4a^2 \\
&= Ma^2\left[\frac{2}{3}+\frac{2}{3}+\frac{2}{3}+\frac{4}{3}+\frac{8}{3}+\frac{2}{3}\right]=\frac{20Ma^2}{3}.
\end{aligned}$$

$$\therefore \quad A = A_0+3M(\bar{y}^2+\bar{z}^2)=A_0+3M\left(a^2+\frac{a^2}{9}\right)=A_0+\frac{10}{3}Ma^2.$$

$$\therefore \quad A_0 = A-\frac{10}{3}Ma^2=\frac{20Ma^2}{3}-\frac{10}{3}Ma^2=\frac{10}{3}Ma^2.$$

$$B = B_0+3M(\bar{z}^2+\bar{x}^2)=B_0+3M\left(\frac{a^2}{9}+\frac{a^2}{9}\right)=B_0+\frac{6Ma^2}{9}.$$

$$\therefore \quad B_0 = B - \frac{2}{3}Ma^2 = \frac{8}{3}Ma^2 - \frac{2}{3}Ma^2 = 2Ma^2.$$

$$C = C_0 + 3M(\bar{x}^2 + \bar{y}^2) = C_0 + 3M\left(\frac{a^2}{9} + a^2\right) = C_0 + \frac{10Ma^2}{3}.$$

$$\therefore \quad C_0 = C - \frac{10Ma^2}{3} = \frac{20Ma^2}{3} - \frac{10Ma^2}{3} = \frac{10}{3}Ma^2.$$

► **Problem 1.18.7** *Prove that a uniform triangular lamina of mass M is equimomental with three particles, each of mass $M/12$ placed at the angular points and a particle of mass $3M/4$ placed at the centre of mass of the triangles.* **[CUH 2000; VUH 2002]**

Solution: Let ABC is the given triangular lamina of mass M. We take Bx and By two perpendicular lines at B as x and y-axis. Let it be required to find the m.i. of the triangular lamina ABC about BD. We produce AC to meet Bx at D. We also assume that ABD be a triangular lamina of same density as that of the given lamina ABC.

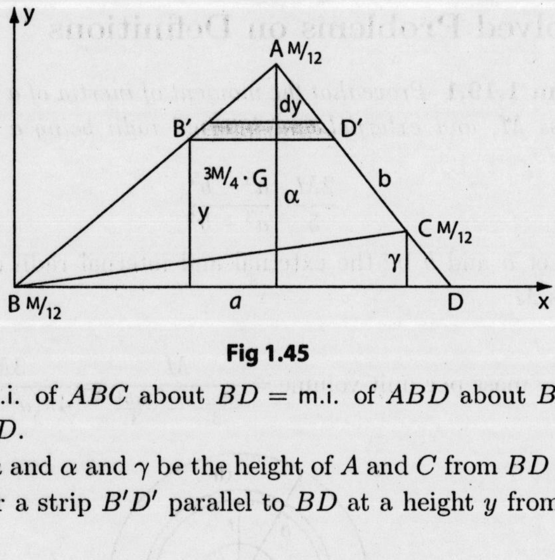

Fig 1.45

Thus the m.i. of ABC about BD = m.i. of ABD about BD − m.i. of BCD about BD.

Let $BD = a$ and α and γ be the height of A and C from BD respectively.

We consider a strip $B'D'$ parallel to BD at a height y from BD and of thickness dy.

Let

$$m = \frac{\text{mass}}{\text{area}} = \frac{M}{\frac{1}{2}a(\alpha - \gamma)} = \frac{2M}{a(\alpha - \gamma)}.$$

$\therefore$ The m.i. of $B'D'$ about BD

$$= m\frac{\alpha - y}{\alpha}a \cdot dy \cdot y^2 \quad \left[\because \frac{B'D'}{\alpha - y} = \frac{BD}{y}, \quad \text{or,} \quad B'D' = \frac{\alpha - y}{\alpha} \cdot a\right].$$

$\therefore$ m.i. of ABD about BD

$$= \int_0^\alpha m\frac{\alpha - y}{\alpha} a \cdot y^2\, dy = \frac{ma}{\alpha}\left[\alpha \cdot \frac{\alpha^3}{3} - \frac{\alpha^4}{4}\right] = \frac{ma}{12}\alpha^3.$$

Similarly the m.i. of BCD about

$$BD = \frac{ma}{12}\gamma^3.$$

Thus the m.i. of ABC about BD

$$= \frac{ma}{12}\alpha^3 - \frac{ma}{12}\gamma^3 = \frac{ma}{12}(\alpha^3 - \gamma^3)$$

$$= \frac{2Ma}{a(\alpha-\gamma)}\frac{(\alpha^3-\gamma^3)}{12} = \frac{M}{6}(\alpha^2 + \alpha\gamma + \gamma^2).$$

Now the m.i. of three particles of mass $M/12$ placed at the vertices about BD + m.i. of the particles $3M/4$ placed the centre of mass about BD

$$= \frac{M}{12}\cdot\alpha^2 + \frac{M}{12}\gamma^2 + \frac{3M}{4}\left(\frac{\alpha+\gamma}{3}\right)^2 = \frac{M}{12}[\alpha^2 + \gamma^2 + \alpha^2 + \gamma^2 + 2\alpha\gamma]$$

$$= \frac{M}{6}(\alpha^2 + \gamma^2 + \alpha\gamma).$$

1.19 Solved Problems on Definitions

► **Problem 1.19.1** *Prove that the moment of inertia of a hallow spherical shell of mass M, and external and internal radii being a and b, about a diameter is*

$$\frac{2M}{5}\cdot\frac{a^5 - b^5}{a^3 - b^3}.$$

Solution: Let a and b be the external and internal radii of the spherical shell of mass M.

Let

$$m = \text{mass per unit volume} = \frac{M}{\frac{4\pi a^3}{3} - \frac{4\pi b^3}{3}} = \frac{3M}{4\pi(a^3 - b^3)}.$$

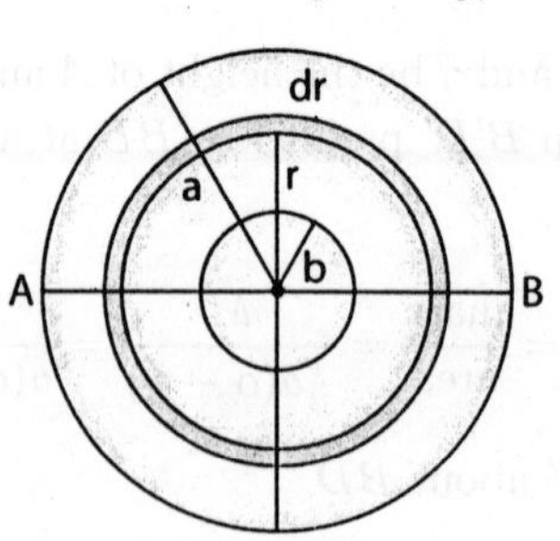

Fig 1.46

Let us consider an elementary spherical shell of radius r and of thickness dr.

$\therefore$ The mass of the spherical shell $= 4\pi r^2\,dr\cdot m$.

The m.i. of the spherical shell about diameter AB

$$= 4\pi r^2 m\cdot dr\cdot\frac{2}{3}r^2 \quad \text{[by (1.4.25)]}$$

$$= \frac{8}{3}\pi m\cdot r^4\,dr.$$

$\therefore$ m.i. of the whole hallow sphere

$$= \frac{8}{3}\pi m\int_b^a r^4\,dr = \frac{8}{3}\pi m\cdot\frac{a^5-b^5}{5}$$

$$= \frac{8}{3}\pi\cdot\frac{3M}{4\pi(a^3-b^3)}\cdot\frac{a^5-b^5}{5} = \frac{2M}{5}\cdot\frac{a^5-b^5}{a^3-b^3}.$$

▶ **Problem 1.19.2** *Prove that the moment of inertia of a truncated cone of mass M about its axis, the radius of its ends being a and b, is*

$$\frac{3M}{10}\cdot\frac{a^5-b^5}{a^3-b^3}.$$

Solution: Let M be the mass of the truncated cone, having a and b radii of the ends and α is semi-vertical angle. The volume of the truncated cone

$$= \tfrac{1}{3}\pi a^2\cdot a\cot\alpha - \tfrac{1}{3}\pi b^2\cdot b\cot\alpha = \tfrac{1}{3}\pi\cot\alpha(a^3-b^3).$$

Let

$$m = \frac{\text{mass}}{\text{vol.}} = \frac{M}{\frac{1}{3}\pi\cot\alpha(a^3-b^3)} = \frac{3M}{\pi\cot\alpha(a^3-b^3)}.$$

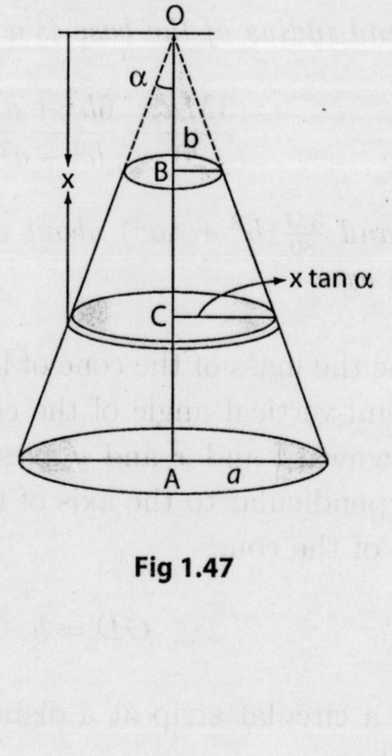

Fig 1.47

Let us consider a circular strip at a distance x from the vertex O and of thickness dx.

$\therefore$ The mass of the circular strip

$$= m \cdot \pi (x \tan \alpha)^2 \, dx = \frac{3M}{\pi(a^3 - b^3) \cot \alpha} \pi x^2 \tan^2 \alpha \, dx$$

$$= \frac{3M}{a^3 - b^3} \cdot \tan^3 \alpha \cdot x^2 \, dx.$$

$\therefore$ m.i. of the strip about the axis OA

$$= \frac{3M}{a^3 - b^3} \cdot \tan^3 \alpha \cdot x^2 \, dx \cdot \frac{x^2 \tan^2 \alpha}{2} \quad \text{[by (1.4.11)]}$$

$$= \frac{3M}{2(a^3 - b^3)} \cdot \tan^5 \alpha \cdot x^4 \, dx$$

$\therefore$ m.i. of the whole truncated cone about OA

$$= \frac{3M}{2(a^3 - b^3)} \cdot \tan^5 \alpha \int_{b \cot \alpha}^{a \cot \alpha} x^4 \, dx$$

$$= \frac{3M}{2(a^3 - b^3)} \cdot \tan^5 \alpha \cdot \frac{a^5 - b^5}{5} \cot^5 \alpha$$

$$= \frac{3M}{10} \cdot \frac{a^5 - b^5}{a^3 - b^3}.$$

► **Problem 1.19.3** *Show that the m.i. of a right circular solid cone, of mass M, height h and radius of the base is a is*

$$\frac{3Ma^2}{20} \cdot \frac{6h^2 + a^2}{h^2 + a^2}$$

about a slant side and $\frac{3M}{80}(h^2 + 4a^2)$ about a line through c.g. of the cone perpendicular to its axis.

KUH '01

Solution: Let M be the mass of the cone of height h, and radius of the base is a and α is the semi-vertical angle of the cone. We take z-axes along the axis of the cone downward and x and y axes perpendicular to it at O. Let Gx' be any line perpendicular to the axis of the cone and parallel to x-axis, where G is the c.g. of the cone.

$$\therefore \; OD = h.$$

Let us consider a circular strip at a distance z and of thickness dz and parallel to circular base.

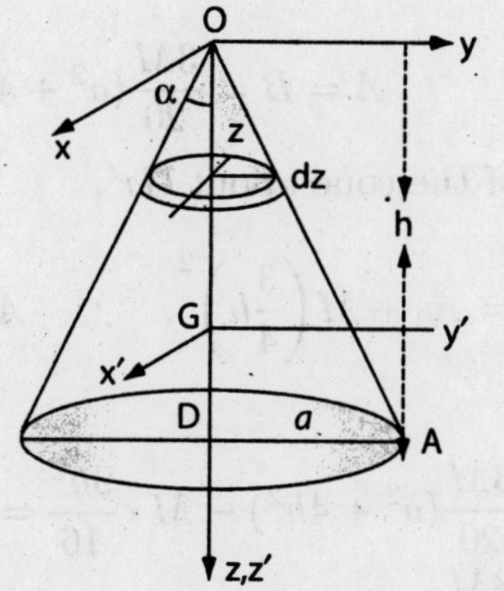

Fig 1.48

Let

$$m = \frac{\text{mass}}{\text{volume}} = \frac{M}{\frac{1}{3}\pi a^2 \cdot OD} = \frac{3M}{\pi h^2 \tan^2\alpha} \cdot \frac{1}{h} = \frac{3M}{\pi h^3 \tan^2\alpha}.$$

$\therefore$ mass of the circular strip $= m\pi z^2 \tan^2\alpha \cdot dz$.

$\therefore$ The m.i. of the strip about any line through c.g. of the strip and parallel to x-axes

$$= m\pi z^2 \tan^2\alpha\, dz \cdot \frac{z^2\tan^2\alpha}{4} \quad \text{[by (1.4.9)]}$$

Thus m.i. of the strip about x-axis

$$\begin{aligned} &= m\pi z^2 \tan^2\alpha\, dz \cdot \frac{z^2\tan^2\alpha}{4} + m\pi z^2\tan^2\alpha\, dz \cdot z^2 \\ &= m\pi \tan^2\alpha \left(\frac{\tan^2\alpha}{4} + 1\right) z^4\, dz. \end{aligned}$$

$\therefore$ $A =$ m.i. of the cone about x-axis.

$$\begin{aligned} &= \int_0^h m\pi \tan^2\alpha \left[\frac{\tan^2\alpha}{4} + 1\right] z^4\, dz \\ &= \frac{3M}{\pi h^3 \tan^2\alpha} \cdot \pi\tan^2\alpha \left(\frac{\tan^2\alpha}{4} + 1\right) \cdot \frac{h^5}{5} \\ &= \frac{3}{5}Mh^2 \left(\frac{\tan^2\alpha}{4} + 1\right) = \frac{3}{5}Mh^2\left(\frac{a^2}{4h^2} + 1\right) \\ &= \frac{3}{5}Mh^2\left(\frac{a^2 + 4h^2}{4h^2}\right) = \frac{3M}{20}(a^2 + 4h^2). \end{aligned}$$

By symmetry

$$A = B = \frac{3M}{20}(a^2 + 4h^2).$$

Let A_0 = m.i. of the cone about Gx'.

$$\therefore \quad A = A_0 + M\left(\frac{3}{4}h\right)^2. \quad \therefore \quad A_0 = A - M \cdot \frac{9h^2}{16}.$$

$$\begin{aligned}\therefore \quad A_0 &= \frac{3M}{20}(a^2 + 4h^2) - M \cdot \frac{9h^2}{16} = \frac{3M}{80}[4a^2 + 16h^2 - 15h^2] \\ &= \frac{3M}{80}(4a^2 + h^2).\end{aligned}$$

Now the moment of inertia of the circular strip about z-axis

$$= m\pi z^2 \tan^2\alpha\, dz \cdot \frac{z^2 \tan^2\alpha}{2} \quad \text{[by (1.4.11)]}$$

$\therefore$ m.i. of the cone about z-axis

$$\begin{aligned}C &= \int_0^h \pi m z^2 \tan^2\alpha \cdot dz \cdot \frac{z^2\tan^2\alpha}{2} = \pi m \cdot \frac{\tan^4\alpha}{2}\int_0^h z^4\, dz \\ &= \pi m \cdot \frac{\tan^4\alpha}{2} \cdot \frac{h^5}{5} = \frac{3M}{\pi h^3 \tan^2\alpha} \cdot \pi \cdot \frac{\tan^4\alpha}{2} \cdot \frac{h^5}{5} \\ &= \frac{3M}{10}h^2\tan^2\alpha = \frac{3M}{10}h^2 \cdot \frac{a^2}{h^2} = \frac{3Ma^2}{10}.\end{aligned}$$

Now let (l, m, n) be the d.c's of the generator OA.

$$\therefore \quad n = \cos\alpha = \frac{h}{\sqrt{h^2 + a^2}}.$$

$$\begin{aligned}\therefore \quad I &= \text{m.i. of the cone about the generator } OA \\ &= Al^2 + Bm^2 + Cn^2 \\ &= A(l^2 + m^2) + Cn^2 \quad [\text{since } D = E = F = 0 \text{ and } A = B] \\ &= A(1 - n^2) + Cn^2 \\ &= \frac{3M}{20}(a^2 + 4h^2)\left(1 - \frac{h^2}{h^2 + a^2}\right) + C\left(\frac{h^2}{a^2 + h^2}\right) \\ &= \frac{3M}{20}(a^2 + 4h^2)\frac{a^2}{h^2 + a^2} + \frac{3Ma^2}{10} \cdot \frac{h^2}{h^2 + a^2} \\ &= \frac{3M}{20} \cdot a^2\left[\frac{a^2 + 4h^2}{h^2 + a^2} + \frac{2h^2}{a^2 + h^2}\right] = \frac{3M}{20}a^2\left[\frac{a^2 + 6h^2}{a^2 + h^2}\right].\end{aligned}$$

► **Problem 1.19.4** *Show that the moment of inertia of a thin homogeneous ellipsoidal shell (bounded by similar, similarly situated and concentric ellipsoid) about an axis say x-axis is* $M\frac{b^2+c^2}{3}$, *where M is the mass of the shell.*

Solution: Let the length of semi-axes of the inner and outer ellipsoid are respectively a, b, c and ka, kb, kc.

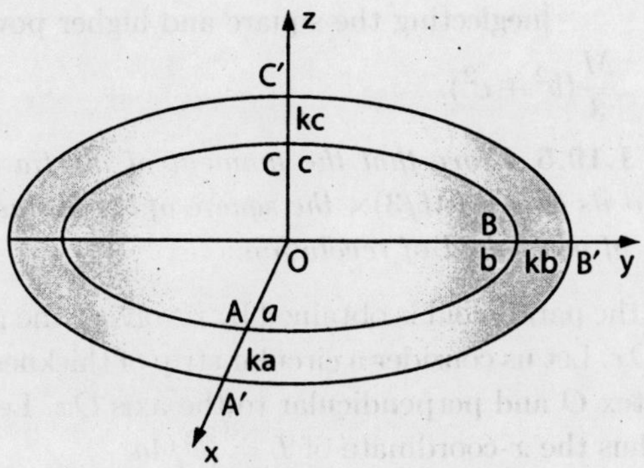

Fig 1.49

Let

$$m = \frac{\text{mass}}{\text{volume}} = \frac{M}{\frac{4}{3}\pi abck^3 - \frac{4}{3}\pi abc} = \frac{3M}{4\pi abc(k^3-1)}.$$

$\therefore$ The m.i. of the outer ellipsoid about x-axis

$$= \frac{4}{3}\pi abck^3 \cdot m \cdot \frac{k^2b^2 + k^2c^2}{5} = \frac{4}{3}\pi abc \cdot m \cdot \frac{b^2+c^2}{5}k^5 \quad \text{[by (1.4.35)]}$$

m.i. of the inner ellipsoid about x-axis

$$= \frac{4}{3}\pi abc \cdot m \cdot \frac{b^2+c^2}{5}$$

$\therefore$ The moment of inertia of the required ellipsoidal shell

$$\begin{aligned}
&= \frac{4}{3}\pi abc \cdot m \cdot \frac{b^2+c^2}{5} \cdot k^5 - \frac{4}{3}\pi abc \cdot m \cdot \frac{b^2+c^2}{5} \\
&= \frac{4}{3}\pi abc \cdot m \cdot \frac{b^2+c^2}{5} \cdot (k^5-1) \\
&= \frac{4}{3}\pi abc \cdot \frac{3M}{4\pi abc(k^3-1)} \cdot \frac{b^2+c^2}{5} \cdot (k^5-1) \\
&= \frac{M}{5}(b^2+c^2) \cdot \frac{k^5-1}{k^3-1}.
\end{aligned}$$

Let us put $k = K' + 1$, where K' is very small, for shell.

$$\begin{aligned}\therefore\quad I &= \frac{M}{5}(b^2+c^2)\frac{(K'+1)^5-1}{(K'+1)^3-1}\\ &= \frac{M}{5}(b^2+c^2)\frac{K'^5+5K'^4+10K'^3+10K'^2+5K'+1-1}{K'^3+3K'^2+3K'+1-1}\\ &= \frac{M}{5}(b^2+c^2)\cdot\frac{5K'}{3K'} = \frac{M}{5}(b^2+c^2)\cdot\frac{5}{3}\end{aligned}$$

[neglecting the square and higher power of K']

$$= \frac{M}{3}(b^2+c^2).$$

► **Problem 1.19.5** *Prove that the moment of inertia of a paraboloid of revolution about its axis is $(M/3)\times$ the square of the radius of its base, where M is the mass of paraboloid of revolution.*

Solution: Let the paraboloid is obtained by revolving the parabola $y^2 = 4ax$ about its axis Ox. Let us consider a circular strip of thickness dx at a distance x from the vertex O and perpendicular to the axis Ox. Let R be the radius of the base. Thus the x-coordinate of $L = R^2/4a$.

$\therefore$ Volume of the paraboloid

$$= \int_0^{R^2/4a} \pi y^2\cdot dx = \pi\int_0^{\frac{R^2}{4a}} 4ax\,dx = \frac{4a\pi}{2}\left(\frac{R^2}{4a}\right)^2 = \frac{\pi R^4}{8a}$$

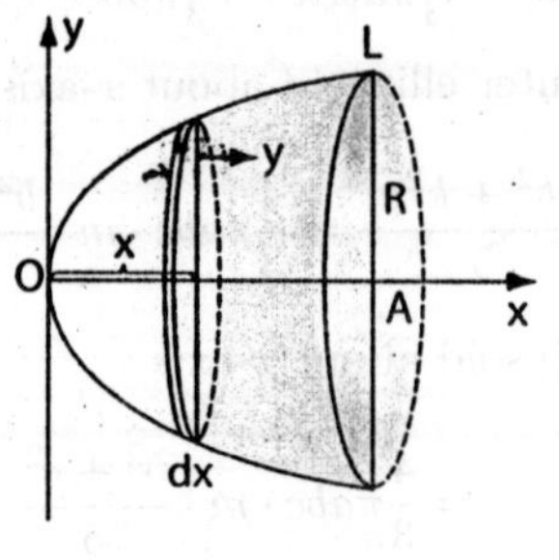

Fig 1.50

Let

$$m = \frac{\text{mass}}{\text{volume}} = \frac{M\times 8a}{\pi R^4} = \frac{8Ma}{\pi R^4}.$$

Now the m.i. of the circular strip about Ox

$$\begin{aligned} &= m\cdot\pi y^2\,dx\cdot\frac{y^2}{2}\quad\text{[by (1.4.11)]}\\ &= \frac{\pi m}{2}\cdot y^4\,dx = \frac{\pi m}{2}\cdot(4ax)^2\,dx\\ &= 8a^2\pi m\cdot x^2\,dx.\end{aligned}$$

Thus the m.i. of the whole paraboloid about Ox

$$= 8a^2\pi m \int_0^{R^2/4a} x^2\,dx = \frac{8}{3}a^2\pi m \,|\, x^3 \,|_0^{R^2/4a} = \frac{8}{3}a^2\pi m \cdot \frac{R^6}{64a^3}$$

$$= \frac{8}{3}a^2\pi \cdot \frac{8Ma}{\pi R^4} \cdot \frac{R^6}{64a^3} = \frac{1}{3}MR^2 = \frac{1}{3}M \times \text{ the square of the radius of the base.}$$

► **Problem 1.19.6** *Prove that the moment of inertia of a parabolic area of latus rectum 4a cut off by an ordinates at distance h from the vertex is* $\frac{3}{7}Mh^2$ *about the tangent at vertex and* $\frac{4}{5}Mah$ *about the axis.*

Solution: Let the equation of the parabola be $y^2 = 4ax$. We consider a strip of thickness dx at a distance x from the vertex O and perpendicular to the axis of the parabola Ox.

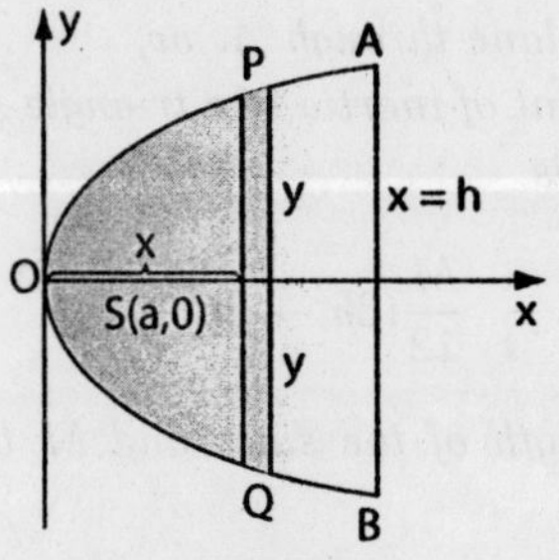

Fig 1.51

Now area

$$AOB = \int_{x=0}^{h} 2y dx = \int_{x=0}^{h} 2 \cdot 2\sqrt{a} \cdot \sqrt{x} \cdot dx = 4\sqrt{a} \left| \frac{x^{3/2}}{3/2} \right|_0^h = \frac{8\sqrt{a}}{3} \cdot h^{3/2}.$$

$$\therefore\ m = \frac{\text{mass}}{\text{area}} = \frac{M}{\frac{8\sqrt{a}}{3} \cdot h^{3/2}} = \frac{3M}{8\sqrt{a}\, h^{3/2}}.$$

$\therefore$ mass of the strip $= m \cdot 2ydx = 2m \cdot 2\sqrt{a} \cdot \sqrt{x}dx.$

$\therefore$ m.i. of the strip about $Oy = 4m\sqrt{a} \cdot \sqrt{x} \cdot x^2\,dx.$

Thus m.i. of the whole area

$$= 4m\sqrt{a} \int_0^h x^{5/2} dx = 4m\sqrt{a} \cdot \frac{h^{7/2}}{7/2} = \frac{8m\sqrt{a}}{7} \cdot h^{7/2}$$

$$= \frac{8\sqrt{a}}{7} \cdot \frac{3M}{8\sqrt{a}} \cdot \frac{1}{h^{3/2}} \cdot h^{7/2} = \frac{3M}{7} \cdot h^2.$$

Now the m.i. of the strip about the axis of the parabola

$$= m \cdot 2ydx \cdot \frac{y^2}{3} \quad \text{[by (1.4.3)]}$$
$$= \frac{2}{3} \cdot m \cdot y^3\, dx.$$

$\therefore$ m.i. of the whole area about the axis of the parabola

$$= \frac{2}{3}m \int_{x=0}^{h} y^3\, dx = \frac{2}{3}m \int_0^h \left(\sqrt{4a} \cdot \sqrt{x}\right)^3 dx = \frac{16}{3}m \cdot a^{3/2} \cdot \frac{h^{5/2}}{5/2}$$
$$= \frac{32}{15} \cdot m \cdot a^{3/2} \cdot h^{5/2} = \frac{32}{15} \cdot \frac{3M}{8\sqrt{a} \cdot h^{3/2}} \cdot a^{3/2} \cdot h^{5/2}$$
$$= \frac{4}{5}ahM.$$

▶ **Problem 1.19.7** *Find the moment of inertia of a triangle ABC, about a perpendicular to the plane through A. or,*

Prove that the moment of inertia of a triangle ABC about a perpendicular to the plane through A is

$$\frac{M}{12}(3b^2 + 3c^2 - a^2),$$

where a, b, c be the length of the sides and M be the mass of the triangle ABC. **[VUH 2006]**

Solution: Let ABC be the triangle of mass M. $AD = h =$ perpendicular distance of A from BC. $AE = d =$ median.

Let us consider a strip PQ of thickness dx parallel to BC and at a distance x from A.

Let AO be the line perpendicular to the plane of the triangle through A.

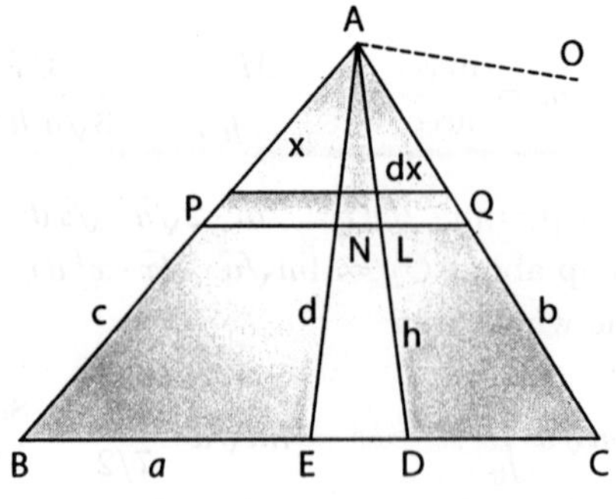

Fig 1.52

Let
$$m = \frac{\text{mass}}{\text{area}} = \frac{M}{\frac{1}{2}ah} = \frac{2M}{ah}.$$

Thus mass of the strip $= mPQ \cdot dx$.

$\therefore$ m.i. of the strip PQ about AO

$$\begin{aligned} &= \text{m.i. of } PQ \text{ about a line through mid-point } N \text{ parallel to } AO \\ &\quad + \text{mass of the strip} \times AN^2 \\ &= \frac{m \cdot PQ \cdot dx}{3}\left(\frac{PQ}{2}\right)^2 + m \cdot PQ \cdot dx \cdot AN^2. \end{aligned}$$

Now from geometry we have,

$$\frac{PQ}{BC} = \frac{x}{h} = \frac{AL}{AD} = \frac{AN}{AE} = \frac{AN}{d}, \quad \text{or,} \quad \frac{PQ}{a} = \frac{x}{h} = \frac{AN}{d}.$$

$$\therefore \quad PQ = \frac{ax}{h} \quad \text{and} \quad AN = \frac{dx}{h}.$$

$\therefore$ The m.i. of the strip PQ about AO

$$\begin{aligned} &= \frac{mdx}{12} \cdot \left(\frac{ax}{h}\right)^3 + mdx \cdot \frac{ax}{h} \cdot \frac{x^2\,d^2}{h^2} = \frac{m}{12h^3}[a^3x^3 + 12ad^2x^3]dx \\ &= \frac{2M}{12ah^4} \cdot (a^3 + 12ad^2)x^3\,dx = \frac{M}{6h^4}(a^2 + 12d^2)x^3\,dx. \end{aligned}$$

$\therefore$ The m.i. of the triangle about AO

$$I = \frac{M}{6h^4}(a^2 + 12d^2)\int_0^h x^3\,dx = \frac{M}{24}(a^2 + 12d^2). \tag{1.19.1}$$

Now

$$\begin{aligned} d^2 &= AE^2 = AD^2 + ED^2 = AD^2 + (EC - CD)^2 \\ &= AD^2 + EC^2 + CD^2 - 2EC \cdot CD \\ &= AC^2 + EC^2 - 2EC \cdot CD = b^2 + \left(\frac{a}{2}\right)^2 - 2 \cdot \frac{a}{2} \cdot b\cos C \\ &= b^2 + \frac{a^2}{4} - ab \cdot \frac{a^2 + b^2 - c^2}{2ab} = b^2 + \frac{a^2}{4} - \frac{a^2 + b^2 - c^2}{2} \\ &= \frac{4b^2 + a^2 - 2a^2 - 2b^2 + 2c^2}{4} = \frac{2b^2 + 2c^2 - a^2}{4}. \end{aligned}$$

$\therefore$ From (1.19.1) we get,

$$\begin{aligned} I &= \frac{M}{24}\left[a^2 + 12 \cdot \frac{2b^2 + 2c^2 - a^2}{4}\right] \\ &= \frac{M}{24}[a^2 + 6b^2 + 6c^2 - 3a^2] \\ &= \frac{M}{12}(3b^2 + 3c^2 - a^2). \end{aligned}$$

▶ **Problem 1.19.8** *Prove that the m.i. of the area bounded by* $r^2 = a^2 \cos 2\theta$ *about its axis is*

$$\frac{Ma^2}{16}(\pi - 8/3),$$

where M is the mass of the area.

Solution: Let Ox is the axis of the area of mass M bounded by the curve $r^2 = a^2 \cos 2\theta$. The loop is formed between $\theta = \pi/4$ and $\theta = -\pi/4$. Let us consider an elementary area $rd\theta\, dr$ at $P(r, \theta)$. Now the area bounded by the curve

$$= 4\int_{\theta=0}^{\pi/4}\int_{r=0}^{a\sqrt{\cos 2\theta}} rd\theta\, dr = 4\int_{\theta=0}^{\pi/4} d\theta \cdot \frac{r^2}{2}\Big|_0^{a\sqrt{\cos 2\theta}}$$

$$= 2\int_{\theta=0}^{\pi/4} a^2 \cos 2\theta\, d\theta = 2a^2 \left|\frac{\sin 2\theta}{2}\right|_0^{\pi/4} = a^2.$$

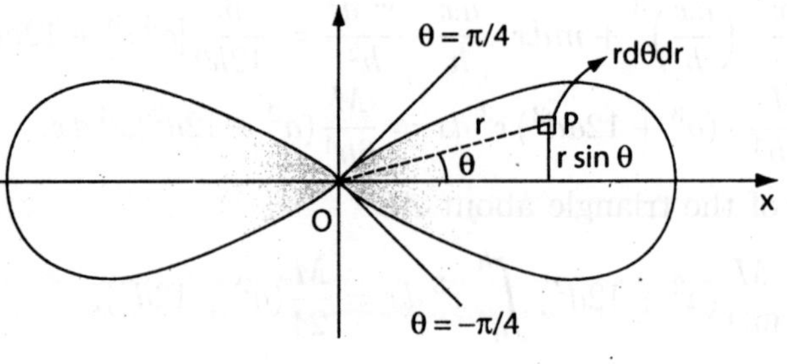

Fig 1.53

Let

$$m = \frac{\text{mass}}{\text{area}} = \frac{M}{a^2}.$$

∴ mass of the elementary area

$$= m \cdot rd\theta\, dr = \frac{M}{a^2} \cdot rd\theta\, dr.$$

Thus the m.i. of the elementary area about the axis Ox

$$= \frac{M}{a^2} \cdot rd\theta\, dr \cdot (r \sin\theta)^2 = \frac{M}{a^2} \cdot r^3 \sin^2\theta dr\, d\theta.$$

$\therefore$ The m.i. of the whole area about the axis Ox

$$= 4 \cdot \frac{M}{a^2} \cdot \int_{\theta=0}^{\pi/4} \int_{r=0}^{a\sqrt{\cos 2\theta}} r^3 \sin^2 \theta dr\, d\theta = \frac{4M}{a^2} \int_{\theta=0}^{\pi/4} \frac{a^4 \cos^2 2\theta}{4} \cdot \sin^2 \theta\, d\theta$$

$$= Ma^2 \int_{\theta=0}^{\pi/4} \cos^2 2\theta \left(\frac{1-\cos 2\theta}{2}\right) d\theta, \quad \text{putting } 2\theta = \phi, 2d\theta = d\phi$$

$$= \frac{Ma^2}{4} \int_{\phi=0}^{\pi/2} \cos^2 \phi (1 - \cos \phi)\, d\phi = \frac{Ma^2}{4}\left[\int_{\phi=0}^{\pi/2} \cos^2 \phi d\phi - \int_{\phi=0}^{\pi/2} \cos^3 \phi\, d\phi\right]$$

$$= \frac{Ma^2}{4}\left[\frac{1}{2}\cdot\frac{\pi}{2} - \frac{2}{3}\right] = \frac{Ma^2}{16}(\pi - 8/3).$$

► **Problem 1.19.9** *A solid body, of density ρ, is in the shape of a solid formed by the revolution of the cardioid $r = a(1 + \cos\theta)$ about the initial line, show that its moment of inertia about a straight line through the pole perpendicular to the initial line is $\frac{352}{105}\pi\rho a^5$.*

Solution: Let us consider an elementary area $rd\theta\, dr$ at $P(r, \theta)$. Let us also consider a circular ring obtained by revolving the elementary area about the axis Ox having radius $r \sin\theta$.

$\therefore$ The mass of the ring $= 2\pi r \sin\theta \cdot \rho \cdot r d\theta\, dr$.

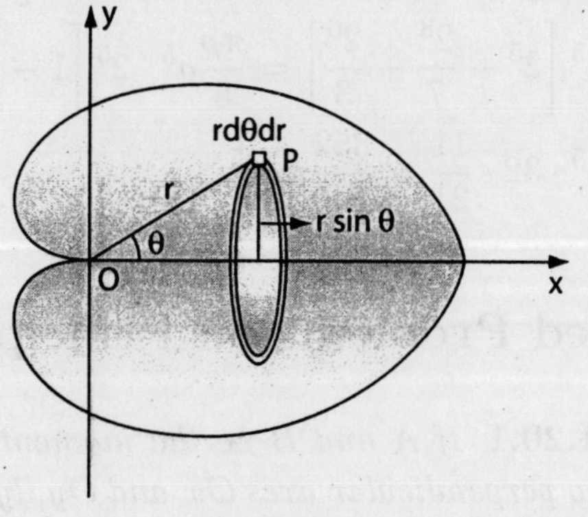

Fig 1.54

$\therefore$ The m.i. of the ring about the axis through its centre and parallel to y-axis

$$= \text{mass of the ring} \times \frac{(\text{radius})^2}{2} \quad [\text{by } (1.4.12)]$$

$$= 2\pi r \sin\theta \cdot r d\theta\, dr \rho \cdot \frac{r^2 \sin^2\theta}{2}.$$

$\therefore$ The m.i. of the ring about y-axis

$$
\begin{aligned}
&= 2\pi r\sin\theta\cdot r d\theta\, dr\rho\cdot\frac{r^2\sin^2\theta}{2} + 2\pi r\sin\theta\cdot r d\theta\, dr\rho\cdot r^2\cos^2\theta \\
&= 2\pi\rho\sin\theta\, r^4\left(\frac{\sin^2\theta}{2}+\cos^2\theta\right) d\theta\, dr \\
&= \pi\rho\sin\theta\cdot r^4(1+\cos^2\theta)\, dr\, d\theta.
\end{aligned}
$$

$\therefore$ The required m.i.

$$
\begin{aligned}
&= \pi\rho\int_{\theta=0}^{\pi}\int_{r=0}^{a(1+\cos\theta)}\sin\theta\cdot r^4\cdot(1+\cos^2\theta)\, dr\, d\theta \\
&= \pi\rho\int_{\theta=0}^{\pi}\sin\theta(1+\cos^2\theta)\frac{a^5(1+\cos\theta)^5}{5}d\theta. \\
&= -\pi\rho\int_2^0\frac{a^5}{5}z^5\{1+(z-1)^2\}dz \quad [\text{Putting } 1+\cos\theta = z \\
&\qquad\qquad\qquad\qquad\qquad\qquad -\sin\theta\, d\theta = dz] \\
&= \frac{\pi\rho}{5}a^5\int_0^2(z^7-2z^6+2z^5)\, dz = \frac{\pi\rho}{5}a^5\left[\frac{2^8}{8}-2\cdot\frac{2^7}{7}+2\cdot\frac{2^6}{6}\right] \\
&= \frac{\pi\rho}{5}a^5\left[2^5-\frac{2^8}{7}+\frac{2^6}{3}\right] = \frac{\pi\rho}{5}a^5\cdot 2^5\left[1-\frac{8}{7}+\frac{2}{3}\right] \\
&= \frac{\pi\rho}{5}a^5\cdot 32\cdot\frac{11}{21} = \frac{352}{105}\pi\rho a^5.
\end{aligned}
$$

1.20 Solved Problems on Principal Axes

► **Problem 1.20.1** *If A and B be the moments of inertia of a uniform lamina about two perpendicular axes Ox and Oy, lying in its plane, and F be the product of inertia of the lamina about these axes, show that the principal moments of inertia at O are equal to*

$$\frac{1}{2}\left[A+B\pm\sqrt{(A-B)^2+4F^2}\right].$$

Solution: Let Ox and Oy are two perpendicular lines are taken as reference axes. Let Ox', Oy' are the principal axes of moment of inertia and Ox' makes an angle θ with Ox. Let A, B and F are the moments and products of inertia of the lamina about the axes Ox and Oy. Let A', B' and F' are the moments and products of inertia about the principal axes Ox' and Oy'

having d.c's $(\cos\theta, \sin\theta, 0)$ and $(-\sin\theta, \cos\theta, 0)$ respectively.

$$\begin{aligned} \therefore \quad A' &= A\cos^2\theta + B\sin^2\theta - 2F\sin\theta\cos\theta \\ &= \frac{A+B}{2} + \frac{A-B}{2}\cos 2\theta - F\sin 2\theta \qquad (1.20.1) \\ B' &= A\sin^2\theta + B\cos^2\theta + 2F\sin\theta\cos\theta \\ &= \frac{A+B}{2} - \frac{A-B}{2}\cos 2\theta + F\sin 2\theta. \qquad (1.20.2) \end{aligned}$$

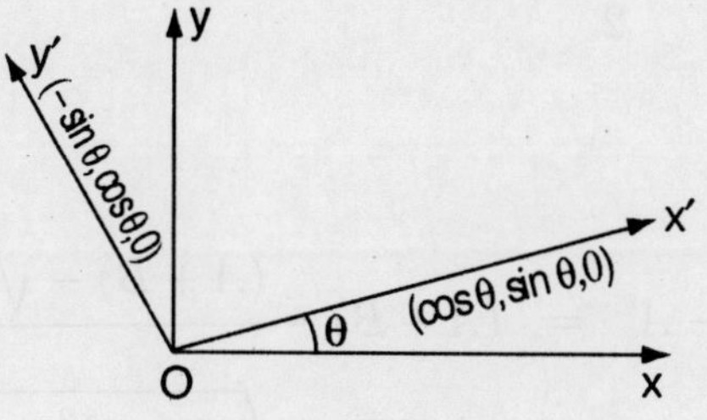

Fig 1.55

and

$$\begin{aligned} F' &= -[A\cos\theta(-\sin\theta) + B\sin\theta\cos\theta] + F(-\sin^2\theta + \cos^2\theta) \\ &= \frac{A}{2}\sin 2\theta - \frac{B}{2}\sin 2\theta + F\cos 2\theta \\ &= \frac{A-B}{2}\sin 2\theta + F\cos 2\theta. \qquad (1.20.3) \end{aligned}$$

$\therefore$ Adding (1.20.1) and (1.20.2) we get

$$A' + B' = A + B. \qquad (1.20.4)$$

Again as Ox' and Oy' are principal axes, then $F' = 0$.

$$\therefore \quad \frac{A-B}{2}\sin 2\theta + F\cos 2\theta = 0,$$

$$\text{or,} \quad (A-B)\sin 2\theta + 2F\cos 2\theta = 0,$$

$$\text{or,} \quad \frac{\sin 2\theta}{2F} = \frac{\cos 2\theta}{B-A} = \frac{1}{\sqrt{(A-B)^2 + 4F^2}}.$$

$$\therefore \quad \sin 2\theta = \frac{2F}{\sqrt{(A-B)^2 + 4F^2}}$$

$$\text{and} \quad \cos 2\theta = \frac{B-A}{\sqrt{(A-B)^2 + 4F^2}}. \qquad (1.20.5)$$

Thus from (1.20.1),

$$\begin{aligned}
A' &= \frac{A+B}{2}+\frac{A-B}{2}\cdot\frac{B-A}{\sqrt{(A-B)^2+4F^2}}-F\cdot\frac{2F}{\sqrt{(A-B)^2+4F^2}}\\
&= \frac{A+B}{2}-\frac{1}{2}\frac{(A-B)^2+4F^2}{\sqrt{(A-B)^2+4F^2}}=\frac{A+B}{2}-\frac{1}{2}\left[\sqrt{(A-B)^2+4F^2}\right]\\
&= \frac{(A+B)-\sqrt{(A-B)^2+4F^2}}{2}
\end{aligned}$$

From (1.20.4),

$$\begin{aligned}
B'=A+B-A' &= (A+B)-\frac{(A+B)-\sqrt{(A-B)^2+4F^2}}{2}\\
&= \frac{A+B}{2}+\frac{\sqrt{(A-B)^2+4F^2}}{2}\\
&= \frac{1}{2}\left[(A+B)+\sqrt{(A-B)^2+4F^2}\right].
\end{aligned}$$

► **Problem 1.20.2** *The lengths AB and AD of the sides of rectangle ABCD are* $2a, 2b$; *show that the inclination of AB of one of the principal axis at A is*

$$\frac{1}{2}\tan^{-1}\frac{3ab}{2(a^2-b^2)}.$$

[KUH '03]

Solution: Let M be the mass of the rectangle of sides $2a$ and $2b$. We take the axes Ax and Ay along the sides AB and AD. We take Gx_0, Gy_0 axes at the centre of gravity G of the rectangle $ABCD$, parallel to Ax and Ay. Let A, B and F be the moments and product of inertia about Ax and Ay and A_0, B_0, F_0 are the moments and product of inertia about Gx_0 and Gy_o.

$$\therefore\quad A_0=\frac{Mb^2}{3},\quad B_0=\frac{Ma^2}{3},\quad F_0=0.$$

$$\begin{aligned}
\therefore\quad A &= A_0+Mb^2=\frac{Mb^2}{3}+Mb^2=\frac{4M}{3}b^2\\
B &= B_0+Ma^2=\frac{Ma^2}{3}+Ma^2=\frac{4M}{3}a^2\\
F &= F_o+Mab=0+Mab=Mab.
\end{aligned}$$

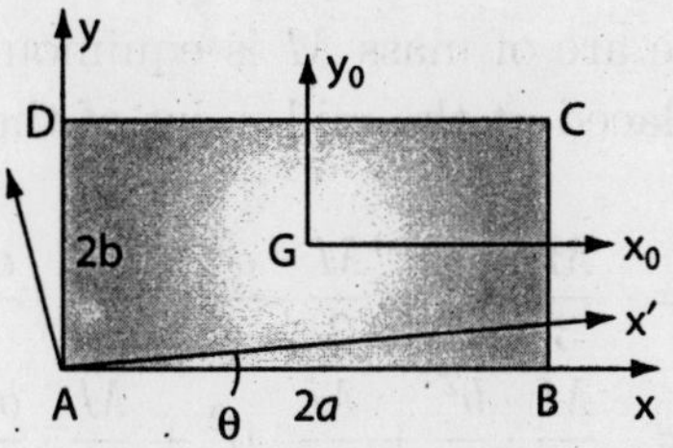

Fig 1.56

Let Ax' and Ay' are the principal axes at A where Ax' makes an angle θ with Ax.

Thus

$$\tan 2\theta = \frac{2F}{B-A} = \frac{2Mab}{\frac{4M}{3}a^2 - \frac{4M}{3}b^2} = \frac{3ab}{2(a^2-b^2)}.$$

$$\therefore \quad \theta = \frac{1}{2}\tan^{-1}\frac{3ab}{2(a^2-b^2)}.$$

▶ **Problem 1.20.3** *At the vertex C of a triangle ABC, which is right angled at C; the principal axes are, a perpendicular to the plane and two others inclined to the sides at an angle*

$$\frac{1}{2}\tan^{-1}\frac{ab}{a^2-b^2}.$$

[BU H '02; KUH '05]

Solution: We take CA and CB as axes, where ABC is the triangle and $\angle ACB = 90°$. Let M be the mass of the triangle. Let Cx', Cy' and Cz' are principal axis and Cz' is perpendicular to ABC at C and Cx' makes an angle θ with CA. Let $CA = b$ and $CB = a$ and D, E, F are the mid-points of BA, CA and CB respectively. Thus the coordinates of D, E and F are

$$\left(\frac{b}{2},\frac{a}{2}\right), \quad \left(\frac{b}{2},0\right) \quad \text{and} \quad \left(0,\frac{a}{2}\right).$$

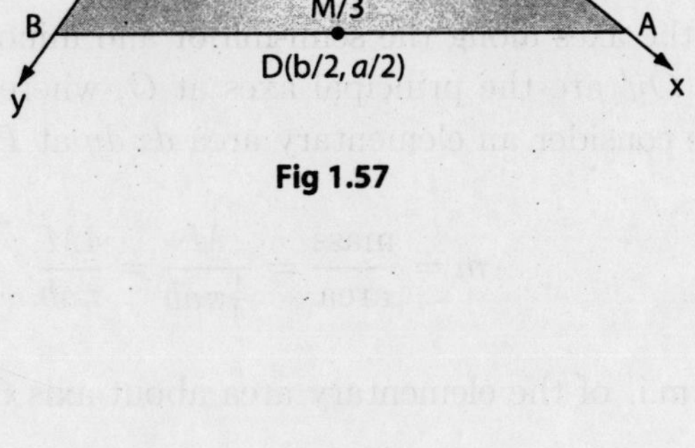

Fig 1.57

As the triangle are of mass M is equimomental with the three particles each mass $M/3$ placed at the mid-point of the sides; we have.

$$\begin{aligned} A &= \frac{M}{3}\cdot 0^2 + \frac{M}{3}\cdot\frac{a^2}{4} + \frac{M}{3}\cdot\frac{a^2}{4} = \frac{Ma^2}{6} \\ B &= \frac{M}{3}\cdot\frac{b^2}{4} + \frac{M}{3}\cdot 0^2 + \frac{M}{3}\cdot\frac{b^2}{4} = \frac{Mb^2}{6} \\ F &= \frac{M}{3}\cdot\frac{a}{2}\cdot 0 + \frac{M}{3}\cdot\frac{b}{2}\cdot 0 + \frac{M}{3}\cdot\frac{a}{2}\cdot\frac{b}{2} = \frac{Mab}{12}, \end{aligned}$$

where A, B and F are the moments of product of inertia with respect to CA and CB as axes.

As θ be the angle made by the principal axis Cx' with the axis Cx,

$$\tan 2\theta = \frac{2F}{B-A} = \frac{2\cdot\frac{Mab}{12}}{\frac{M}{6}(b^2-a^2)} = \frac{ab}{a^2-b^2}.$$

$$\therefore\quad \theta = \frac{1}{2}\tan^{-1}\frac{ab}{b^2-a^2}.$$

$$\therefore\quad |\theta| = \left|-\frac{1}{2}\tan^{-1}\frac{ab}{a^2-b^2}\right| = \frac{1}{2}\tan^{-1}\frac{ab}{a^2-b^2}.$$

► **Problem 1.20.4** *Show that at the centre of the quadrant of an ellipse the principal axes in its plane are inclined at an angle*

$$\frac{1}{2}\tan^{-1}\frac{4ab}{\pi(a^2-b^2)}.$$

Solution: Let M be the mass of the quadrant of the ellipse

$$\frac{x^2}{a^2} + \frac{y^2}{b^2} = 1.$$

We take the axes along the semi-major and minor axes at the centre O. Let Ox' and Oy' are the principal axes at O, where Ox' makes an angle θ with Ox. We consider an elementary area $dx\,dy$ at $P(x, y)$.

Let

$$m = \frac{\text{mass}}{\text{area}} = \frac{M}{\frac{1}{4}\pi ab} = \frac{4M}{\pi ab}.$$

Now the m.i. of the elementary area about axis $Ox = m dx\,dy\cdot y^2$.

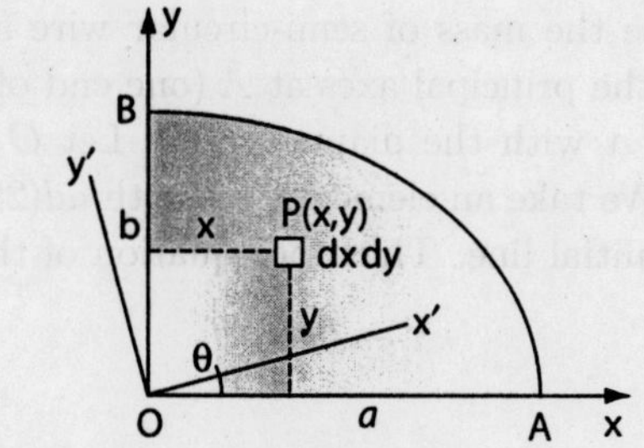

Fig 1.58

Thus the moment of inertia of the quadrant about Ox

$$
\begin{aligned}
A &= \int_{x=0}^{a}\int_{y=0}^{b\sqrt{1-x^2/a^2}} m\,dx\,dy\cdot y^2 = m\int_0^a dx\cdot\left|\frac{y^3}{3}\right|_0^{\frac{b}{a}\sqrt{a^2-x^2}} \\
&= m\int_0^a dx\cdot\frac{b^3}{3a^3}(a^2-x^2)^{3/2} = \frac{mb^3}{3a^3}\int_0^a (a^2-x^2)^{3/2}\,dx \\
&\qquad [\text{putting } x = a\sin\theta, dx = a\cos\theta\,d\theta] \\
&= \frac{mb^3}{3a^3}\int_0^{\pi/2} a^3\cos^3\theta\cdot\cos\theta\,d\theta = \frac{mb^3a}{3}\int_0^{\pi/2}\cos^4\theta\,d\theta \\
&= \frac{mab^3}{3}\cdot\frac{3}{4}\cdot\frac{1}{2}\cdot\frac{\pi}{2} = \frac{mab^3}{16}\pi = \frac{4M}{\pi ab}\cdot\frac{ab^3}{16}\pi = \frac{Mb^2}{4}.
\end{aligned}
$$

Similarly, m.i. of the quadrant about Oy

$$B = \int_{x=0}^{a}\int_{y}^{\frac{b}{a}\sqrt{a^2-x^2}} m\,dx\,dy\cdot x^2 = \frac{Ma^2}{4}$$

and the product of inertia about Ox and Oy

$$B = \int_{x=0}^{a}\int_{y=0}^{\frac{b}{a}\sqrt{a^2-x^2}} m\,dx\,dy\cdot xy = \frac{Mab}{2\pi}.$$

If θ be angle made by the principal axes Ox' with Ox, we have

$$
\begin{aligned}
\tan 2\theta &= \frac{2F}{B-A} = \frac{\frac{2Mab}{2\pi}}{\frac{M}{4}(b^2-a^2)} = \frac{4ab}{\pi(a^2-b^2)}. \\
\therefore\quad \theta &= \frac{1}{2}\tan^{-1}\left(\frac{4}{\pi}\cdot\frac{ab}{a^2-b^2}\right).
\end{aligned}
$$

▶ **Problem 1.20.5** *A wire is in the form of a semi-circle of radius a, show that at an end of its diameter the principal axes in its plane are inclined to the diameter at angles*

$$\frac{1}{2}\tan^{-1}\frac{4}{\pi}\quad \textit{and}\quad \pi/2+\frac{1}{2}\tan^{-1}\frac{4}{\pi}.$$

Solution: Let M be the mass of semi-circular wire and AB its diameter. Let Ax and Ay are the principal axes at A (one end of the diameter) where Ax makes an angle α with the diameter AB. Let O be the centre of the semi-circular wire. We take an elementary length $ad(2\theta)$ at $P(r,\theta)$ taking A as pole and AB as initial line. Thus the equation of the circular wire is

$$r = 2a\cos\theta.$$

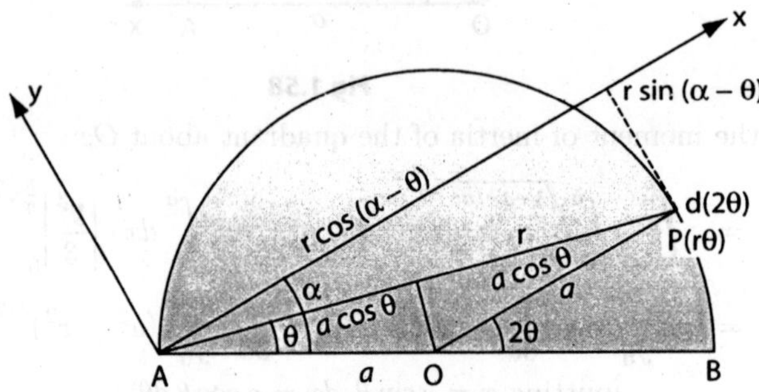

Fig 1.59

Let

$$m = \frac{\text{mass}}{\text{length}} = \frac{M}{\pi a}.$$

$\therefore$ The mass of the elementary length $= mad(2\theta)$. As Ax and Ay are the principal axes at A, the product of inertia of the circular wire with respect to Ax and Ay is zero.

$$\therefore \quad F = 0 = \int_0^{\pi/2} ma\,d(2\theta)\cdot r\cos(\alpha-\theta)\cdot[-r\sin(\alpha-\theta)]$$

$$= -2ma\int_0^{\pi/2} r^2\sin(\alpha-\theta)\cos(\alpha-\theta)\,d\theta.$$

$$\therefore \quad \int_0^{\pi/2} r^2\sin(\alpha-\theta)\cos(\alpha-\theta)\,d\theta = 0,$$

$$\text{or,} \quad \int_0^{\pi/2} 4a^2\cos^2\theta\sin(\alpha-\theta)\cos(\alpha-\theta)\,d\theta = 0,$$

$$\text{or,} \quad \int_0^{\pi/2} \cos^2\theta\sin(2\alpha-2\theta)\,d\theta = 0,$$

$$\text{or,} \quad \sin 2\alpha\int_0^{\pi/2}\cos^2\theta\cos 2\theta\,d\theta - \cos 2\alpha\int_0^{\pi/2}\cos^2\theta\sin 2\theta\,d\theta = 0,$$

$$\text{or,} \quad \sin 2\alpha\int_0^{\pi/2}(2\cos^4\theta-\cos^2\theta)\,d\theta - \cos 2\alpha\int_0^{\pi/2} 2\sin\theta\cos^3\theta\,d\theta = 0,$$

$$\text{or,}\quad \sin 2\alpha\left[2\cdot\frac{3}{4}\cdot\frac{1}{2}\cdot\frac{\pi}{2}-\frac{1}{2}\cdot\frac{\pi}{2}\right]-2\cos 2\alpha\int_0^1 z(1-z^2)\,dz=0$$

[putting $\sin\theta = z$],

$$\text{or,}\quad \sin 2\alpha\left[\frac{3\pi}{8}-\frac{\pi}{4}\right]-2\cos 2\alpha\left[\frac{1}{2}-\frac{1}{4}\right]=0,$$

$$\text{or,}\quad \sin 2\alpha\cdot\frac{\pi}{8}-2\cos 2\alpha\cdot\frac{1}{4}=0.$$

$$\therefore\quad \tan 2\alpha=\frac{4}{\pi};\quad \alpha=\frac{1}{2}\tan^{-1}\frac{4}{\pi}.$$

$$\therefore\quad \text{The other angle}=\frac{\pi}{2}+\alpha=\frac{\pi}{2}+\frac{1}{2}\tan^{-1}\frac{4}{\pi}.$$

▶ **Problem 1.20.6** *Show that the principal axes at the node of a half-loop of the lemniscate $r^2 = a^2\cos 2\theta$ are inclined to the initial line at angles*

$$\frac{1}{2}\tan^{-1}\frac{1}{2}\quad \textit{and}\quad \frac{\pi}{2}+\frac{1}{2}\tan^{-1}\frac{1}{2}.$$

Solution: We consider Ox and Oy as axes at the node O (pole) and Ox as initial line and Ox' and Oy' as principal axes where Ox' makes an angle α with Ox. Let m = mass/area of the lemniscate. We take an elementary area $rd\theta\,dr$ at $P(r,\theta)$. As Ox' and Oy' are principal axes, the product of inertia with respect to Ox' and Oy' is zero. The product of inertia with respect to Ox' and Oy'

$$\begin{aligned}F=0\ &= \int_{\theta=0}^{\pi/4}\int_{r=0}^{a\sqrt{\cos 2\theta}} mr\,d\theta\,dr\cdot r\cos(\alpha-\theta)[-r\sin(\alpha-\theta)]\\ &= -m\int_{\theta=0}^{\pi/4}\int_{r=0}^{a\sqrt{\cos 2\theta}} r^3\sin(\alpha-\theta)\cos(\alpha-\theta)\,dr\,d\theta\\ &= -\frac{m}{2}\int_{\theta=0}^{\pi/4}\int_{r=0}^{a\sqrt{\cos 2\theta}} r^3\sin(2\alpha-2\theta)\,dr\,d\theta,\end{aligned}$$

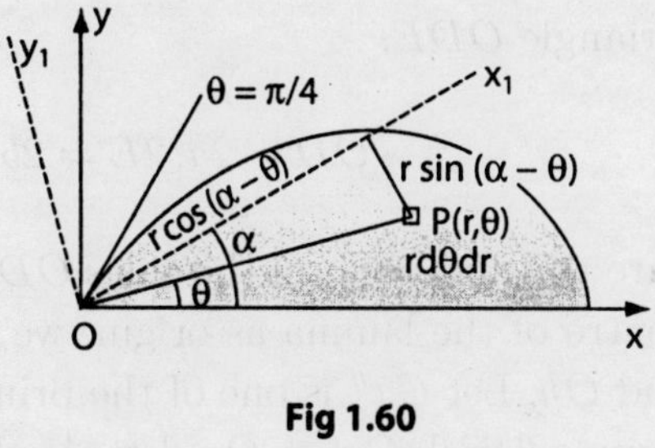

Fig 1.60

$$\text{or,}\quad \int_0^{\pi/4} \sin(2\alpha - 2\theta)\, d\theta \int_{r=0}^{a\sqrt{\cos 2\theta}} r^3\, dr = 0,$$

$$\text{or,}\quad \int_0^{\pi/4} \sin(2\alpha - 2\theta)\, d\theta \cdot \frac{a^4 \cos^2 2\theta}{4} = 0,$$

$$\text{or,}\quad \int_0^{\pi/4} (\sin 2\alpha \cos 2\theta - \cos 2\alpha \sin 2\theta) \cos^2 2\theta\, d\theta = 0,$$

$$\text{or,}\quad \int_0^{\pi/2} (\sin 2\alpha \cos z - \cos 2\alpha \sin z) \cos^2 z\, dz = 0 \quad [\text{putting } 2\theta = z],$$

$$\text{or,}\quad \sin 2\alpha \int_0^{\pi/2} \cos^3 z\, dz - \cos 2\alpha \int_0^{\pi/2} \cos^2 z \sin z\, dz = 0,$$

$$\text{or,}\quad \sin 2\alpha \cdot \tfrac{2}{3} + \cos 2\alpha \int_1^0 z^2\, dz = 0,$$

$$\text{or,}\quad \sin 2\alpha \cdot \tfrac{2}{3} - \cos 2\alpha \cdot \tfrac{1}{3} = 0.$$

$$\therefore\quad \tan 2\alpha = \tfrac{1}{2} \quad \therefore\quad \alpha = \tfrac{1}{2} \tan^{-1} \frac{1}{2}.$$

$$\therefore\quad \text{other angle} = \frac{\pi}{2} + \tfrac{1}{2} \tan^{-1} \frac{1}{2}.$$

▶ **Problem 1.20.7** *A uniform square lamina is bounded by the axes of x and y and the lines $x = 2c$, $y = 2c$ and a corner is cut off by the line $\frac{x}{a} + \frac{y}{b} = 2$, show that the principal axes at the centre of the square are inclined to the axis of x at an angle given by*

$$\tan 2\theta = \frac{ab - 2(a+b)c + 3c^2}{(a-b)(a+b-2c)}. \qquad \textbf{[BUH 1996]}$$

Solution: Let M be the mass of the square lamina $OABC$ and M' be the mass of cut-off triangle ODE.

$$\therefore\quad OD = 2a, 2E = 2b.$$

Let P, Q, R are the mid-point of the sides OD, DE and EO respectively. Taking G, the centre of the lamina as origin, we consider Gx' and Gy' axes parallel to Ox and Oy. Let Gx'' is one of the principal axes at the centre G, which makes an angle θ with Gx' or Ox. Let A_1, B_1 and F_1 are the moments and product of inertia of the whole lamina about Gx' and Gy'.

$$\therefore\quad A_1 = B_1 = \frac{M}{3} c^2 \quad \text{and} \quad F_1 = 0.$$

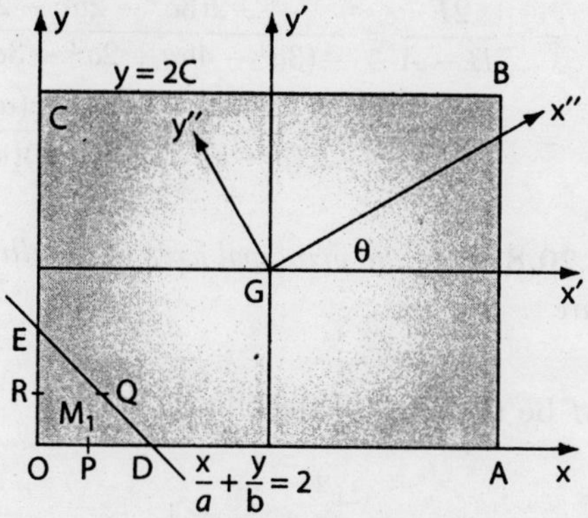

Fig 1.61

Let A, B and F are the m.i. and product of inertia of the cut off lamina about Gx' and Gy'.

Let A', B', F' be the moments and product of inertia of the cut-off triangle ODE about Gx' and Gy'. Thus the coordinates of mid-points are $P[-(c-a), -c]$, $Q[-(c-a), -(c-b)]$, $R[-c, -(c-b)]$ with respect to Gx' and Gy' as axes.

As triangular lamina ODE of mass M' equimomental with three particles each of mass $M'/3$ is placed at mid-point of the sides we have

$$\begin{aligned} A' &= \frac{M'}{3}\left[(c-b)^2+c^2+(c-b)^2\right] = \frac{M'}{3}\left[3c^2-4bc+2b^2\right] \\ B' &= \frac{M'}{3}\left[(c-a)^2+(c-a)^2+c^2\right] = \frac{M'}{3}\left[3c^2-4ac+2a^2\right] \\ F' &= \frac{M'}{3}\left[(c-b)(c-a)+c(c-a)+c(c-b)\right] \\ &= \frac{M'}{3}\left[3c^2-2ca-2bc+ab\right] \end{aligned}$$

Now the m.i. of the cut-off lamina about Gx' and Gy' are

A = (m.i. of the square lamina $OABC$ about Gx') – (m.i. of the triangular lamina ODE about Gx')

$$= A_1 - A' = \frac{M}{3}c^2 - \frac{M'}{3}[3c^2-4bc+2b^2].$$

Similarly, moment of inertia of the cut-off lamina about Gy'

$$\begin{aligned} B &= B_1 - B' = \frac{M}{3}c^2 - \frac{M'}{3}[3c^2-4ac+2a^2] \\ F &= 0 - F' = -\frac{M'}{3}[3c^2-2ac-2bc+ab]. \end{aligned}$$

$$\therefore \quad \tan 2\theta = \frac{2F}{B-A} = \frac{-2(3c^2 - 2ac - 2bc + ab)}{-(3c^2 - 4ac + 2a^2 - 3c^2 + 4bc - 2b^2)}$$
$$= \frac{3c^2 - 2c(a+b) + ab}{a^2 - b^2 - 2c(a-b)} = \frac{ab - 2c(a+b) + 3c^2}{(a-b)(a+b-2c)}.$$

► **Problem 1.20.8** *Find the principal axes of an elliptic area at any point of its bounding arc.*

Solution: Let M be the mass of the elliptic area

$$\frac{x^2}{a^2} + \frac{y^2}{b^2} = 1$$

and $P(a\cos\theta, b\sin\theta)$ be a point on the bounding arc. We take Px' and Py' as axes parallel to the axes Ox and Oy. Let Px'' and Py'' be the principal axes at P and Px'' makes an angle α with Px' or Ox. Let A_0, B_0 and F_0 are the moments and product of inertia with respect to Ox and Oy and A, B and F be the moments and product of inertia with respect to Px' and Py'.

$$\therefore \quad A_0 = \frac{Mb^2}{4} \quad \text{and} \quad B_0 = \frac{Ma^2}{4} \quad \text{and} \quad F = 0.$$

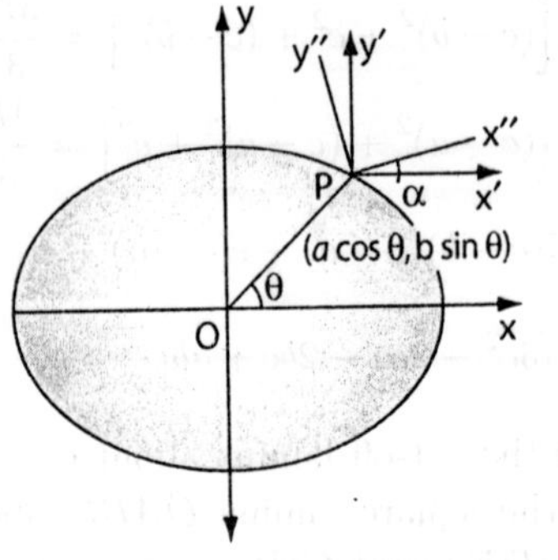

Fig 1.62

Thus

$$A = A_0 + Mb^2\sin^2\theta = \frac{Mb^2}{4} + Mb^2\sin^2\theta = \frac{Mb^2}{4}(1 + 4\sin^2\theta)$$
$$B = B_0 + Ma^2\cos^2\theta = \frac{Ma^2}{4} + Ma^2\cos^2\theta = \frac{Ma^2}{4}(1 + 4\cos^2\theta)$$
$$F = F_0 + Mab\sin\theta\cos\theta = 0 + Mab\sin\theta\cos\theta = Mab\sin\theta\cos\theta.$$

$$\therefore \tan 2\alpha = \frac{2F}{B-A} = \frac{2Mab\sin\theta\cos\theta}{\frac{M}{4}[a^2+4a^2\cos^2\theta - b^2 - 4b^2\sin^2\theta]}$$

$$= \frac{8ab\sin\theta\cos\theta}{a^2-b^2+4a^2\cos^2\theta-4b^2\sin^2\theta}$$

$$= \frac{4ab\sin 2\theta}{(a^2-b^2)(\sin^2\theta+\cos^2\theta)+4a^2\cos^2\theta-4b^2\sin^2\theta}$$

$$= \frac{4ab\sin 2\theta}{(5a^2-b^2)\cos^2\theta+(a^2-5b^2)\sin^2\theta}.$$

▶ **Problem 1.20.9** *Show that at an extremity of the bounding diameter of a semi-circle lamina, the principal axis makes an angle* $\frac{1}{2}\tan^{-1}\frac{8}{3\pi}$ *to the diameter.* **[KUH '06]**

Solution: Let $r = 2a\cos\theta$ be the equation of the circle, where O is the pole and $OD(=2a)$ as initial line. Let us consider an elementary are $r\delta\theta\delta r$ at the point $P(r,\theta)$.

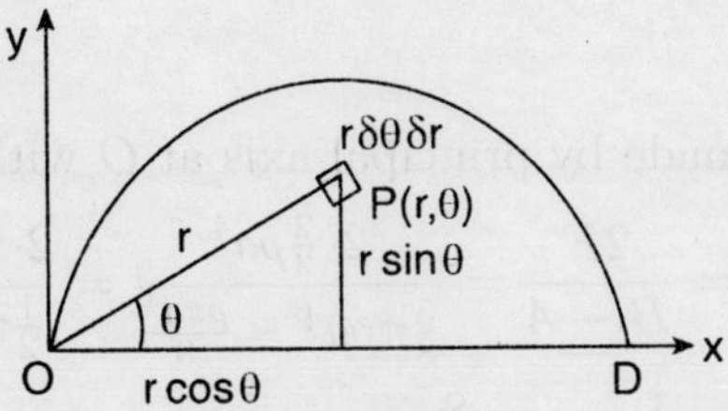

Fig 1.63

Therefore, the moment of inertia about ox

$$A = \rho\int_0^{\pi/2}\int_0^{2a\cos\theta} rd\theta dr(r^2\sin^2\theta), \quad \text{where } \rho \text{ is the density}$$

$$= \rho\int_0^{\pi/2}\int_0^{2a\cos\theta} r^3\sin^2\theta\cdot dr\,d\theta$$

$$= \rho\int_0^{\pi/2}\left.\frac{r^4}{4}\right|_0^{2a\cos\theta}\cdot\sin^2\theta\cdot d\theta$$

$$= \rho\int_0^{\pi/2}\frac{(2a)^4}{4}\cdot\cos^4\theta\sin^2\theta\,d\theta$$

$$= 4\rho a^4\int_0^{\pi/2}\cos^4\theta\sin^2\theta\,d\theta = 4\rho a^4\cdot\frac{3}{4+2}\cdot\frac{1}{2+2}\cdot\frac{1}{2}\cdot\frac{\pi}{2}$$

$$= 4\rho\cdot a^4\frac{3}{6\cdot 4\cdot 4}\pi = \frac{\rho a^4\pi}{8}$$

B = moment of inertia about Oy

$$= \rho \int_0^{\pi/2} \int_0^{2a\cos\theta} r d\theta\, dr (r^2 \cos^2\theta) = \rho \int_0^{\pi/2} \int_0^{2a\cos\theta} r^3 \cos^2\theta dr\, d\theta$$

$$= \rho \int_0^{\pi/2} \frac{r^4}{4}\bigg|_0^{2a\cos\theta} \cos^2\theta d\theta = \rho \int_0^{\pi/2} \frac{(2a)^4}{4} \cos^4\theta \cdot \cos^2\theta d\theta$$

$$= 4\rho a^4 \int_0^{\pi/2} \cos^6\theta d\theta = 4\rho a^4 \cdot \frac{5}{6} \cdot \frac{3}{4} \cdot \frac{1}{2} \cdot \frac{\pi}{2}$$

$$= \frac{5}{8} \pi \rho a^4$$

F = product of inertia about Ox and Oy

$$= \rho \int_0^{\pi/2} \int_0^{2a\cos\theta} r\cos\theta \cdot r\sin\theta \cdot r dr\, d\theta = \rho \int_0^{\pi/2} \int_0^{2a\cos\theta} r^3 \sin\theta\cos\theta dr\, d\theta$$

$$= \rho \int_0^{\pi/2} \frac{r^4}{4}\bigg|_0^{2a\cos\theta} \sin\theta\cos\theta\, d\theta = \rho \frac{(2a)^4}{4} \int_0^{\pi/2} \cos^5\theta \cdot \sin\theta\, d\theta$$

$$= 4\rho a^4 \cdot \frac{1}{6} = \frac{2}{3} \rho a^4$$

If ψ be the angle made by principal axis at O with Ox, then

$$\tan 2\psi = \frac{2F}{B - A} = \frac{2 \cdot \frac{2}{3}\rho a^4}{\frac{5}{8}\pi\rho a^4 - \frac{\rho\pi a^4}{8}} = \frac{2 \cdot \frac{2}{3}}{\frac{1}{2}\pi} = \frac{8}{3\pi}$$

$$\psi = \frac{1}{2} \tan^{-1} \frac{8}{3\pi}.$$

► **Problem 1.20.10** *Show that the principle axes at a corner of rectangular lamina of sides* $2a, 2b$ *makes angles with the two sides* θ *and* $\frac{\pi}{2} + \theta$, *where* $\tan 2\theta = \frac{3ab}{2(a^2+b^2)}$. **[KUH '03]**

Solution: Let $ABCD$ be the rectangle of length $2a$ and $2b$ of mass M. Let O be the c.g. of the rectangle. Let Ax' and Ay' are the principle axes at the corner A. Let Ax' makes an angle θ with AB.

Let Ox and Oy are the axes parallel to AB and AD.

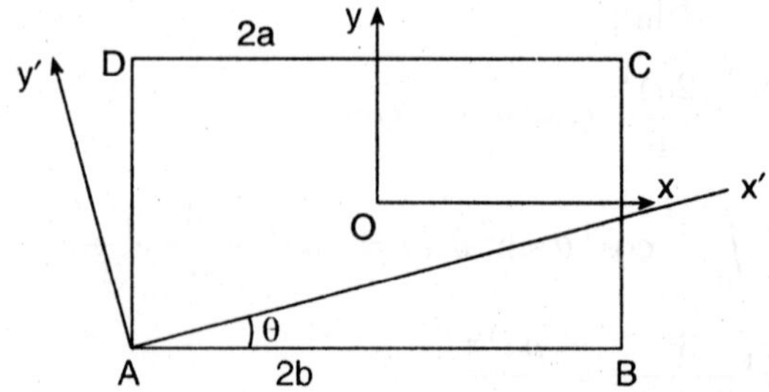

Fig 1.64

Now $\quad A$ = moment of inertia of the rectangle about AB

= moment of inertia of the rectangle about Ox + moment of inertia of the mass M placed at O about AD

$$= \frac{M}{3}b^2 + Mb^2 = \frac{4Mb^2}{3}$$

B = moment of inertia of the rectangle about AD

$$= \frac{M}{3}a^2 + Ma^2 = \frac{4Ma^2}{3}$$

F = product of inertia about AB and AD

= product of inertia of the rectangle about Ox and Oy + product of inertia of the mass M placed at O about AB and AD

$= 0 + Mab = Mab.$

Therefore
$$\tan 2\theta = \frac{2F}{B - A} = \frac{Mab}{\frac{4Ma^2}{3} - \frac{4Mb^2}{3}} = \frac{3}{4}\frac{ab}{(a^2 - b^2)}.$$

EXERCISES

1. Show that the product of inertia of the first quadrant of an elliptic lamina with respect to the major and minor axes of length 2a, 2b is $M\frac{ab}{2\pi}$, where M is the mass of the ellipse.

2. Show that the product of inertia of an octant of an ellipsoid is $\frac{2Mab}{5\pi}$, where M is the mass of the ellipsoid.

3. Show that the moment of inertia of a semi-circular lamina about a tangent parallel to the bounding diameter is $Ma^2\left(\frac{5}{4} - \frac{8}{3\pi}\right)$ where M is the mass of the lamina.

4. Show that the moment of inertia of a rectangle of sides 2a and 2b about a line through its centre and in its plane making an angle θ with the side 2a is $\frac{1}{3}M(b^2\cos^2\theta + a^2\sin^2\theta)$, where M is the mass of the rectangle.

5. Find the moment of inertia of an elliptic area about a line CP inclined at θ to the major axis and about a tangent parallel to CP, C being the centre and P is a point on the ellipse. [CUH 1987]

   [*Ans.* $-\frac{M}{4}(b^2\cos^2\theta + a^2\sin^2\theta)$, $\frac{5M}{4}(b^2\cos^2\theta + a^2\sin^2\theta)$]

Chapter 2

General Equations of Motion of Material System

2.1 Definitions

(i) **Impressed force:** The external force acting on a body or a particles or a system as particles, exerted on it by some external agency is called the *external impressed force.*

(ii) **Effective force:** The rate of change of momentum of a moving particles is called the *effective force on the particles.*

Thus it f be the acceleration of a particles of man m, then mf is called the *effective force* on the particles and $-mf$ is called the *reversed effective force* or *inertia force* on the body.

2.2 D'Alembert's Principle

The D'Alembert's principle states that the reversed effective forces acting on each particle of a body and the external forces acting on the body are in equilibrium.

[The D'Alembert's principle assume that the internal forces of the body in motion are in equilibrium among themselves.]

Verification: If (x, y, z) be the position of a particle of mass m at time t and X, Y, Z be the components of external force, along parallel to the axes, acting on it, then by Newton's Second law of forces, we have

$$m\frac{d^2x}{dt^2} = X; \qquad m\frac{d^2y}{dt^2} = Y; \qquad m\frac{d^2z}{dt^2} = Z,$$

$$\text{i.e.,} \quad X - m\frac{d^2x}{dt^2} = 0; \quad Y - m\frac{d^2y}{dt^2} = 0; \quad Z - m\frac{d^2z}{dt^2} = 0.$$

Thus the particle is in equilibrium.

If however,the system consists of a system of particles or a rigid body, we must include the internal forces acting on the particles which are due to the action (attraction or repulsion) of the rest of the particles of the system on it. If X_1, Y_1, Z_1 be the algebraic sum of the components of the internal forces acting on it parallel to the coordinate axes, we have

$$m\frac{d^2x}{dt^2} = X + X_1, \qquad m\frac{d^2y}{dt^2} = Y + Y_1, \qquad m\frac{d^2z}{dt^2} = Z + Z_1,$$

$$\text{i.e.,} \quad X + X_1 - m\frac{d^2x}{dt^2} = 0; \quad Y + Y_1 - m\frac{d^2y}{dt^2} = 0, \quad Z + Z_1 - m\frac{d^2z}{dt^2} = 0.$$

Hence the reversed effective force, the external force and the internal forces, acting on a particle of mans m of the system are in equilibrium. This is true for every particle of the system.

Therefore for the whole system, reversed effective forces acting on all particles, the external forces and the internal forces for the body are in equilibrium. Now by Newton's third low, to every action, there is an equal and opposite reaction, the internal forces among themselves are in equilibrium. Hence the principle.

2.3 Equations of Motion of a Rigid Body Due to D'Alembert's Principle [K. U. '05]

Let (X, Y, Z) be the components of the external forces, parallel to the coordinate axes, acting on a particle of mass m at (x, y, z) at time t. Then by D'Alembert's principle the forces

$$X - m\frac{d^2x}{dt^2}, \quad Y - m\frac{d^2y}{dt^2}, \quad Z - m\frac{d^2z}{dt^2}$$

at (x, y, z) together with similar forces acting on each particles of the body are in equilibrium.

Therefore, the algebraic sum of the forces parallel to the axes are equal to zero.

$$\text{i.e.,}\quad \sum\left(X - m\frac{d^2x}{dt^2}\right) = 0 \quad \text{or,} \quad \sum m\frac{d^2x}{dt^2} = \sum X \qquad (2.3.1)$$

$$\sum\left(Y - m\frac{d^2y}{dt^2}\right) = 0 \quad \text{or,} \quad \sum m\frac{d^2y}{dt^2} = \sum Y \qquad (2.3.2)$$

$$\sum\left(Z - m\frac{d^2z}{dt^2}\right) = 0 \quad \text{or,} \quad \sum m\frac{d^2z}{dt^2} = \sum Z. \qquad (2.3.3)$$

And the algebraic sums of moment of the forces about the axes are also equal to zero.

$$\text{i.e.,}\quad \sum\left\{y\left(Z - m\frac{d^2z}{dt^2}\right) - z\left(Y - m\frac{d^2y}{dt^2}\right)\right\} = 0$$

$$\sum\left\{z\left(X - m\frac{d^2x}{dt^2}\right) - x\left(Z - m\frac{d^2z}{dt^2}\right)\right\} = 0$$

$$\sum\left\{x\left(Y - m\frac{d^2y}{dt^2}\right) - y\left(X - m\frac{d^2x}{dt^2}\right)\right\} = 0.$$

$$\left[\because \sum\begin{vmatrix} i & j & k \\ x & y & z \\ X - m\ddot{x} & Y - m\ddot{y} & Z - m\ddot{z}\end{vmatrix}\right]$$

$$\text{or,}\quad \sum m\left(y\frac{d^2z}{dt^2} - z\frac{d^2y}{dt^2}\right) = \sum(yY - zY) = L \quad (\text{say}) \qquad (2.3.4)$$

$$\sum m\left(z\frac{d^2x}{dt^2} - x\frac{d^2z}{dt^2}\right) = \sum(zX - xZ) = M \quad (\text{say}) \qquad (2.3.5)$$

$$\sum m\left(x\frac{d^2y}{dt^2} - y\frac{d^2x}{dt^2}\right) = \sum(xY - yX) = N \quad (\text{say}), \qquad (2.3.6)$$

where L, M, N are the algebraic sum of the moment of the external forces about the coordinates axes.

These six equations namely (2.3.1) to (2.3.6) are the equations of motion for any rigid body. The equations (2.3.1), (2.3.2)and (2.3.3) states that the sums of the effective forces are respectively equal to the algebraic sums of the components of the external forces, parallel to the axes. The equation (2.3.4), (2.3.5) and (2.3.6) states that the sum of the moments of the effective forces about the coordinate axes are respectively equal to the algebraic sum of the moments of the external forces about the same axes.

[The equation (2.3.4), (2.3.5) and (2.3.6) can be written as the scheme

$$\text{LHS} = \sum\begin{vmatrix} i & j & k \\ x & y & z \\ m\ddot{x} & m\ddot{y} & m\ddot{z}\end{vmatrix} \quad \text{and} \quad \text{RHS} = \sum\begin{vmatrix} i & j & k \\ x & y & z \\ X & Y & Z\end{vmatrix}\,]$$

2.4 Motion of Centre of Inertia and Equations of Motion Relative to the Centre of Inertia

[CUH '02, '04, VUH '01, '02, '04, '05; KUH '01, '04 '06, '08]

Let $P(x, y, z)$ be the position of an elementary mass m at time t, referred to the axes Ox, Oy and Oz, fixed in space. Let $(\bar{x}, \bar{y}, \bar{z})$ be the coordinates of centre of inertia G of the body at that instant referred to axes Ox, Oy and Oz and (x', y', z') be the coordinates of the same elementary mass m at the same instant referred to the parallel axes Gx', Gy' and Gz' at centre of inertia G.

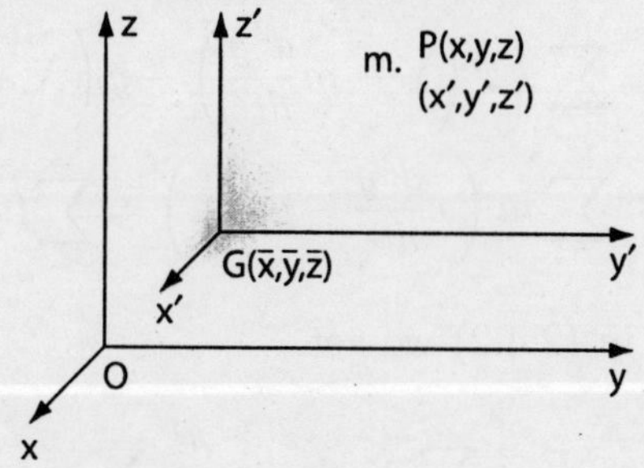

Fig 2.1

Then

$$\left.\begin{array}{l} x = \bar{x} + x', \quad y = \bar{y} + y' \quad \text{and} \quad z = \bar{z} + z' \\ \text{and} \quad M\bar{x} = \sum mx, \quad M\bar{y} = \sum my \quad \text{and} \quad M\bar{z} = \sum mz \\ \sum mx' = \sum my' = \sum mz' = 0 \end{array}\right\}, \tag{2.4.1}$$

where M is the mass of the body.

If (X, Y, Z) be the components of the external forces acting on the elementary mass m at $P(x, y, z)$ at time t, then by D'Alembert's principle, we have

$$\sum\left(X - m\frac{d^2x}{dt^2}\right) = 0 \quad \text{or,} \quad \sum m\frac{d^2x}{dt^2} = \sum X \tag{2.4.2}$$

$$\sum\left(Y - m\frac{d^2y}{dt^2}\right) = 0 \quad \text{or,} \quad \sum m\frac{d^2y}{dt^2} = \sum Y \tag{2.4.3}$$

$$\sum\left(Z - m\frac{d^2z}{dt^2}\right) = 0 \quad \text{or,} \quad \sum m\frac{d^2z}{dt^2} = \sum Z \tag{2.4.4}$$

and

$$\sum\left\{y\left(Z-m\frac{d^2z}{dt^2}\right)-z\left(Y-m\frac{d^2y}{dt^2}\right)\right\}=0$$

$$\text{or,}\quad \sum m\left(y\frac{d^2z}{dt^2}-z\frac{d^2y}{dt^2}\right)=\sum(yZ-zY). \tag{2.4.5}$$

$$\sum\left\{z\left(X-m\frac{d^2x}{dt^2}\right)-x\left(Z-m\frac{d^2z}{dt^2}\right)\right\}=0$$

$$\text{or,}\quad \sum m\left(z\frac{d^2x}{dt^2}-x\frac{d^2z}{dt^2}\right)=\sum(zX-xZ) \tag{2.4.6}$$

$$\sum\left\{x\left(Y-m\frac{d^2y}{dt^2}\right)-y\left(X-m\frac{d^2x}{dt^2}\right)\right\}=0$$

$$\text{or,}\quad \sum m\left(x\frac{d^2y}{dt^2}-y\frac{d^2x}{dt^2}\right)=\sum(xY-yX). \tag{2.4.7}$$

Using (2.4.1) in (2.4.2) we get,

$$\sum m\frac{d^2}{dt^2}(\bar{x}+x')=\sum X \quad \text{or,} \quad \frac{d^2}{dt^2}\sum m(\bar{x}+x')=\sum X$$

$$\text{or,}\quad \frac{d^2}{dt^2}\left[\sum m\bar{x}+\sum mx'\right]=\sum X$$

$$\text{or,}\quad \frac{d^2}{dt^2}\sum m\bar{x}=\sum X \quad [\text{by (2.4.1)}]$$

$$\text{or,}\quad \frac{d^2\bar{x}}{dt^2}\sum m=\sum X.$$

$$\therefore \quad M\frac{d^2\bar{x}}{dt^2}=\sum X. \tag{2.4.8}$$

Similarly

$$M\frac{d^2\bar{y}}{dt^2}=\sum Y. \tag{2.4.9}$$

$$M\frac{d^2\bar{z}}{dt^2}=\sum Z. \tag{2.4.10}$$

Thus equation (2.4.8) is the equation of motion of a particle of mass M placed at centre of inertia (c.g.) of the body and acted on by a force ΣX parallel and equal to the sum of the components of the external forces in the direction of x-axis acting on the particles of the body. The interpretation of equations (2.4.9) and (2.4.10) will be same. Thus *the centre of inertia of a body moves as if whole mass of the body moves as if all the mass of the body were collected at it and as if all the external forces were acting at it in the directions parallel to these in which they act.*

Now using (2.4.1), we get from (2.4.5)

$$\sum m\left[(\bar{y}+y')\frac{d^2}{dt^2}(\bar{z}+z')-(\bar{z}+z')\frac{d^2}{dt^2}(\bar{y}+y')\right]$$

$$=\sum\left[(\bar{y}+y')Z-(\bar{z}+z')Y\right].$$

$$\text{or,}\quad \sum m\left[(\bar{y}+y')\left(\frac{d^2\bar{z}}{dt^2}+\frac{d^2z'}{dt^2}\right)-(\bar{z}+z')\left(\frac{d^2\bar{y}}{dt^2}+\frac{d^2y'}{dt^2}\right)\right]$$

$$=\bar{y}\sum Z-\bar{z}\sum Y+\sum(y'Z-z'Y)$$

$$\text{or,}\quad \sum m\left[\bar{y}\frac{d^2\bar{z}}{dt^2}-\bar{z}\frac{d^2\bar{y}}{dt^2}\right]+\sum m\left(\bar{y}\frac{d^2z'}{dt^2}-\bar{z}\frac{d^2y'}{dt^2}\right)$$

$$+\sum m\left[y'\frac{d^2\bar{z}}{dt^2}-z'\frac{d^2\bar{y}}{dt^2}\right]+\sum m\left[y'\frac{d^2z'}{dt^2}-z'\frac{d^2y'}{dt^2}\right]$$

$$=\bar{y}\sum Z-\bar{z}\sum Y+\sum(y'Z-z'Y)$$

$$\text{or,}\quad \bar{y}\sum m\frac{d^2\bar{z}}{dt^2}-\bar{z}\sum m\frac{d^2\bar{y}}{dt^2}+\bar{y}\frac{d^2}{dt^2}\left(\sum mz'\right)-\bar{z}\frac{d^2}{dt^2}\left(\sum my'\right)$$

$$+\frac{d^2\bar{z}}{dt^2}\sum my'-\frac{d^2\bar{y}}{dt^2}\sum mz'+\sum m\left[y'\frac{d^2z'}{dt^2}-z'\frac{d^2y'}{dt^2}\right]$$

$$=\bar{y}\sum Z-\bar{z}\sum Y+\sum(y'Z-z'Y)$$

$$\text{or,}\quad \bar{y}\sum Z-\bar{z}\sum Y+\sum m\left[y'\frac{d^2z'}{dt^2}-z'\frac{d^2y'}{dt^2}\right]$$

$$=\bar{y}\sum Z-\bar{z}\sum Y+\sum(y'Z-z'Y)$$

[by (2.4.1), (2.4.9) and (2.4.10)]

$$\text{or,}\quad \sum m\left[y'\frac{d^2z'}{dt^2}-z'\frac{d^2y'}{dt^2}\right]=\sum(y'Z-z'Y). \tag{2.4.11}$$

Similarly from (2.3.5) and (2.3.6), we get

$$\sum m\left[z'\frac{d^2x'}{dt^2}-x'\frac{d^2z'}{dt^2}\right]=\sum(z'X-x'Z) \tag{2.4.12}$$

$$\text{and}\quad \sum m\left[x'\frac{d^2y'}{dt^2}-y'\frac{d^2x'}{dt^2}\right]=\sum(x'Y-y'X). \tag{2.4.13}$$

The result (2.4.8), (2.4.9), (2.4.10) and (2.4.11), (2.4.12) and (2.4.13) shows that the motion of translation of the body can be considered independently of the motion of rotation by the first three. We also see that the motion of *centre of inertia* is to be found by the methods of dynamics of particles (neglecting the motion of rotation of the body). By the last three equations we see that the motion of rotation is reduced to finding that of a body about a fixed point.

Remark 2.4.1 *The equation (2.4.8), (2.4.9) and (2.4.10) can be written in the form*

$$\frac{d}{dt}\left[\sum m\frac{dx}{dt}\right]=\sum X,\quad \frac{d}{dt}\left[\sum m\frac{dy}{dt}\right]=\sum Y,\quad \frac{d}{dt}\left[\sum m\frac{dz}{dt}\right]=\sum Z$$

i. e., $\frac{d}{dt}$ *[sum of the* $\overrightarrow{\textit{momentums}}$ *along a direction] = Algebraic sum of the components of the* $\overrightarrow{\textit{applied forces}}$ *along the same direction.*

The equation (2.4.11), (2.4.12) and (2.4.13) can be written as

$$\frac{d}{dt}\left[\sum m\left(y\frac{dz}{dt}-z\frac{dy}{dt}\right)\right]=\sum(yZ-zY);$$
$$\frac{d}{dt}\left[\sum m\left(z\frac{dx}{dt}-x\frac{dz}{dt}\right)\right]=\sum(yX-xY);$$
$$\text{and}\quad \frac{d}{dt}\left[\sum m\left(x\frac{dy}{dt}-y\frac{dx}{dt}\right)\right]=\sum(xY-yX).$$

Thus $\frac{d}{dt}$ *[sum of moment of* $\overrightarrow{\textit{momentum}}$*] = Algebraic sums of the moments of the* $\overrightarrow{\textit{applied forces}}$*.*

The term moment of momentum sometime called angular momentum.

2.5 Solved Problems

► **Problem 2.5.1** *A plank, of mass M, is initially at rest along a line of greatest slope of a smooth plane inclined at an angle α to the horizon, and a man, of mass M', starting from the upper end walks down the plank so that it does not move, shew that he gets to the other end in time $\sqrt{\frac{2M'a}{(M+M')\,g\sin\alpha}}$, where a is the length of the plank.* **[CUH 1988, 1992, 2002, 2004]**

Solution: We take the axes Ox and Oy along and perpendicular at O. Let at any time t, P is the position of the man at a distance x from O. Let G be the c.g. of the plank OA of mass M and the weight Mg acts vertically downward through G. So the man exerts a force F (say) down the plane on the plank and therefore the plank also exerts a force F on the man up the plane.

Considering the motion of the plank (as the plank is at rest), we get

$$\text{M}.0 = Mg\sin\alpha + F. \tag{2.5.1}$$

Again considering the motion of the man, we get

$$M'.\frac{d^2x}{dt^2} = M'g\sin\alpha - F. \tag{2.5.2}$$

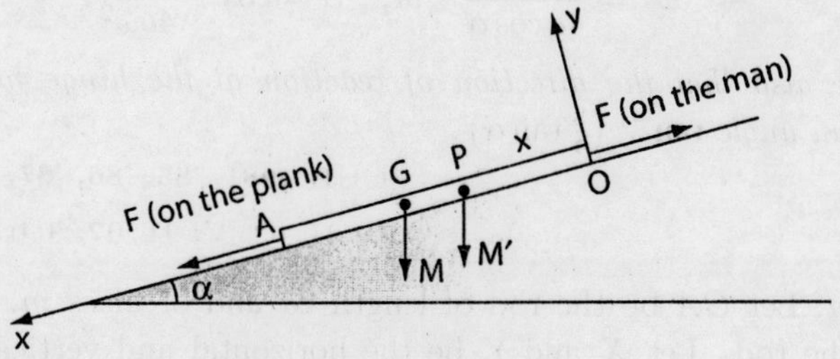

Fig 2.2

Adding (2.5.1) and (2.5.2), we get

$$\begin{aligned} M'.\frac{d^2x}{dt^2} &= (M+M')\,g\sin\alpha \\ \text{or,}\quad \frac{d^2x}{dt^2} &= \frac{M+M'}{M'}\,g\sin\alpha. \end{aligned}$$

$\therefore$ Integrating we get,

$$\frac{dx}{dt} = \frac{M+M'}{M'}\,g\sin\alpha\cdot t + C_1$$

Now $\frac{dx}{dt} = 0$, when $t = 0$. $\quad\therefore\ C_1 = 0$.

$$\therefore\quad \frac{dx}{dt} = \frac{M+M'}{M'}\cdot g\sin\alpha.t.$$

Again integrating, we have

$$x = \frac{M+M'}{M'}\cdot g\sin\alpha\cdot\frac{t^2}{2} + C_2,$$

As $x = 0$, when $t = 0$. $\quad\therefore\ C_2 = 0$.
Thus

$$x = \frac{M+M'}{M'}\cdot g\sin\alpha\cdot\frac{t^2}{2}.$$

Let $t = t_1$, when $x = a$.

$$\therefore\quad a = \frac{M+M'}{M'}\cdot g\sin\alpha\cdot\frac{t_1^2}{2}.$$

Thus

$$t_1 = \sqrt{\frac{2aM'}{(M+M')\,g\sin\alpha}}.$$

▶ **Problem 2.5.2** *A rod of length 2a, revolves with uniform angular velocity ω about a vertical axes through a smooth joint at one extremity of the rod so that it describe a cone of semi-vertical angle α, show that*

$$\omega^2 = \frac{3g}{4a\cos\alpha} \quad or, \quad \alpha = \cos^{-1}\frac{3g}{4a\omega^2}.$$

Prove also that the direction of reaction at the hinge makes with the vertical an angle $\tan^{-1}\left(\frac{3}{4}\tan\alpha\right)$.

[CUH 1981, '85, '86, '87, '93, 2001, '07; BUH '03, VUH '02, '04; KUH '03, '08]

Solution: Let OA be the rod of length $2a$ and of mass m. Let G be the c.g. of the rod. Let X and Y be the horizontal and vertical components of reaction at the hinge O and α is the semi-vertical angle of the cone. We consider an elementary length dr at P at a distance r from O. Thus the mass of the elementary length dr at P is $\frac{m}{2a}.dr$.

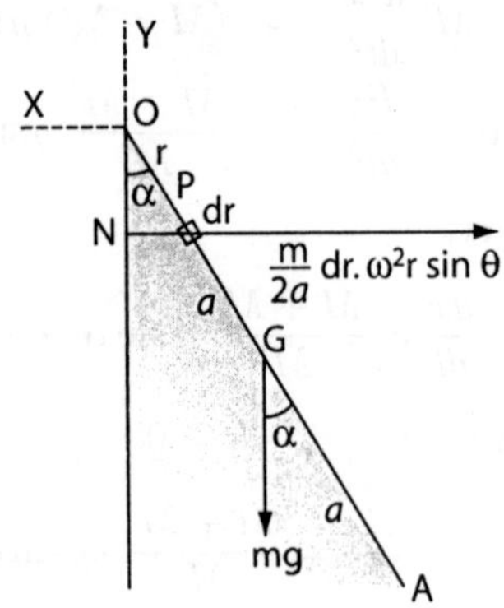

Fig 2.3

Thus the effective force on the elementary mass is

$$\frac{m}{2a}dr\cdot\omega^2\cdot PN = \frac{m}{2a}\omega^2\cdot r\sin\alpha\cdot dr.$$

Now $OG = GA = a$.
Now taking moment about O, we get

$$\begin{aligned} mga\sin\alpha &= \int_0^{2a}\frac{m}{2a}\omega^2\cdot r\sin\alpha\cdot r\cos\alpha = \frac{m}{2a}\omega^2\sin\alpha\cos\alpha\int_0^{2a} r^2\,dr \\ &= \frac{m\omega^2}{2a}\sin\alpha\cos\alpha\cdot\frac{8a^3}{3} = m\omega^2\sin\alpha\cos\alpha\cdot\frac{4a^2}{3}. \end{aligned}$$

$$\text{or,}\quad ma\sin\alpha\left[\frac{4\omega^2 a}{3}\cos\alpha - g\right] = 0.$$

$\therefore$ either, $\sin\alpha = 0$, i. e., $\alpha = 0$, i. e., the rod is vertical.

$$\text{or,}\quad \cos\alpha = \frac{3g}{4\omega^2 a} \quad \text{or,}\quad \alpha = \cos^{-1}\frac{3g}{4\omega^2 a} \quad \text{or,}\quad \omega^2 = \frac{3g}{4a\cos\alpha} \qquad (2.5.3)$$

provided $3g < 4\omega^2 a$.

Again we have

$$\begin{aligned} Y &= mg. \\ X &= \int_0^{2a} \frac{m}{2a} dr \omega^2 r \sin\alpha = \frac{m\omega^2}{2a}\sin\alpha \int_0^{2a} r dr \\ &= m\omega^2 a \sin\alpha = m.\frac{3g}{4a\cos\alpha}\cdot a\sin\alpha \quad \text{[by (2.5.3)]} \\ &= \frac{3}{4} mg\tan\alpha. \end{aligned}$$

If ϕ be the angle made by the direction of the reaction with the vertical,

$$\begin{aligned} \tan\phi &= \frac{X}{Y} = \frac{\frac{3}{4}mg\tan\alpha}{mg} = \frac{3}{4}\tan\alpha. \\ \therefore\quad \phi &= \tan^{-1}\left(\frac{3}{4}\tan\alpha\right). \end{aligned}$$

► **Problem 2.5.3** *A non-homogeneous rod AB of length 2a whose density at any point is directly proportional to the distance of the point A, is rotating with a uniform angular velocity ω about a vertical axes through A. If the rod is inclined at an angle α to the vertical, show that the value of α is either 0 or* $\cos^{-1}\left(\frac{2g}{3l\omega^2}\right)$. **[CUH 2000, 2003; VUH 2005]**

Solution: Let AB be the rod of length $2l$. Let the rod is rotating with uniform angular velocity ω about the vertical making an angle α. Let ρ be the density of the rod at the end A. We consider an elementary length dr at P at a distance r from A. Thus the mass of the element dr at P is $\rho r dr$. Thus the effective force on element dr and perpendicular to vertical $AC = \rho r dr \cdot \omega^2 \cdot r\sin\alpha$.

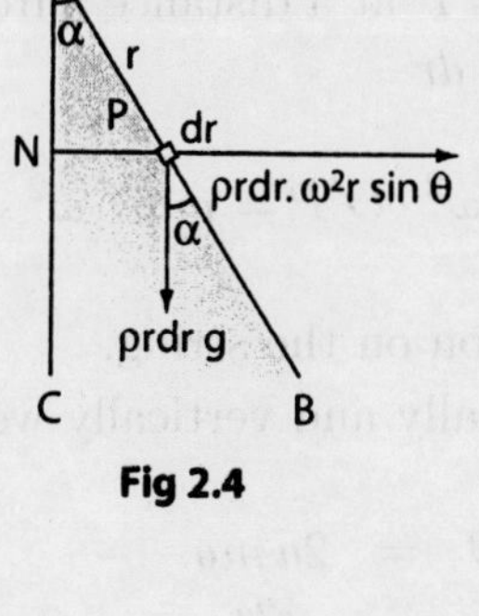

Fig 2.4

Now taking moment about A, we get

$$\int_0^{2l} \rho r dr g \cdot r \sin\alpha = \int_0^{2l} \rho r dr \cdot \omega^2 r \sin\alpha \cdot r\cos\alpha$$

$$\text{or,}\quad g\sin\alpha \int_0^{2l} r^2\,dr = \omega^2 \sin\alpha\cos\alpha \int_0^{2l} r^3\,dr$$

$$\text{or,}\quad g\sin\alpha \cdot \frac{8l^3}{3} = \omega^2 \sin\alpha\cos\alpha \cdot \frac{16l^4}{4}$$

$$\text{or,}\quad g\sin\alpha = 3\omega^2 \sin\alpha\cos\alpha \cdot \frac{l}{2}$$

$$\text{or,}\quad \sin\alpha\left[\frac{3l\omega^2}{2g}\cos\alpha - 1\right] = 0.$$

$$\therefore\quad \text{either}\quad \sin\alpha = 0 \quad \text{i.e.,}\quad \alpha = 0$$

$$\text{or,}\quad \cos\alpha = \frac{2g}{3l\omega^2},\quad \text{i.e.,}\quad \alpha = \cos^{-1}\left(\frac{2g}{3l\omega^2}\right).$$

► **Problem 2.5.4** *A rod, of length 2a, is suspended by a string, of length l, attached to one end; if the string and rod revolve about the vertical with uniform angular velocity and their inclination to the vertical be θ and φ respectively, show that*

KUH '02, '04

$$\frac{3l}{a} = \frac{(4\tan\theta - 3\tan\phi)\sin\phi}{(\tan\phi - \tan\theta)\sin\theta}.$$

Solution: Let OA be string of length l and AB be the rod of length $2a$ and G be its c.g. so $AG = BG = a$. Let m is the line density of the rod. Let ω be the angular velocity with which the system rotates. We take an elementary length dr of the rod at P at a distance r from A. Thus the effective force on the elementary length dr

$$= m dr \omega^2 \cdot O'P = m dr \cdot \omega^2 \cdot (l\sin\theta + r\sin\phi).$$

Let T be the tension on the string.
Resolving horizontally and vertically we get

$$T\cos\theta = 2a\,mg. \tag{2.5.4}$$

$$\begin{aligned} T\sin\theta &= \int_0^{2a} m dr \cdot \omega^2 \cdot (l\sin\theta + r\sin\phi) \\ &= m\omega^2[2al\sin\theta + 2a^2\sin\phi] \end{aligned} \tag{2.5.5}$$

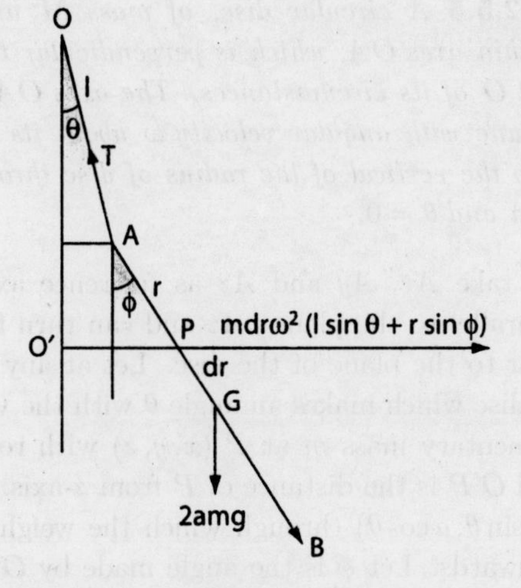

Fig 2.5

and taking moment about A, we get,

$$\int_0^{2a} mdr \cdot \omega^2(l\sin\theta + r\sin\phi)\cdot r\cos\phi = 2a\,mg\cdot a\sin\phi$$

$$\text{or,}\quad m\omega^2\cos\phi\left(\frac{4a^2}{2}\cdot l\sin\theta + \frac{8a^3}{3}\sin\phi\right) = 2a^2mg\cdot\sin\phi$$

$$\text{or,}\quad \omega^2\cos\phi\left(l\sin\theta + \frac{4a}{3}\sin\phi\right) = g\sin\phi. \qquad (2.5.6)$$

Dividing (2.5.5) by (2.5.4), we get,

$$\tan\theta = \frac{m\omega^2[2al\sin\theta + 2a^2\sin\phi]}{2a\,mg} = \frac{\omega^2(l\sin\theta + a\sin\phi)}{g}$$

$$\text{or,}\quad g\tan\theta = \omega^2(l\sin\theta + a\sin\phi). \qquad (2.5.7)$$

Dividing (2.5.7) by (2.5.6) we get

$$\frac{(l\sin\theta + a\sin\phi)}{(l\sin\theta + \frac{4a}{3}\sin\phi)\cos\phi} = \frac{\tan\theta}{\sin\phi}$$

$$\text{or,}\quad \tan\phi(l\sin\theta + a\sin\phi) = \tan\theta\left(l\sin\theta + \frac{4a}{3}\sin\phi\right)$$

$$\text{or,}\quad l(\tan\phi - \tan\theta)\sin\theta = a\sin\phi\left(\frac{4}{3}\tan\theta - \tan\phi\right)$$

$$\text{or,}\quad \frac{3l}{a} = \frac{4\tan\theta - 3\tan\phi}{\tan\phi - \tan\theta}\cdot\frac{\sin\phi}{\sin\theta}.$$

▶ **Problem 2.5.5** *A circular disc, of mass M and radius a, can turn freely about a thin axes OA, which is perpendicular to its plane and passes through a point O of its circumstances. The axes OA is compelled to move a horizontal plane with angular velocity ω about its end A. Show that the inclination θ to the vertical of the radius of disc through O is $\cos^{-1}\left(\frac{g}{a\omega^2}\right)$, unless $\omega^2 < g/a$ and $\theta = 0$.* **[CUH 1993]**

Solution: We take Ax, Ay and Az as reference axes. Let plane of the circular disc parallel to the plane yAz and can turn freely about Ax which is perpendicular to the plane of the disc. Let at any time t, G' be the c.g. of the circular disc which makes an angle θ with the vertical Az or OB. Let us take an elementary mass m at P (x, y, z) with respect to the axes Ax, Ay and Az and $O'P$ is the distance of P from z-axis. Thus the coordinates of G' are $(x, a\sin\theta, a\cos\theta)$ through which the weight of the disc Mg acts vertically downwards. Let ϕ is the angle made by $O'P$ with Ax. Thus the effective force force on the mass m at $P(x, y, z)$ is $m\omega^2 \cdot O'P$.

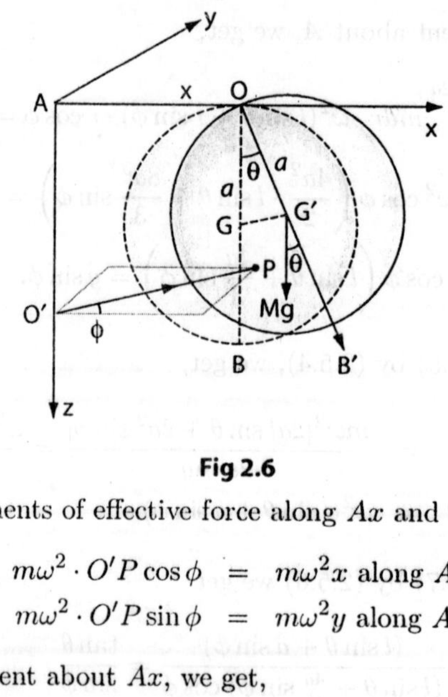

Fig 2.6

Thus the components of effective force along Ax and Ay are

$$
\begin{aligned}
m\omega^2 \cdot O'P\cos\phi &= m\omega^2 x \text{ along } Ax \\
\text{and} \quad m\omega^2 \cdot O'P\sin\phi &= m\omega^2 y \text{ along } Ay.
\end{aligned}
$$

Now taking moment about Ax, we get,

$$-\sum m\omega^2 yz = -Mga\sin\theta \quad (\text{as } \theta \text{ in decreasing sense})$$

$$\text{or,} \quad -\omega^2 D + Mga\sin\theta = 0.$$

$$\left[\text{LHS} = \begin{vmatrix} i & j & k \\ x & y & z \\ m\omega^2 x & m\omega^2 y & 0 \end{vmatrix} \text{ and RHS} = \begin{vmatrix} i & j & k \\ x & a\sin\theta & a\cos\theta \\ 0 & 0 & Mg \end{vmatrix}\right]$$

where D = Product of inertia of the circular disc about x-axis

= Product of inertia of the disc about parallel axes through c.g. G' parallel to x-axes + Product of inertia of the mass M at c.g. G' about the x-axes

$= O + M(a \sin\theta)(a\cos\theta) = Ma^2 \sin\theta\cos\theta.$

$$\therefore \quad -\omega^2 Ma^2 \sin\theta\cos\theta + Mga\sin\theta = 0 \quad \text{or,} \quad \sin\theta(a\omega^2\cos\theta - g) = 0.$$

$$\therefore \quad \text{either} \quad \sin\theta = 0 \quad \text{or,} \quad \cos\theta = \frac{g}{a\omega^2}.$$

Thus either, $\theta = 0$, or, $\theta = \cos^{-1}\frac{g}{a\omega^2}$ unless $a\omega^2 < g$, or, $\omega^2 < g/a$.

► **Problem 2.5.6** *A thin heavy rectangular disc can turn freely about an axis in its own plane, and this axis revolves horizontally with a uniform angular velocity ω about a fixed point on itself. Show that the inclination θ of the plane of the disc to the vertical is $\cos^{-1}\left(\frac{gh}{K^2\omega^2}\right)$, where h is the distance of the centre of inertia of the rectangular disc from the axis and K is the radius of gyration of the disc about the axis. If $\omega^2 < \frac{gh}{K^2}$, then the plane of the disc is vertical.* (CUH 1987, '89, '94, VH '92; KUH '05)

Solution: Let $ABCD$ be the heavy rectangular disc of mass M. Let h be the depth of c.g. G below Ox and any instant θ be the inclination of the plane with vertical. We take Ox, Oy and Oz as reference axes at O. Let at any instant $P(\alpha, \beta, \gamma)$ be the potion of an elementary mass m and R be the distance of P from Ox. Thus the effective force on the elementary mass m at $P = m\omega^2 \cdot O'P$, where $O'P$ is the distance of P from Oz. Thus the components of effective force $m\omega^2 O'P$ are $m\omega^2 O'P \cdot \cos\phi$ along Ox and $m\omega^2 O'P \sin\phi$ along Oy, i.e., $m\omega^2\alpha$ along Ox and $m\omega^2\beta$ along Oy.

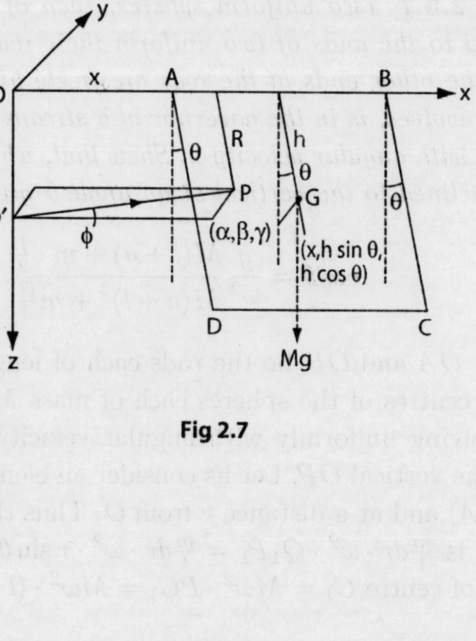

Fig 2.7

Taking moment about Ox we get

$$\sum -m\omega^2\beta\gamma = -Mgh\sin\theta \quad \text{[as } \theta \text{ in decreasing sense]}$$

$$\left[\text{LHS} = \begin{vmatrix} i & j & k \\ \alpha & \beta & \gamma \\ m\omega^2\alpha & m\omega^2\beta & 0 \end{vmatrix}\right.$$

$$\left.\text{and RHS} = \begin{vmatrix} i & j & k \\ x & h\sin\theta & h\cos\theta \\ 0 & 0 & Mg \end{vmatrix}\right]$$

$$\begin{aligned} \text{or,} \quad \sum m\omega^2 \cdot R\sin\theta \cdot R\cos\theta &= Mgh\sin\theta \\ \omega^2\sin\theta\cos\theta \sum mR^2 &= Mgh\sin\theta \\ \text{or,} \quad \omega^2\sin\theta\cos\theta \cdot MK^2 &= Mgh\sin\theta, \end{aligned}$$

where K is the radius of gyration of the disc.

or, $\sin\theta(\omega^2K^2\cos\theta - gh) = 0$

either

$$\cos\theta = \frac{gh}{\omega^2K^2} \quad \text{or,} \quad \theta = \cos^{-1}\frac{gh}{\omega^2K^2},$$

provided

$$\omega^2K^2 > gh \quad \text{or,} \quad \omega^2 > \frac{gh}{K^2}.$$

If $\omega^2 < \frac{gh}{K^2}$, then $\sin\theta = 0$, i. e., $\theta = 0$, i. e., the disc will be vertical.

▶ **Problem 2.5.7** *Two uniform spheres, each of mass M and radius a, are firmly fixed to the ends of two uniform their rods, each of mass m and length l, and the other ends of the rods are freely hinged to a point O. The whole system revolves, is in the governor of a stream-engine, about a vertical line through O with angular velocity ω. Shew that, when the motion is steady, the rods are inclined to the vertical at an angle θ given by the equation*

$$\cos\theta = \frac{g}{\omega^2}\frac{M(l+a) + m\cdot\frac{l}{2}}{M(a+l)^2 + m\frac{l^2}{3}}.$$

Solution: Let OA and OB be the rods each of length l and mass m. And C_1 and C_2 be centres of the spheres each of mass M and radius a. Let the system is revolving uniformly with angular velocity ω and the rod makes angle θ with the vertical OP. Let us consider an elementary length dr at Q_1 (on the rod OA) and at a distance r from O. Thus the effective force on the element at Q_1 is $\frac{m}{l}dr \cdot \omega^2 \cdot Q_1P_1 = \frac{m}{l}dr \cdot \omega^2 \cdot r\sin\theta$ and the effective force on the sphere of centre $C_1 = M\omega^2 \cdot PC_1 = M\omega^2 \cdot (l+a)\sin\theta$.

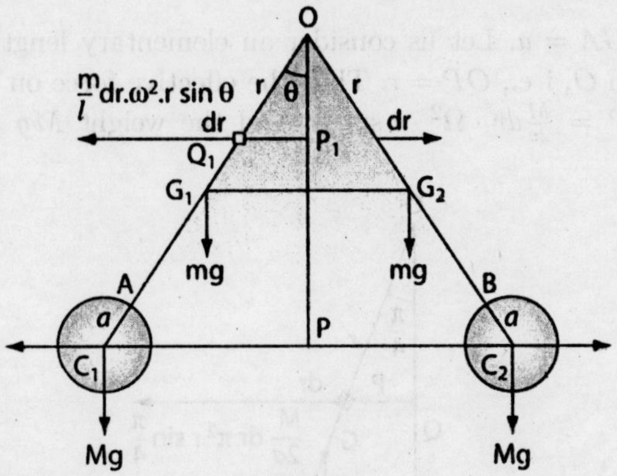

Fig 2.8

Thus taking moment about O, we get

$$M\omega^2(l+a)\sin\theta\cdot OP+\int_0^l\frac{m}{l}dr\cdot\omega^2\cdot r\sin\theta\cdot OP_1$$

$$=mg\cdot\frac{l}{2}\sin\theta+Mg\cdot C_1P$$

$$\text{or,}\quad M\omega^2(l+a)\sin\theta\cdot(l+a)\cos\theta+\frac{m}{l}\omega^2\cdot\int_0^l r\,dr\cdot\sin\theta\cdot r\cos\theta$$

$$=mg\cdot\frac{l}{2}\sin\theta+Mg\cdot(l+a)\sin\theta$$

$$\text{or,}\quad M\omega^2(l+a)^2\sin\theta\cos\theta+\frac{m}{l}\omega^2\sin\theta\cos\theta\cdot\frac{l^3}{3}$$

$$-g\left[\frac{l}{2}m+M(l+a)\right]\sin\theta=0$$

$$\text{or,}\quad \sin\theta\left[\omega^2\left\{M(l+a)^2+\frac{ml^2}{3}\right\}\cos\theta-g\left\{\frac{lm}{2}+M(l+a)\right\}\right]=0$$

$$\therefore\quad \text{either}\quad \cos\theta=\frac{g}{\omega^2}\left[\frac{\frac{lm}{2}+M(l+a)}{M(l+a)^2+\frac{ml^2}{3}}\right]$$

$$\text{or,}\quad \sin\theta=0,\quad \text{i.e.,}\quad \theta=0.$$

▶ **Problem 2.5.8** *A heavy uniform rigid straight rod OA of length 2a is free to turn about its end O. If the rod revolves uniformly with angular velocity Ω about the vertical through O making a constant angle $\frac{\pi}{4}$ with the vertical, show that*

$$\Omega^2=\frac{3g}{2\sqrt{2}a}.$$

[CUH 1998, 2002]

Solution: Let G be the c.g. of the uniform rod OA of mass M and length $2a$.

$\therefore$ $OG = GA = a$. Let us consider an elementary length dr at P at a distance r from O, i.e., $OP = r$. Thus the effective force on the elementary length dr at $P = \frac{M}{2a}dr \cdot \Omega^2 \cdot r\sin\frac{\pi}{4}$. And the weight Mg acts at c.g. G downward.

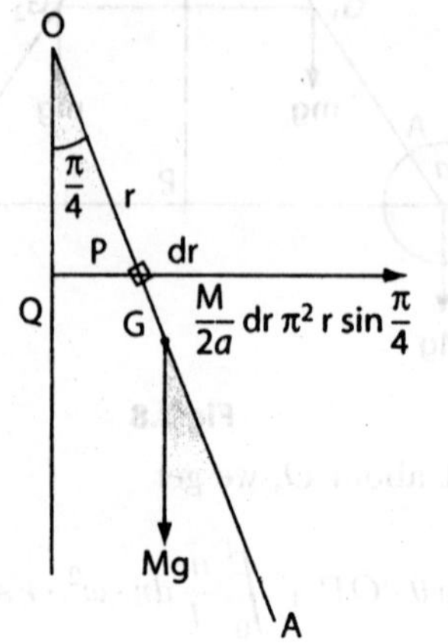

Fig 2.9

Therefore, taking moment about O, we get

$$\int_0^{2a} \frac{M}{2a} dr \cdot \Omega^2 \cdot r\sin\frac{\pi}{4} \cdot r\cos\frac{\pi}{4} = Mg \cdot a\sin\frac{\pi}{4}$$

$$\text{or,}\quad \frac{M}{2a}\Omega^2 \int_0^{2a} r^2 \cdot \frac{1}{\sqrt{2}} \cdot \frac{1}{\sqrt{2}} \cdot dr = \frac{Mga}{\sqrt{2}}$$

$$\text{or,}\quad \frac{M}{4a}\Omega^2 \cdot \frac{8a^3}{3} = \frac{Mga}{\sqrt{2}}.$$

$$\frac{2a^2}{3}\Omega^2 = \frac{ga}{\sqrt{2}} \quad \text{or,} \quad \Omega^2 = \frac{3g}{2\sqrt{2}a}.$$

► **Problem 2.5.9** *A boat of mass m_B and length l is floating stationary on the still water of a lake. A man of mass m_M initially standing at one end of the boat walks to the other end of it and stops. Show that the distance through which the boat will move in the opposite direction is $\frac{m_M l}{m_M + m_B}$, assuming the water offers no resistance to the boat.*

[CUH 2005; VUH 2001]

Solution: Let B be the one end of the boat and the man starts from A, the other end. Let $AB = l$. Let $\bar{x}$ be the distance of c.g. G of the system from the end A.

$$\therefore \quad (m_M + m_B)\bar{x} = m_B \cdot \frac{l}{2} + m_M \times 0 = \frac{l}{2} m_B \tag{2.5.8}$$

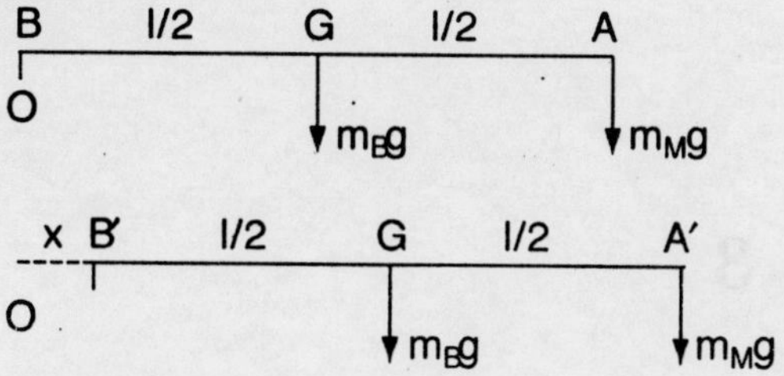

Fig 2.10

Let the boat describes a distance x, when the man reaches the other end B', as the boat is floating stationary, the c.g. remain at rest, then

$$\begin{aligned}(m_M + m_B)\bar{x} &= \left(\frac{l}{2} - x\right) m_B + (l - x) m_M \\ &= \frac{l}{2}(m_B + 2m_M) - x(m_M + m_B).\end{aligned}$$

$$\therefore \quad \frac{l}{2}(m_B + 2m_M) - x(m_M + m_B) = \frac{l}{2} m_B \quad \text{[by (2.5.8)]}$$

$$\text{or,} \quad x = \frac{\frac{l}{2} m_B - \frac{l}{2} m_B + l m_M}{m_M + m_B} = \frac{l m_M}{m_M + m_B}.$$

Particular Case I. Putting $l = 2a$, $m_M = M$ and $m_B = m$, we get $x = \frac{2aM}{m+M}$. **[VUH 2001]**

Chapter 3

Motion of a Rigid Body about a Fixed Axis

3.1 To investigate the motion of the Rigid Body about the axis, if a Rigid Body can turn freely about an axis fixed in Space

Let the body rotates about an axis Oz fixed in space. Let $OABz, OCDz$ and $OEFz$ be three planes each is passing through the fixed axis Oz, such that the plane $OABz$ is fixed in space. The plane $OCDz$ is fixed in the body and the plane $OEFz$ passes through any point P of the body. The lines OA, OC and OE are taken perpendicular to Oz. Let $\angle AOC = \theta$, $\angle AOE = \phi$. The planes $OCDz$ and $OEFz$ are fixed relative to the body, therefore the angle between them is constant, i. e., $\angle EOC =$ constant $= \alpha$ (say).

$$\therefore \quad \phi = \theta - \alpha \quad \text{or,} \quad \dot{\phi} = \dot{\theta} \quad \text{and} \quad \ddot{\phi} = \ddot{\theta}.$$

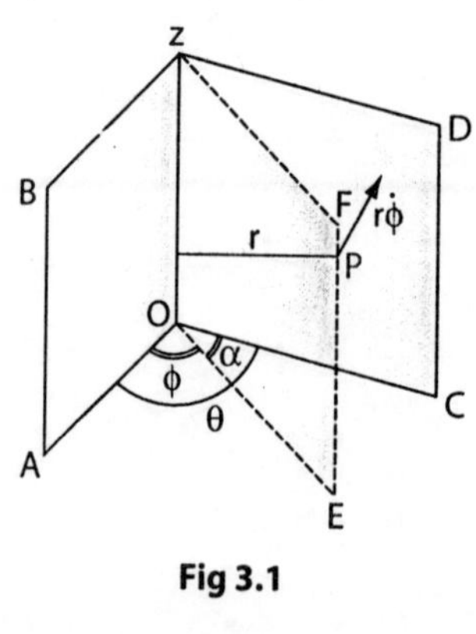

Fig 3.1

Thus any plane through Oz-axis and a point in the body moves with same angular velocity as that of the plane $OCDz$ fixed in the body. Let P be any point in the body, where perpendicular distance from fixed axis Oz is r. Thus the velocity of P perpendicular to the plane $OEFz$ in a direction ϕ increasing $= r\dot{\phi} = r\dot{\theta}$.

Thus the momentum of elementary mass m at $P = m \cdot r\dot{\theta}$. Therefore, the moment of momentum of the elementary mass m at $P = m \cdot r\dot{\theta} \cdot r = mr^2\dot{\theta}$.

By D'Alembert's principle we have

$$\frac{d}{dt}[\text{Total } \overrightarrow{\text{moment of}} \text{ momentum}] = L = \text{Algebraic sum of the } \overrightarrow{\text{moment of}} \text{ applied forces.}]$$

Thus

$$\frac{d}{dt}\left[\sum mr^2\dot{\theta}\right] = L = \text{Algebraic sum of the moments of the applied forces}$$

$$\text{or,} \quad \frac{d}{dt}\left[\dot{\theta}\sum mr^2\right] = L$$

$$\text{or,} \quad \frac{d}{dt}[I\dot{\theta}] = L \quad \text{or,} \quad I\ddot{\theta} = L \quad \text{or,} \quad MK^2\ddot{\theta} = L,$$

where I = moment of inertia of the body about the line about which the body rotates and K = radius of gyration of the body about the line about which the body rotates.

Thus

$$I\ddot{\theta} = L, \quad \text{or,} \quad MK^2\ddot{\theta} = L \tag{3.1.1}$$

is the equation of motion of the body about an axis fixed in space.

$$\begin{aligned} \text{Again Kinetic Energy} &= \text{K.E.} = \tfrac{1}{2}\sum m(r\dot{\theta})^2 \\ &= \tfrac{1}{2}\sum mr^2\dot{\theta}^2 = \tfrac{1}{2}\dot{\theta}^2\sum mr^2 \\ &= \tfrac{1}{2}I\dot{\theta}^2, \end{aligned}$$

where I = moment of inertia of the body about the fixed axis Oz.

3.2 Motion of a Compound Pendulum

(CUH 2004, '07; VUH '03, '05; KUH '03, '05

If a rigid body of any size and shape is so suspended from a horizontal axis, fixed in space, rigidly attached to the body that it can have free swing (there is no friction of the axis and no resistance or they are negligible) under gravity about the axis, then the body is called a *compound pendulum*.

Let the vertical plane be taken as the plane through the centre of gravity of the body G and perpendicular to the axis of rotation. Let OA be the line through c.g. of the body G and perpendicular to the axis of rotation PQ and let PQ meet OA at O. Thus O is the point of suspension. Let $OG = h$. Let K and K_1 are the radii of gyration of the body about the axis of rotation PQ and about an axis through G parallel to PQ respectively. Let at any time t, OG makes an angle θ with the vertical and M is the mass of the body. Thus Mg the weight of the body acts vertically downward through G. Now the velocity of a particle at a distance r from O perpendicular to $OA = r\dot{\theta}$, as the radial velocity is zero. Thus the Kinetic Energy of the body

$$= \tfrac{1}{2}\sum m(r\dot{\theta})^2 = \tfrac{1}{2}\dot{\theta}^2\sum mr^2 = \tfrac{1}{2}I\dot{\theta}^2, \tag{3.2.1}$$

where I is the moment of inertia of the body about the axis of rotation PQ through the point of suspension O,

$$\text{i.e.,}\quad I = MK^2 = M(K_1^2 + h^2) \tag{3.2.2}$$

and the potential energy $= -Mgh\cos\theta$.

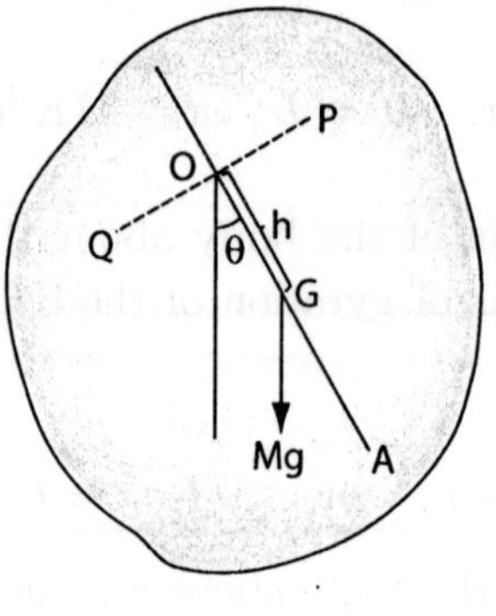

Fig 3.2

Now by principle of energy, K.E. + P.E. = constant = initial value.
We get,

$$\tfrac{1}{2}I\dot{\theta}^2 - Mgh\cos\theta = \text{constant} = -Mgh\sin\alpha$$

assuming initially, i.e., when $t = 0$, $\dot{\theta} = 0$ and $\theta = \alpha$.

Thus

$$\tfrac{1}{2}I\dot{\theta}^2 = Mgh(\cos\theta - \cos\alpha) \quad \text{or,} \quad \tfrac{1}{2}M(K_1^2 + h^2)\dot{\theta}^2 = Mgh(\cos\theta - \cos\alpha)$$

$$\text{or,}\quad \dot{\theta}^2 = \frac{2gh(\cos\theta - \cos\alpha)}{K_1^2 + h^2}. \tag{3.2.3}$$

To find the Time of Small Oscillation

Let α is small, θ is also small, then from (3.2.3) we get,

$$\dot{\theta}^2 = \frac{2gh}{K_1^2+h^2}\left[\left(1-\frac{\theta^2}{2}+\ldots\right)-\left(1-\frac{\alpha^2}{2}+\ldots\right)\right]$$

[Neglecting the higher power of α and θ as α and θ are small]

$$= \frac{2gh}{K_1^2+h^2}\left[\frac{\alpha^2-\theta^2}{2}\right] = \frac{gh}{K_1^2+h^2}\left(\alpha^2-\theta^2\right)$$

$$\text{or,}\quad \dot{\theta} = -\sqrt{\frac{gh}{K_1^2+h^2}}\sqrt{\alpha^2-\theta^2}$$

[Negative sign is taken as θ tends to decrease]

$$\text{or,}\quad -\frac{d\theta}{\sqrt{\alpha^2-\theta^2}} = \sqrt{\frac{gh}{K_1^2+h^2}}dt.$$

Integrating,

$$\cos^{-1}\frac{\theta}{\alpha} = \sqrt{\frac{gh}{K_1^2+h^2}}\,t + C \quad \text{(integration constant).}$$

Since $\theta = \alpha$, when $t = 0$, $C = 0$.

$$\therefore\quad \cos^{-1}\frac{\theta}{\alpha} = \sqrt{\frac{gh}{K_1^2+h^2}}\cdot t, \quad \text{or,}\quad \theta = \alpha\cos\left[\sqrt{\frac{gh}{K_1^2+h^2}}\cdot t\right].$$

Thus the time period of oscillation

$$T = \frac{2\pi}{\sqrt{\dfrac{gh}{K_1^2+h^2}}} = 2\pi\sqrt{\frac{K_1^2+h^2}{gh}}. \tag{3.2.4}$$

Comparing (3.2.4) with the period of simple pendulum $2\pi\sqrt{l/g}$, we get

$$l = \frac{K_1^2+h^2}{h}.$$

Thus the length of the equivalent simple pendulum

$$l = \frac{K^2}{h} = \frac{K_1^2+h^2}{h} = h + \frac{K_1^2}{h} > h, \tag{3.2.5}$$

where K and K_1 are the radii of gyration of the body about an axis through the point of suspension and about an parallel axis through c.g. G of the body respectively and h is the distance of G from the point of suspension.

Let $OO' = l =$ length of equivalent simple pendulum. Let O' is called the *centre of oscillation.*

We shall now prove that *the point of suspension O and centre of oscillation of a compound pendulum are convertable, i. e., if the body is suspended from O', then the centre of oscillation will be O.* [VUH '03, KUH '07]

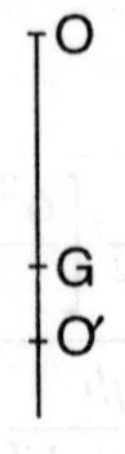

Fig 3.3

Now

$$OO' = l = \frac{K^2}{h} = \frac{K_1^2 + h^2}{h} = \frac{K_1^2}{h} + h = \frac{K_1^2}{h} + OG,$$

$$\text{or,} \quad OO' - OG = \frac{K_1^2}{h}, \quad \text{or,} \quad O'G = \frac{K_1^2}{h} = \frac{K_1^2}{OG},$$

$$\text{or,} \quad O'G \cdot OG = K_1^2. \tag{3.2.6}$$

Suppose the body is suspended from O' and O_1 be the corresponding centre of oscillation

$$\therefore \quad O'G \cdot O_1G = K_1^2. \tag{3.2.7}$$

Thus from (3.2.6) and (3.2.7) we get

$$O'G \cdot OG = O'G \cdot O_1G. \qquad \therefore \quad OG = O_1G.$$

Thus O_1 coincides with O.

Minimum Period of Oscillation

[KUH 2003, '05]

From (3.2.4), we get,

$$T = 2\pi\sqrt{\frac{1}{g}\left(\frac{K_1^2}{h} + h\right)} = 2\pi\sqrt{\frac{1}{g}\left\{\left(\frac{K_1}{\sqrt{h}} - \sqrt{h}\right)^2 + 2K_1\right\}}.$$

Thus T will be minimum when $\frac{K_1}{\sqrt{h}} - \sqrt{h} = 0$ or, $K_1 = h$.

$$\therefore \quad T_{\min} = 2\pi\sqrt{\frac{2K_1}{g}} = 2\pi\sqrt{\frac{2h}{g}}. \tag{3.2.8}$$

Remark 3.2.1 *A pendulum is suspended at O and O' is the centre of oscillation. Show that if an additional weight is rigidly attached at O', the period of oscillation is unaltered.*

Let O be the point of suspension of a body of mass M forming a compound pendulum and h be the depth of c.g. G from the point of suspension O. Let O' be the centre of oscillation.

Thus from (3.2.5)we have

$$OO' = \frac{K^2}{h} \tag{3.2.9}$$

and the period of oscillation

$$= 2\pi\sqrt{\frac{K^2}{gh}}. \tag{3.2.10}$$

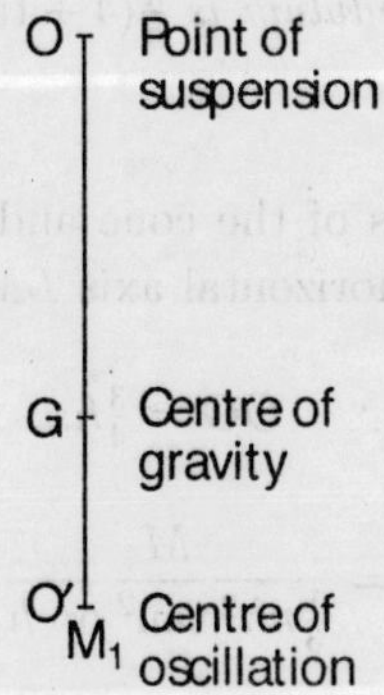

Fig 3.4

Let an additional mass M_1 be attached rigidly at O' and then we assume that K' be the radius of gyration about the horizontal axis through O.

Thus

$$\begin{aligned}
(M+M_1)K'^2 &= MK^2 + M_1 \cdot OO'^2 \\
\text{or,}\quad (M+M_1)K'^2 &= MK^2 + M_1\frac{K^4}{h^2} \\
\text{or,}\quad (M+M_1)K'^2 &= K^2\left[M + \frac{M_1K^2}{h^2}\right]. \qquad (3.2.11)
\end{aligned}$$

Again

$$\begin{aligned}
(M+M_1)h_1 &= M\cdot h + OO'M_1 = Mh + M_1\frac{K^2}{h} \\
(M+M_1)h_1 &= h\left[M + M_1\frac{K^2}{h^2}\right], \qquad (3.2.12)
\end{aligned}$$

where h_1 is the depth of c.g. when the mass M_1 attached at O'. Now dividing (3.2.11) by (3.2.12) we get

$$\frac{K'^2}{h_1} = \frac{K^2}{h}.$$

If T' be the period of oscillation when the mass M_1 added at O', we get

$$T' = 2\pi\sqrt{\frac{K'^2}{gh'}} = \sqrt{\frac{K^2}{gh}}.$$

Thus the period of oscillation is unaltered.

3.3 Solved Problems on Simple Equivalent Pendulum

► **Problem 3.3.1** *A solid homogeneous cone, of height h and vertical angle 2α, oscillates about a horizontal axis through its vertex, show that the length of simple equivalent pendulum is $\frac{h}{5}(4 + \tan^2\alpha)$.*

[CUH 2005; VUH 2006]

Solution: Let M be the mass of the cone and G be the c.g. of the cone. Let the cone oscillates about horizontal axis LM through the vertex O.

$$\therefore \quad OG = \tfrac{3}{4}h.$$

$$\text{Let } m = \text{mass/volume} = \frac{M}{\frac{1}{3}\pi h^2 \tan^2\alpha \cdot h} = \frac{3M}{\pi h^3 \tan^2\alpha}. \tag{3.3.1}$$

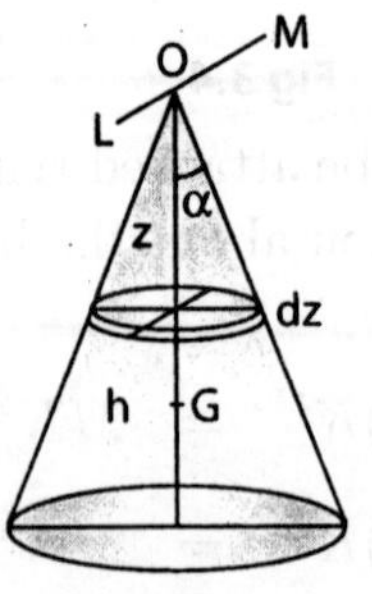

Fig 3.5

We consider a circular strip of width dz at a distance z from the vertex O and parallel to the axis of rotation.

∴ The m.i. of the strip about an axis through the centre of the strip and parallel to axis of rotation

$$= m\pi \cdot z^2 \tan^2\alpha\, dz \cdot \frac{z^2 \tan^2\alpha}{4} = \frac{m\pi \tan^4\alpha}{4} \cdot z^4 \cdot dz.$$

Thus the m.i. of the strip about the axis of rotation

$$\begin{aligned} &= \frac{1}{4} m\pi \tan^4\alpha \cdot z^4\, dz + m\pi z^2 \tan^2\alpha\, dz \cdot z^2 \\ &= \frac{1}{4} m\pi \tan^4\alpha \cdot z^4\, dz + m\pi \tan^2\alpha \frac{z^4}{4} dz \\ &= m\pi \tan^2\alpha \left(\frac{\tan^2\alpha}{4} + 1\right) \cdot z^4 dz. \end{aligned}$$

Thus the m.i. of the cone about the axis of rotation

$$\begin{aligned} &= m\pi \tan^2\alpha \left(\frac{\tan^2\alpha}{4} + 1\right) \cdot \int_0^h z^4\, dz \\ &= \frac{3M}{h^3}\left(\frac{\tan^2\alpha}{4} + 1\right)\frac{h^5}{5} \quad \text{[by (3.3.1)]} \\ &= \frac{3Mh^2}{5}\left(\frac{\tan^2\alpha}{4} + 1\right). \end{aligned}$$

Thus the radius of gyration K of the cone about the axis of rotation is given by

$$K^2 = \frac{3h^2}{5}\left(\frac{\tan^2\alpha}{4} + 1\right) = \frac{3}{20} h^2 (\tan^2\alpha + 4).$$

Again the distance of c.g. G of the cone from the vertex $h_1 = \frac{3}{4}h$.
Thus the length of simple equivalent pendulum

$$= \frac{K^2}{h_1} = \frac{\frac{3h^2}{20}(\tan^2\alpha + 4)}{\frac{3}{4}h} = \frac{h}{5}(4 + \tan^2\alpha).$$

► **Problem 3.3.2** *A bent lever, whose arms of length a and b, angle between them being α, makes small oscillations in its own plane about the fulcrum, show that the length of the corresponding simple pendulum is*

$$\frac{2}{3}\frac{a^3 + b^3}{\sqrt{a^4 + 2a^2b^2\cos\alpha + b^4}}.$$

Solution: Let OA and OB be the arms of lengths a and b and G_1, G_2 be their c.g. respectively. We take the x-axis along OA and y-axis perpendicular to OA at O and $\angle AOB = \alpha$. Let m = mass/length of the arms. Thus the coordinates of G_1 and G_2 respectively are $(\frac{a}{2}, 0)$ and $(\frac{b}{2}\cos\alpha, \frac{b}{2}\sin\alpha)$. If $(\bar{x}, \bar{y})$ be the coordinates of c.g. G of the system, we get

$$\begin{aligned}(ma+mb)\bar{x} &= ma\cdot\frac{a}{2}+mb\cdot\frac{b}{2}\cos\alpha=\frac{1}{2}m(a^2+b^2\cos\alpha)\\ \therefore\quad \bar{x} &= \frac{a^2+b^2\cos\alpha}{2(a+b)}\end{aligned}$$

and

$$(ma+mb)\bar{y}=ma\cdot 0+mb\cdot\frac{b}{2}\sin\alpha,\quad \text{or,}\quad \bar{y}=\frac{1}{2}\frac{b^2\sin\alpha}{a+b}.$$

$$\begin{aligned}\therefore\quad OG=h &= \sqrt{\bar{x}^2+\bar{y}^2}\\ &= \frac{1}{2(a+b)}\sqrt{(a^2+b^2\cos\alpha)^2+(b^2\sin\alpha)^2}\\ &= \frac{1}{2(a+b)}\sqrt{a^4+b^4+2a^2b^2\cos\alpha}.\end{aligned}$$

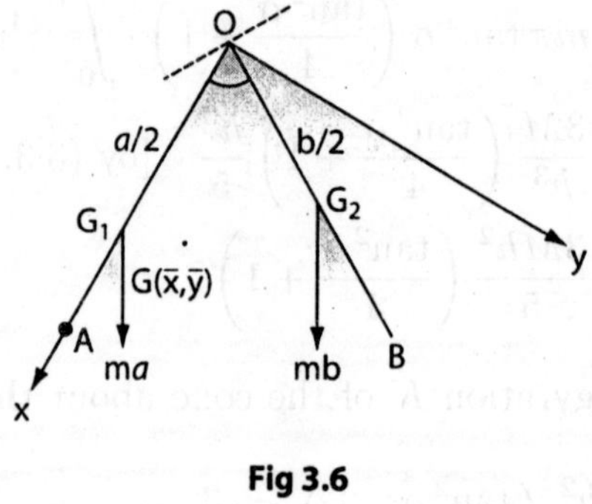

Fig 3.6

If K be the radius of gyration of the system about O, we have

$$\begin{aligned}(ma+mb)K^2 &= \frac{ma}{3}\cdot\frac{a^2}{4}+ma\cdot\frac{a^2}{4}+\frac{mb}{3}\cdot\frac{b^2}{4}+mb\cdot\frac{b^2}{4}\\ &= \frac{ma^3}{3}+\frac{mb^3}{3}=\frac{m(a^3+b^3)}{3}.\\ \therefore\quad K^2 &= \frac{a^3+b^3}{3(a+b)}.\end{aligned}$$

$\therefore$ The length of equivalent simple pendulum

$$=\frac{K^2}{h}=\frac{\dfrac{(a^3+b^3)}{3(a+b)}}{\dfrac{\sqrt{a^4+b^4+2a^2b^2\cos\alpha}}{2(a+b)}}=\frac{2}{3}\cdot\frac{a^3+b^3}{\sqrt{a^4+b^4+2a^2b^2\cos\alpha}}.$$

► **Problem 3.3.3** *A weightless straight rod ABC, of length 2a, is movable about the end A which is fixed and carries two particles of same mass, one fastend to the middle point B and the other to the end C of the rod. If the rod be held in a horizontal portion and be then let go, show that its angular velocity when vertical is* $\sqrt{\frac{6g}{5a}}$ *and that* $\frac{5a}{3}$ *is the length of the simple equivalent pendulum.*

Solution: Let $AC(=2a)$ be the weightless rod, and the end A be fixed such that $AB = BC = a$. Let two equal masses M each fastend at B and C. Let G be the c.g. of the system and $AG = h$. Let the rod makes an angle θ with the vertical at any time t. Therefore,

$$(M+M)h = Ma + M\cdot 2a, \quad \text{or,} \quad h = \frac{3a}{2}. \tag{3.3.2}$$

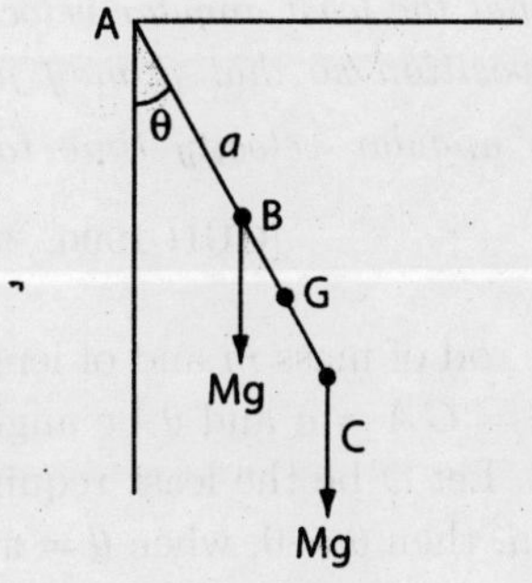

Fig 3.7

If K be the radius of gyration of the system about A, we have

$$2MK^2 = Ma^2 + M\cdot 4a^2, \quad \text{or,} \quad K^2 = \frac{5a^2}{2}. \tag{3.3.3}$$

Thus the length of simple equivalent pendulum

$$l = \frac{K^2}{h} = \frac{\frac{5a^2}{2}}{\frac{3a}{2}} = \frac{5a}{3}.$$

Considering the equation of motion of the system, we have,

$$2MK^2\ddot{\theta} = -Mga\sin\theta - Mg\cdot 2a\sin\theta = -3Mga\sin\theta,$$

$$\text{or,} \quad 2\cdot\frac{5a^2}{2}\ddot{\theta} = -3ga\sin\theta \quad [\text{by } (3.3.3)]$$

$$\text{or,} \quad \ddot{\theta} = -\frac{3}{5}\frac{g}{a}\cdot\sin\theta,$$

$$\text{or,} \quad 2\dot{\theta}\ddot{\theta} = -\frac{6g}{5a}\sin\theta\dot{\theta}.$$

Integrating,

$$\dot{\theta}^2 = \frac{6g}{5a}\cos\theta + C.$$

Initially, $t = 0$, $\dot{\theta} = 0$, when $\theta = \pi/2$.

$$\therefore \quad C = 0, \quad \therefore \quad \dot{\theta}^2 = \frac{6g}{5a}\cos\theta.$$

$$\therefore \quad \dot{\theta}^2\Big|_{\theta=0} = \frac{6g}{5a}.$$

Thus,

$$\dot{\theta}^2\Big|_{\theta=0} = \sqrt{\frac{6g}{5a}}.$$

► **Problem 3.3.4** *A uniform rod of mass m and length 2a, can turn freely about a fixed axis; show that the least angular velocity with which it must be started from the lowest position so that it may just make complete revolution is* $\sqrt{\frac{3g}{a}}$*, with this angular velocity time to describe an angle* θ *is* $2\sqrt{\frac{a}{3g}}\log\tan(\pi/4 + \theta/4)$. **[CUH 1990, '96, '01, '04; KUH '01, '08]**

Solution: Let OA be the rod of mass m and of length $2a$. Let G be the c.g. of the rod, such that $OG = GA = a$ and θ be angle made by the rod with the vertical at any time t. Let ω be the least required angular velocity just to complete the revolution, then $\dot{\theta} = 0$, when $\theta = \pi$.

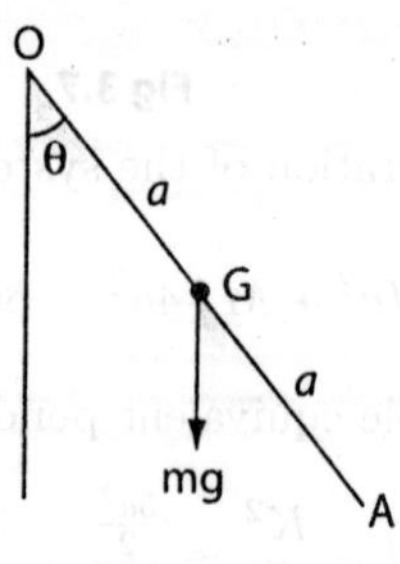

Fig 3.8

$\therefore$ We have the conditions $\dot{\theta} = 0$, when $\theta = \pi$ and $\dot{\theta} = \omega$, when $\theta = 0$. Considering the equation of motion of the system we have

$$mK^2\ddot{\theta} = -mga\sin\theta, \quad \text{or,} \quad \frac{4}{3}a^2\ddot{\theta} = -ga\sin\theta,$$

$$\text{or,} \quad \ddot{\theta} = -\frac{3g}{4a}\sin\theta, \quad \text{or,} \quad 2\dot{\theta}\ddot{\theta} = -\frac{3g}{2a}\sin\theta\cdot\dot{\theta},$$

$$\text{or,} \quad d(\dot{\theta}^2) = -\frac{3g}{2a}\sin\theta\cdot d\theta, \qquad (3.3.4)$$

$$\text{or,} \quad \int_{\omega}^{0} d(\dot{\theta}^2) = -\frac{3g}{2a}\int_0^{\pi} \sin\theta \cdot d\theta,$$

$$\text{or,} \quad -\omega^2 = -\frac{3g}{2a}\left| -\cos\theta \right|_0^{\pi} = -\frac{3g}{2a}(+1+1) = -\frac{3g}{a},$$

$$\text{or,} \quad \omega^2 = \frac{3g}{a}, \quad \text{or,} \quad \omega = \sqrt{\frac{3g}{a}}.$$

Again integrating (3.3.4) we get

$$\int_{\omega=\sqrt{\frac{3g}{a}}}^{\dot{\theta}} d(\dot{\theta}^2) = -\frac{3g}{2a}\int_0^{\theta} \sin\theta\, d\theta = \frac{3g}{2a}\left| \cos\theta \right|_0^{\theta},$$

$$\text{or,} \quad \dot{\theta}^2 - \frac{3g}{a} = \frac{3g}{2a}(\cos\theta - 1), \quad \text{or,} \quad \dot{\theta}^2 = \frac{3g}{2a}\cos\theta + \frac{3g}{2a} = \frac{3g}{a}\cos^2\frac{\theta}{2},$$

$$\text{or,} \quad \dot{\theta} = \sqrt{\frac{3g}{a}}\cos\frac{\theta}{2}, \quad \text{or,} \quad dt = \sqrt{\frac{a}{3g}}\sec\frac{\theta}{2}d\theta, \tag{3.3.5}$$

$$\text{or,} \quad \int_0^t dt = \int_0^{\theta}\sqrt{\frac{a}{3g}}\sec\frac{\theta}{2}d\theta$$

$$\text{or,} \quad t = 2\sqrt{\frac{a}{3g}}\log\tan\left(\frac{\pi}{4}+\frac{\theta}{4}\right)\Big|_0^{\theta} = 2\sqrt{\frac{a}{3g}}\log\tan\left(\frac{\pi}{4}+\frac{\theta}{4}\right).$$

► **Problem 3.3.5** *A sphere, of radius a, is suspended by a fine wire from a fixed point at a distance l from its centre; show that the time of small oscillation is given by*

$$2\pi\sqrt{\frac{5l^2+2a^2}{5lg}}\left[1+\frac{1}{4}\sin^2\frac{\alpha}{2}\right],$$

where α represents the amplitude of the vibration and find the length of the simple equivalent pendulum. **[VUH 1989; KUH '02]**

Solution: Let O' be the centre of the sphere of radius a and mass M. Let O is the point of suspension and $OO' = l$. Let at any time t, the wire makes an angle θ with the vertical, i.e., amplitude of the small oscillation.

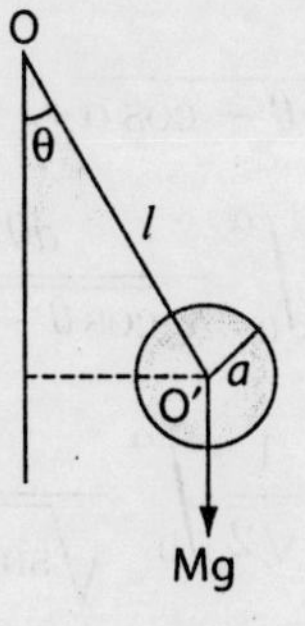

Fig 3.9

Now the m.i. of the sphere about the axis of rotation through O $= MK^2 =$ m.i. of the sphere about an axis through c.g. O' parallel to the axis of rotation + m.i. of the mass M of the sphere at c.g. about the axis of rotation $= \dfrac{2Ma^2}{5} + Ml^2 = \dfrac{M}{5}(5l^2 + 2a^2)$

$$\therefore \quad K^2 = \frac{5l^2 + 2a^2}{5}, \tag{3.3.6}$$

where K is the radius of gyration about O.

Again the distance of c.g. from the centre of rotation $= h = l$. Thus the length of simple equivalent pendulum

$$= \frac{K^2}{h} = \frac{5l^2 + 2a^2}{5l}$$

Now the equation of motion of the system is

$$\begin{aligned} MK^2\ddot{\theta} &= -Mgl\sin\theta \\ \text{or,} \quad K^2\ddot{\theta} &= -gl\sin\theta \\ \text{or,} \quad 2K^2\ddot{\theta}\dot{\theta} &= -2gl\sin\theta\cdot\dot{\theta}, \\ \text{or,} \quad K^2\, d(\dot{\theta}^2) &= -2gl\sin\theta\cdot d\theta. \end{aligned}$$

Integrating

$$K^2\dot{\theta}^2 = 2gl\cos\theta + C.$$

We assume, $\dot{\theta} = 0$, when $\theta = \alpha$.

$$\therefore \quad C = -2gl\cos\alpha.$$

Thus, $K^2\dot{\theta}^2 = 2gl(\cos\theta - \cos\alpha)$

$$\text{or,} \quad \dot{\theta} = \sqrt{\frac{2gl}{K^2}}\sqrt{\cos\theta - \cos\alpha}$$

$$\text{or,} \quad \sqrt{\frac{2gl}{K^2}}\int_0^t dt = \int_0^\alpha \frac{d\theta}{\sqrt{\cos\theta - \cos\alpha}} = \int_0^\alpha \frac{d\theta}{\sqrt{2\sin^2\frac{\alpha}{2} - 2\sin^2\frac{\theta}{2}}}$$

$$= \frac{1}{\sqrt{2}}\int_0^\alpha \frac{d\theta}{\sqrt{\sin^2\frac{\alpha}{2} - \sin^2\frac{\theta}{2}}}.$$

Putting

$$\begin{aligned} \sin\frac{\theta}{2} &= \sin\frac{\alpha}{2}\sin\phi \\ \text{or,} \quad \cos\frac{\theta}{2}d\theta &= 2\sin\frac{\alpha}{2}\cos\phi d\phi, \end{aligned}$$

$\theta \to 0 \Rightarrow \phi \to 0$ and $\theta \to \alpha \Rightarrow \phi \to \frac{\pi}{2}$.

$$\therefore \quad \sqrt{\frac{2gl}{K^2}}t = \frac{1}{\sqrt{2}}\int_0^{\pi/2} \frac{2\sin\frac{\alpha}{2}\cos\phi d\phi}{\sin\frac{\alpha}{2}\cos\phi\cos\frac{\theta}{2}} = \sqrt{2}\int_0^{\pi/2}\frac{d\phi}{\cos\frac{\theta}{2}}$$

$$= \sqrt{2}\int_0^{\pi/2}\frac{d\phi}{\sqrt{1-\sin^2\frac{\alpha}{2}\sin^2\phi}}$$

$$\text{or,}\quad t = \sqrt{\frac{K^2}{gl}}\int_0^{\pi/2}\left(1-\sin^2\frac{\alpha}{2}\sin^2\phi\right)^{-1/2}d\phi$$

$$= \sqrt{\frac{K^2}{gl}}\int_0^{\pi/2}\left(1+\frac{1}{2}\sin^2\frac{\alpha}{2}\sin^2\phi\right)d\phi \quad \text{[neglecting the higher power]}$$

$$= \sqrt{\frac{K^2}{gl}}\left[\frac{\pi}{2}+\frac{1}{2}\sin^2\frac{\alpha}{2}\int_0^{\pi/2}\sin^2\phi d\phi\right]$$

$$= \sqrt{\frac{K^2}{gl}}\left[\frac{\pi}{2}+\frac{1}{2}\sin^2\frac{\alpha}{2}\cdot\frac{1}{2}\cdot\frac{\pi}{2}\right]$$

$$= \sqrt{\frac{K^2}{gl}}\left[\frac{\pi}{2}+\frac{\pi}{8}\sin^2\frac{\alpha}{2}\right]$$

$$= \frac{\pi}{2}\sqrt{\frac{K^2}{gl}}\left[1+\frac{1}{4}\sin^2\frac{\alpha}{2}\right]$$

$$= \frac{\pi}{2}\sqrt{\frac{5l^2+2a^2}{5gl}}\left[1+\frac{1}{4}\sin^2\frac{\alpha}{2}\right].$$

Thus the time of oscillation

$$T = 4t = 2\pi\sqrt{\frac{5l^2+2a^2}{5gl}}\left[1+\frac{1}{4}\sin^2\frac{\alpha}{2}\right].$$

► **Problem 3.3.6** *A solid homogeneous cone of height h and semi-vertical angle α oscillates about a diameter of its base. Show that length of the simple equivalent pendulum is $\frac{1}{5}h(2+3\tan^2\alpha)$.*

[CUH 1990, 2001, 2007; BUH 1996; VUH 1992, 2001]

Solution: Let G be the centre of gravity of the cone of mass M which oscillates about the diameter AB. $\therefore\ CG = \frac{1}{4}h$.

Let us consider a strip of thickness dz at a distance z from the vertex O and parallel to the axis of oscillation ACB.

Let m = mass per volume $= \dfrac{M}{\frac{1}{3}\pi h^3\tan^2\alpha} = \dfrac{3M}{\pi h^3\tan^2\alpha}$.

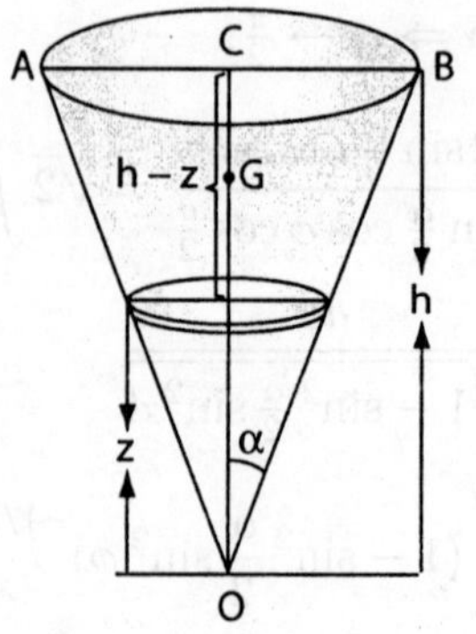

Fig 3.10

Now the moment of inertia of the strip about the axis of oscillation AB

$$= m\pi z^2 \tan^2\alpha\, dz \cdot \frac{z^2\tan^2\alpha}{4} + m\pi z^2\tan^2\alpha\, dz\cdot (h-z)^2$$

$$= m\pi\cdot\tan^2\alpha\left[\frac{1}{4}z^4\tan^2\alpha + z^2(h-z)^2\right]\, dz.$$

Thus the moment of inertia of the cone about AB

$$= m\pi\tan^2\alpha\int_0^h\left[\frac{1}{4}z^4\tan^2\alpha + z^2(h-z)^2\right] dz$$

$$= \frac{3M}{\pi h^3\tan^2\alpha}\cdot\pi\tan^2\alpha\int_0^h\left[\frac{1}{4}\tan^2\alpha\cdot z^4 + h^2z^2 - 2hz^3 + z^4\right] dz$$

$$= \frac{3M}{h^3}\left[\frac{1}{4}\tan^2\alpha\cdot\frac{h^5}{5} + \frac{h^5}{3} - \frac{2h^5}{4} + \frac{h^5}{5}\right]$$

$$= \frac{3M}{h^3}\cdot\frac{h^5}{60}[3\tan^2\alpha + 20 - 30 + 12]$$

$$= \frac{Mh^2}{20}[3\tan^2\alpha + 2].$$

If K be the radius of gyration of the cone about AB

$$\therefore\quad MK^2 = \frac{Mh^2}{20}[3\tan^2\alpha + 2].$$

$$\therefore\quad K^2 = \frac{h^2}{20}(3\tan^2\alpha + 2).$$

$\therefore$ Therefore, the length of simple equivalent pendulum

$$l = \frac{K^2}{CG} = \frac{\frac{h^2}{20}(3\tan^2\alpha + 2)}{\frac{1}{4}h} = \frac{h}{5}(3\tan^2\alpha + 2).$$

► **Problem 3.3.7** *Three equal particles are attached to a weightless rod at equal distances a apart. The system is suspended from, and is free to turn about a point of the rod distance x from the middle particle. Find the time of small oscillation, and show that it is least when $x = 0.82a$ nearly.*

Solution: Let AB be the weightless rod. At A, G, B of the rod three equal particles each of mass m are attached, such that $AG = GB = a$. Let the system is suspended from the point O on the rod such that $OG = x$. Let, at anytime t, θ is the inclination of rod with the vertical. The equation of motion is

$$\begin{aligned} 3mK^2\ddot{\theta} &= mg(a-x)\sin\theta - mgx\sin\theta - mg(a+x)\sin\theta \\ &= -3mgx\sin\theta \\ \text{or,} \quad K^2\ddot{\theta} &= -gx\sin\theta, \end{aligned}$$

where K is the radius of gyration of the system about O.

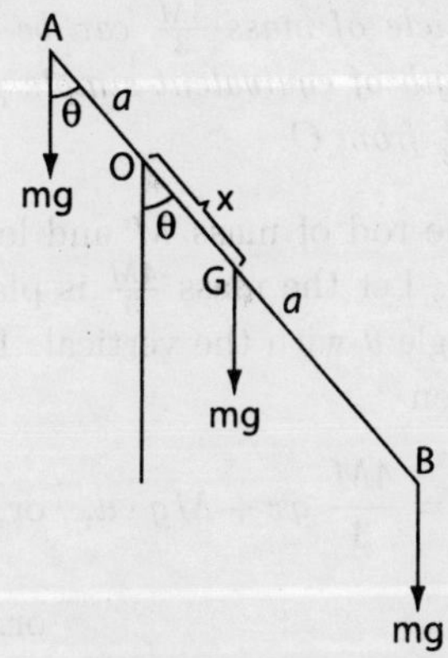

Fig 3.11

Considering the m.i. of the system about an axis through O, we have

$$\begin{aligned} 3mK^2 &= m(a-x)^2 + mx^2 + m(x+a)^2 = 3mx^2 + 2ma^2 \\ \text{or,} \quad K^2 &= \frac{3x^2+2a^2}{3}. \end{aligned}$$

Therefore,

$$\ddot{\theta} = -\frac{gx\sin\theta}{K^2} = -\frac{gx\sin\theta}{\frac{3x^2+2a^2}{3}} = -\frac{3gx\sin\theta}{3x^2+2a^2} = -\frac{3gx\theta}{3x^2+2a^2}$$

[for small oscillation $\sin\theta = \theta$].

The above equation shows that the motion of the system is S.H.M. of time period

$$T = \frac{2\pi}{\sqrt{\frac{3gx}{3x^2+2a^2}}} = \frac{2\pi}{\sqrt{3g}}\sqrt{\frac{3x^2+2a^2}{x}}$$

$$= \frac{2\pi}{\sqrt{3g}}\sqrt{3x+\frac{2a^2}{x}} = \frac{2\pi}{\sqrt{3g}}\sqrt{\left(\sqrt{3x}\right)^2+\left(\frac{\sqrt{2}a}{\sqrt{x}}\right)^2}$$

$$= \frac{2\pi}{\sqrt{3g}}\sqrt{\left(\sqrt{3x}-\frac{\sqrt{2}}{\sqrt{x}}a\right)^2+2\sqrt{6}\,a}.$$

$\therefore$ Time period will be least, when

$$\sqrt{3x}-\frac{\sqrt{2}}{\sqrt{x}}a = 0, \quad \text{or,} \quad x = \frac{\sqrt{2}}{\sqrt{3}}a = \frac{1.429}{1.732}a = 0.82a \quad \text{(Approx)}.$$

▶ **Problem 3.3.8** *A uniform rod of mass M and length 2a is free to turn about one end O. A particle of mass $\frac{4M}{3}$ can be attached to the rod at any point. Prove that the length of equivalent simple pendulum is least when the particle is at a distance $\frac{a}{2}$ from O.*

Solution: Let OA be the rod of mass M and length $2a$ and G be its c.g. such that $OG = GA = a$. Let the mass $\frac{4M}{3}$ is placed at P, where $OP = x$. Let rod OA makes an angle θ with the vertical. If h be the distance of c.g. of the system from O, then

$$\left(\frac{4M}{3}+M\right)gh = \frac{4M}{3}gx + Mg\cdot a, \quad \text{or,} \quad \frac{7}{3}h = \frac{4x+3a}{3},$$

$$\text{or,} \quad h = \frac{4x+3a}{7}.$$

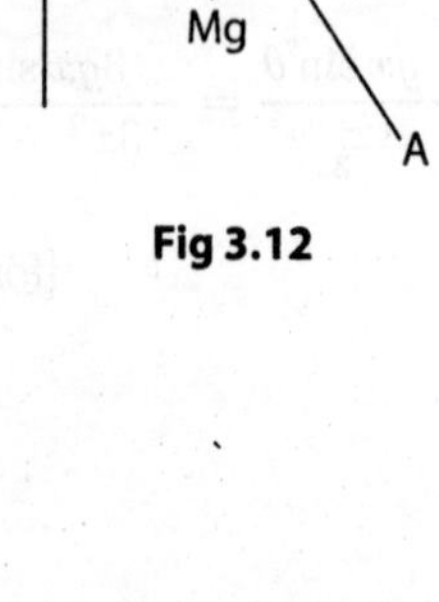

Fig 3.12

If K be the radius of gyration of the system about O, then

$$\left(\frac{4M}{3}+M\right)K^2=\frac{4M}{3}x^2+Mg\cdot a^2+\frac{Ma^2}{3}=M\left(\frac{4x^2+4a^2}{3}\right).$$

$$\therefore \quad K^2=\frac{4x^2+4a^2}{7}=\frac{4(a^2+x^2)}{7}.$$

$\therefore$ The length of simple equivalent pendulum.

$$\begin{aligned} l=\frac{K^2}{h} &= \frac{\frac{4(a^2+x^2)}{7}}{\frac{4x+3a}{7}}=\frac{4x^2+4a^2}{4x+3a} \\ &= \frac{(2x)^2-2\cdot 2x\cdot a+a^2+3a^2+4ax}{3a+4x} \\ &= \frac{a(3a+4x)+(2x-a)^2}{3a+4x} \\ &= a+\frac{(2x-a)^2}{3a+4x}. \end{aligned}$$

The length of equivalent simple pendulum will be minimum when

$$2x-a=0, \quad \text{i.e.,} \quad x=a/2.$$

▶ **Problem 3.3.9** *A uniform rod of mass m and length 2a can oscillates about a fixed horizontal axis through one end. A circular disc of mass 24 m and radius $\frac{1}{3}a$ can have its centre clamped at any point of the rod and its plane contains the axis of rotation, show that for oscillation under gravity the length of the equivalent simple pendulum lies between $\frac{1}{2}a$ and 2a.*

Solution: Let $OA(=2a)$ be the rod and G be the c.g. of the rod so the $OG=a$. Let at $C(OC=x)$, the circular disc of mass 24 m and of radius $\frac{1}{3}a$ be clamped. The distance of c.g. of the system h from O, is given by

$$(24m+m)h=m\cdot a+24m\cdot x.$$

$$\therefore \quad h=\frac{24x+a}{25}.$$

If K be the radius of gyration of the system about O, we have

$$\begin{aligned} (24m+m)K^2 &= \frac{ma^2}{3}+ma^2+\frac{24m\left(\frac{1}{3}a\right)^2}{4}+24mx^2 \\ &= \frac{4ma^2}{3}+\frac{2}{3}ma^2+24mx^2=m(2a^2+24x^2) \\ \text{or,} \quad K^2 &= \frac{2a^2+24x^2}{25}. \end{aligned}$$

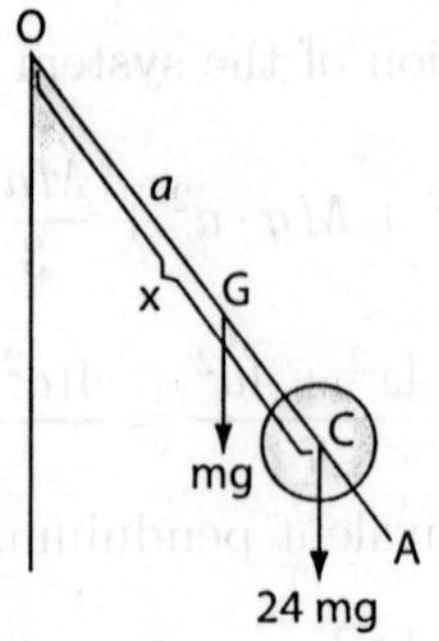

Fig 3.13

$\therefore$ The length of equivalent simple pendulum

$$l = \frac{K^2}{h} = \frac{\frac{2a^2+24x^2}{25}}{\frac{24x+a}{25}} = \frac{2a^2+24x^2}{24x+a} = f(x) \quad \text{(say)}.$$

$$\therefore \quad f'(x) = \frac{48x(24x+a) - 24(2a^2+24x^2)}{(24x+a)^2} = \frac{48\{24x^2+ax-a^2-12x^2\}}{(24x+a)^2}$$

$$= \frac{48\{12x^2+ax-x^2\}}{(24x+a)^2} = \frac{48(4x-a)(3x+a)}{(24x+a)^2}$$

$\therefore$ $f'(x) \gtrless 0$ according as $x > \frac{a}{4}$ or, $x < \frac{a}{4}$.

As x lies between O and $2a$, then $f(x)$ is increasing in $\frac{a}{4} \leq x \leq 2a$ and lies between $f(a/4)$ and $f(2a)$ and in $(0 \leq x \leq a/4)$, $f(x)$ in decreasing and lies between $f(0)$ and $f(a/4)$.

Now

$$f\left(\frac{a}{4}\right) = \frac{2a^2 + 24 \cdot \frac{a^2}{16}}{24 \cdot \frac{a}{4} + a} = \frac{2a^2 + \frac{3}{2}a^2}{7a} = \frac{7a^2}{2 \times 7a} = \frac{a}{2}$$

$$\text{and} \quad f(2a) = \frac{2a^2 + 24 \cdot 4a^2}{48a + a} = \frac{98a^2}{49a} = 2a.$$

Thus $f(x) = l$, lies between $(a/2, 2a)$.

▶ **Problem 3.3.10** *An elliptic lamina is such that when it swings about one latus rectum as a horizontal axis, the other latus rectum passes through the centre of oscillation; prove that the eccentricity of the ellipse is* $1/2$.

If the lamina starts from rest at an inclination $\pi/6$ *to the vertical what is its angular velocity when its plane is vertical.*

Solution: Let the equation of the elliptic lamina be $\frac{x^2}{a^2} + \frac{y^2}{b^2} = 1$ and the lamina swings about the latus rectum AB through $O(ae, 0)$. Since the other latus rectum passes through the centre of oscillation $O'(-ae, 0)$, then the length of simple equivalent pendulum is $OO' = 2ae$.

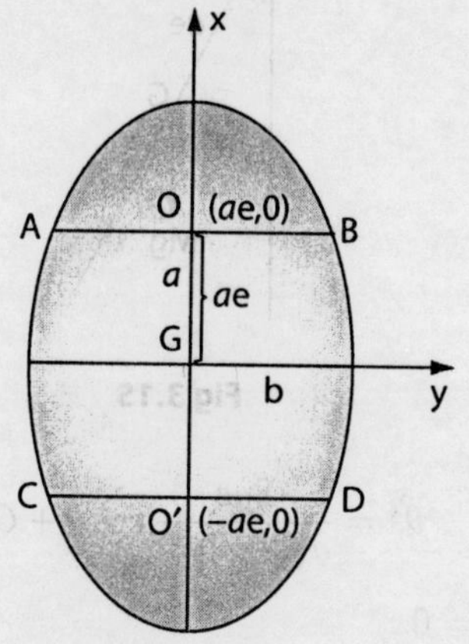

Fig 3.14

Let M be the mass of the lamina and if K be the radius of gyration of the lamina about AB, then

$$MK^2 = \frac{Ma^2}{4} + Ma^2e^2.$$

$$\therefore\ K^2 = \frac{a^2}{4}(1+4e^2) \text{ and } h = OG = ae.$$

$\therefore$ length of equivalent simple pendulum

$$\begin{aligned} 2ae &= \frac{\frac{a^2}{4}(1+4e^2)}{ae} \\ &= \frac{a}{4e}(1+4e^2) \\ \text{or,}\quad 8e^2 &= 1+4e^2, \\ \text{or,}\quad 4e^2 &= 1 \quad \therefore \quad e = \frac{1}{2}. \end{aligned}$$

Let, at any time t, the inclination of the OG with vertical is θ, then the equation of motion $MK^2\ddot{\theta} = L$ gives

$$MK^2\ddot{\theta} = -Mgae\sin\theta, \quad \text{or,} \quad \frac{a^2}{4}(1+4e^2)\ddot{\theta} = -gae\sin\theta,$$

$$\text{or,} \quad \frac{1+4e^2}{4}\cdot a\ddot{\theta} = -ge\sin\theta,$$

$$\text{or,} \quad \ddot{\theta} = -\frac{4ge}{(1+4e^2)a}\sin\theta,$$

$$\text{or,} \quad 2\dot{\theta}\ddot{\theta} = -\frac{8ge}{(1+4e^2)a}\sin\theta\cdot\dot{\theta}.$$

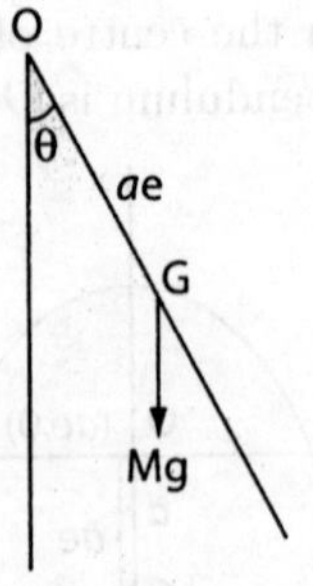

Fig 3.15

Integrating,

$$\dot{\theta}^2 = \frac{8ge}{a(1+4e^2)}\cos\theta + C$$

when $\theta = \pi/6$; then $\dot{\theta} = 0$.

$$\therefore \quad C = -\frac{8ge}{a(1+4e^2)} \cdot \frac{\sqrt{3}}{2} = -\frac{4\sqrt{3}\,ge}{a(1+4e^2)}.$$

$$\therefore \quad \dot{\theta}^2 = \frac{8ge}{a(1+4e^2)}\left[\cos\theta - \frac{\sqrt{3}}{2}\right].$$

$$\therefore \quad \dot{\theta}^2\Big|_{\theta=0} = \frac{8ge}{a(1+4e^2)}\left[1 - \frac{\sqrt{3}}{2}\right]$$

$$= \frac{4ge}{a(1+4e^2)}(2-\sqrt{3}).$$

Putting $e = \frac{1}{2}$, we get

$$\dot{\theta}^2\Big|_{\theta=0} = \frac{2g}{2a}(2-\sqrt{3}) = \frac{g}{a}(2-\sqrt{3})$$

$$\text{or,} \quad \dot{\theta}\Big|_{\theta=0} = \sqrt{\frac{g}{a}(2-\sqrt{3})}.$$

► **Problem 3.3.11** *An elliptic lamina can rotate about a horizontal axis passing through a focus and perpendicular to its plane. If the eccentricity of the ellipse be $\sqrt{2/5}$, prove that the centre of oscillation will be at the other focus.* **[CUH 1997, 2000, 2003; VUH 2004]**

Solution: Let the equation of the ellipse is $\frac{x^2}{a^2} + \frac{y^2}{b^2} = 1$, mass M and G be the c.g. O and O' be the foci. Let the lamina can rotate about an axis through the focus O and perpendicular to the plane of the lamina. Thus the coordinates of O, O' are respectively $(ae, 0)$ and $(-ae, 0)$. Here $OG = ae$ and

$$\begin{aligned} MK^2 &= \frac{M}{4}(a^2+b^2) + Ma^2e^2 = \frac{M}{4}[a^2+b^2+4e^2a^2] \\ &= \frac{M}{4}\left[a^2+a^2(1-e^2)+4e^2a^2\right] = \frac{M}{4}\left[2a^2+3a^2e^2\right]. \end{aligned}$$

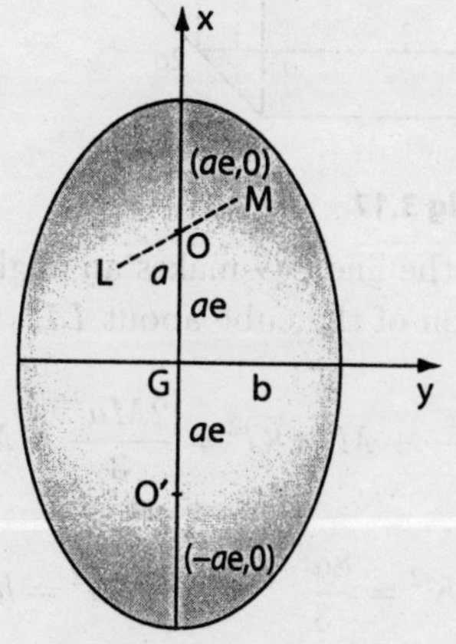

Fig 3.16

Therefore, the length of equivalent simple pendulum

$$\begin{aligned} &= \frac{K^2}{OG} = \frac{\frac{2a^2+3a^2e^2}{4}}{ae} = \frac{a(2+3e^2)}{4e} = \frac{a(2+3\cdot\frac{2}{5})}{4\sqrt{2/5}} \quad \left[\because\ e=\sqrt{2/5}\right] \\ &= \frac{a\cdot\frac{16}{5}}{4\sqrt{2/5}} = \frac{4a}{5\sqrt{2/5}} = \frac{4a\sqrt{2/5}}{5\cdot 2/5} = 2a\sqrt{2/5} = 2ae = OO'. \end{aligned}$$

Thus the centre of oscillation will be at the other focus O'.

► **Problem 3.3.12** *Show that a homogeneous cube of side $2a$ free to turn about of its edge which is horizontal, oscillates in the same time as a simple pendulum of length $\frac{4}{3}\sqrt{2}a$. If this cube moves from rest with one face in a horizontal plane, find its angular velocity, when the face makes an angle $45°$ with the horizontal.*

Solution: Let LL' be one edge of the cube of edge length $2a$ and mass M, where LL' is horizontal. Let G be the c.g. of the cube.

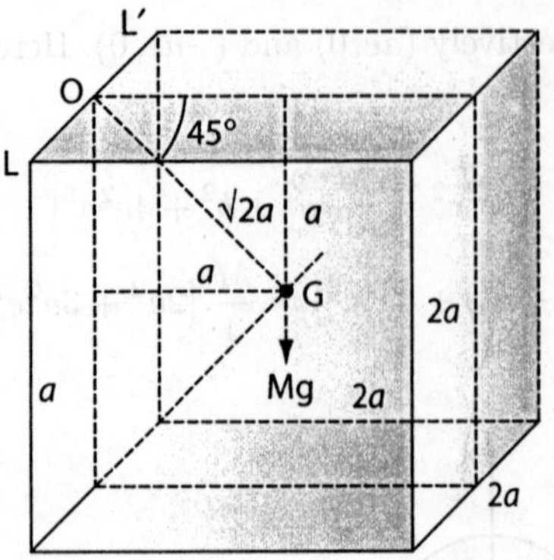

Fig 3.17

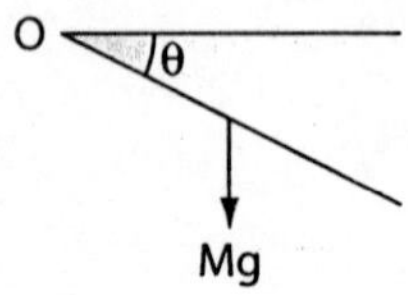

Fig 3.18

Let at any instant, the line OG makes an angle θ with the horizon. If K be the radius of gyration of the cube about LL', we have,

$$MK^2 = \frac{2Ma^2}{3} + M\cdot OG^2 = \frac{2Ma^2}{3} + M\cdot 2a^2 = \frac{8Ma^2}{3}.$$

$$\therefore \quad K^2 = \frac{8a^2}{3} \quad \text{and} \quad OG^2 = h = \sqrt{2}a.$$

Therefore, the length of simple equivalent pendulum

$$= \frac{K^2}{h} = \frac{\frac{8a^2}{3}}{\sqrt{2}a} = \frac{4\sqrt{2}}{3}a.$$

Now the equation of motion is

$$MK^2\ddot{\theta} = Mg\cdot OG\cos\theta = Mg\cdot\sqrt{2}a\cos\theta$$

$$\text{or,} \quad \ddot{\theta} = \frac{\sqrt{2}\,ga}{K^2}\cos\theta = \frac{3\sqrt{2}\,ga}{8a^2}\cos\theta = \frac{3g}{4\sqrt{2}a}\cos\theta,$$

$$\text{or,} \quad 2\dot{\theta}\ddot{\theta} = \frac{3g}{2\sqrt{2}a}\cos\theta\cdot\dot{\theta}.$$

Integrating

$$\dot{\theta}^2 = \frac{3g}{2\sqrt{2}a}\sin\theta + C.$$

Initially $\theta = 45°$, $\dot{\theta} = 0$.

$$\therefore \quad C = -\frac{3g}{4a}.$$

Thus

$$\dot{\theta}^2 = \frac{3g}{2\sqrt{2}a}\sin\theta - \frac{3g}{4a} = \frac{3g}{4a}\left[\sqrt{2}\sin\theta - 1\right]$$

$$\text{or, } \dot{\theta} = \sqrt{\frac{3g}{4a}\left[\sqrt{2}\sin\theta - 1\right]}.$$

When the face makes an angle $45°$ with horizontal, i.e., $\theta = 90°$, we have

$$\dot{\theta}\,|_{\theta=90°} = \sqrt{\frac{3g}{4a}}\,(\sqrt{2}-1).$$

► **Problem 3.3.13** *Find the time of oscillation of a compound pendulum consisting of a rod of mass M and length a, carrying at one end a sphere of mass M_1 and diameter 2b, the other end of the rod being fixed. Find also the length of simple equivalent pendulum.*

Solution: Let OA be the rod of length a and of mass M and O' be the centre of the sphere of mass M_1 and of radius b and the rod fixed at O.

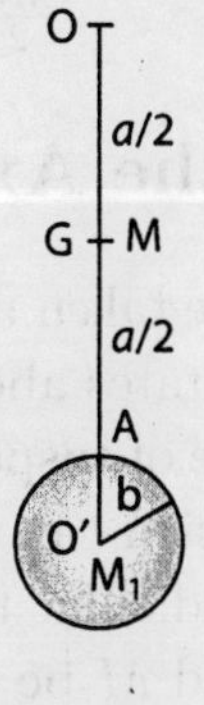

Fig 3.19

Thus the moment of inertia of system about axis of rotation through O

$$\begin{aligned} &= \frac{M}{3}\left(\frac{a^2}{4}\right) + M\frac{a^2}{4} + \frac{2}{5}M_1b^2 + M_1(a+b)^2 \\ &= Ma^2\left(\frac{1}{12}+\frac{1}{4}\right) + M_1\left[\frac{2}{5}b^2 + (a+b)^2\right] \\ &= \frac{Ma^2}{3} + M_1\left[\frac{2}{5}b^2 + (a+b)^2\right]. \end{aligned}$$

Thus the radius of gyration of the system about O, K is given by

$$(M+M_1)K^2 = \frac{Ma^2}{3} + M_1\left[\frac{2}{5}b^2 + (a+b)^2\right].$$

$$\therefore \quad K^2 = \frac{\frac{Ma^2}{3} + M_1\left[\frac{2}{5}b^2 + (a+b)^2\right]}{M+M_1}.$$

Let h be the depth of c.g. from O, then

$$(M+M_1)h = M\cdot\frac{a}{2}+M_1(a+b).$$

$$\therefore\quad h = \frac{\frac{Ma}{2}+M_1(a+b)}{M+M_1}.$$

Thus the time of oscillation

$$= 2\pi\sqrt{\frac{K^2}{gh}} = \frac{2\pi}{\sqrt{g}}\sqrt{\frac{\frac{Ma^2}{2}+M_1\left[\frac{2}{5}b^2+(a+b)^2\right]}{\frac{Ma}{3}+M_1(a+b)}}.$$

Thus the length of simple equivalent pendulum

$$= \frac{K^2}{h} = \frac{\frac{Ma^2}{3}+M_1\left[\frac{2}{5}b^2+(a+b)^2\right]}{\frac{Ma}{2}+M_1(a+b)}.$$

3.4 Reactions of the Axis of Rotation

Let the plane of the paper be taken as the plane through centre of gravity G of the body and the body rotates about *a* horizontal axis and perpendicular to the plane. Let O be centre of suspension and $OG = h$. The reaction of the axis at the centre of suspension can be resolved in two directions one along GO and other along perpendicular to GO. Let OG makes an angle θ with the vertical at any time t and M be the mass of the body. The acceleration along OG (in direction r increasing) $= \ddot{r}-r\dot{\theta}^2 = -h\dot{\theta}^2$ [As $r = h =$ constant] and acceleration along perpendicular OG(in direction θ increasing)

$$= \frac{1}{r}\frac{d}{dt}(r^2\dot{\theta}) = \frac{1}{h}\frac{d}{dt}(h^2\dot{\theta}) = h\ddot{\theta}.$$

Therefore, the equations of motion are

$$M(-h\dot{\theta}^2) = -P+Mg\cos\theta \tag{3.4.1}$$

$$M\cdot h\ddot{\theta} = Q-Mg\sin\theta. \tag{3.4.2}$$

Again the equation of motion about fixed axis, we have

$MK^2\ddot{\theta} = L$ = Algebraic sum of the moment of the external forces about the point of suspension.

$$= -Mgh\sin\theta$$

$$\text{or,}\quad K^2\ddot{\theta} = -gh\sin\theta, \tag{3.4.3}$$

where K is the radius of gyration of the body about O.

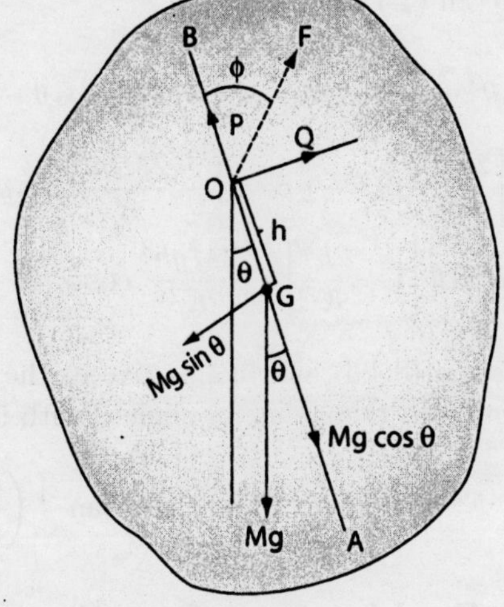

Fig 3.20

From (3.4.2) we get

$$Q = M(h\ddot{\theta} + g\sin\theta) = M\left(-\frac{gh^2\sin\theta}{K^2} + g\sin\theta\right) \quad [\text{by}(3.4.3)]$$

$$= Mg\left(1 - \frac{h^2}{K^2}\right)\sin\theta. \qquad (3.4.4)$$

Again from (3.4.3),

$$\ddot{\theta} = -\frac{gh}{K^2}\sin\theta, \quad \text{or,} \quad 2\dot{\theta}\ddot{\theta} = -\frac{2gh}{K^2}\sin\theta\cdot\dot{\theta},$$

$$\text{or,} \quad \frac{d}{dt}(\dot{\theta}^2) = -\frac{2gh}{K^2}\sin\theta\cdot\dot{\theta}.$$

Integrating,

$$\dot{\theta}^2 = \frac{2gh}{K^2}\cos\theta + C \quad \text{(constant)}.$$

Assuming, $\dot{\theta}^2 = 0$, when $\theta = \alpha$, we get,

$$C = -\frac{2gh}{K^2}\cos\alpha.$$

$$\therefore \quad \dot{\theta}^2 = \frac{2gh}{K^2}(\cos\theta - \cos\alpha). \qquad (3.4.5)$$

Substituting $\dot{\theta}^2$ in (3.4.1) we get,

$$\begin{aligned} P &= Mh\dot{\theta}^2 + Mg\cos\theta = M\left[h\cdot\frac{2gh}{K^2}(\cos\theta - \cos\alpha) + g\cos\theta\right] \\ &= M\left[\frac{2gh^2}{K^2}\cos\theta + g\cos\theta\right] - \frac{2Mgh^2}{K^2}\cos\alpha \\ &= Mg\cos\theta\left[1 + \frac{2h^2}{K^2}\right] - \frac{2Mgh^2}{K^2}\cos\alpha. \qquad (3.4.6) \end{aligned}$$

Thus the equation (3.4.6) and (3.4.3) give us the reaction at O. If F be the resultant reaction at O making an angle ϕ with the GO, we have,

$$F^2 = P^2 + Q^2 \quad \text{and} \quad \phi = \tan^{-1}\left(\frac{Q}{P}\right).$$

3.5 Solved Problems on Reaction

► **Problem 3.5.1** *A circular area can turn freely about a horizontal axis which passes through a point O of its circumference and is perpendicular to its plane. If the motion commences when the diameter through O is vertically above O; the components of strains at O along and perpendicular to this diameter are respectively $\frac{W}{3}(7\cos\theta - 4)$ and $\frac{W}{3}\sin\theta$, where W is the weight of the circular area.* **[CUH 1979]**

Solution: Let G be the c.g. of the circular area of mass M and radius a. Let the motion starts when OG is vertical. We take Ox horizontal. Let at any instant OG makes an angle θ with vertical OZ. Let P and Q are the components of strains along and perpendicular to GO.

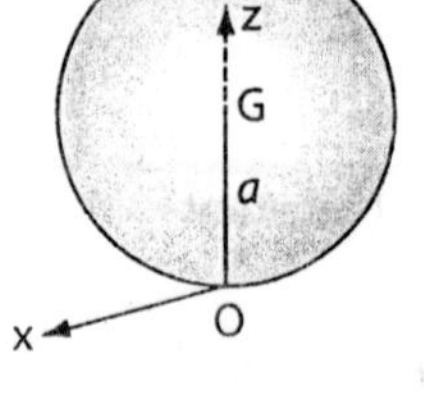

Fig 3.21

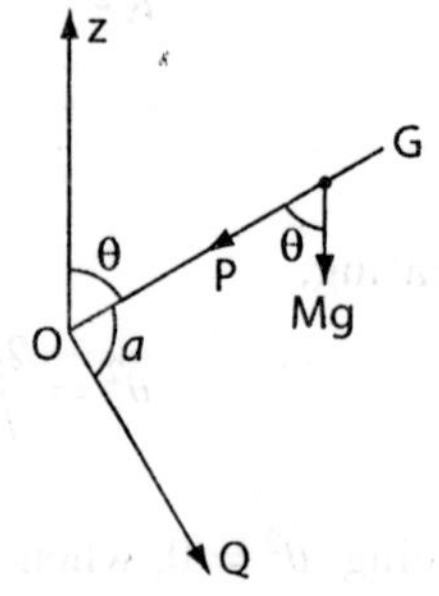

Fig 3.22

The distance of c.g. G from O

$$h = a. \qquad (3.5.1)$$

The m.i. of the circular area about horizontal axis Ox is

$$M \cdot \frac{a^2+a^2}{4} + Ma^2 = \frac{Ma^2}{2} + Ma^2 = \frac{3Ma^2}{2}.$$

$$\therefore \quad K^2 = \frac{3a^2}{2}, \tag{3.5.2}$$

where K is the radius of gyration.

Now the equations of motion along and perpendicular to OG are

$$M(-a\dot{\theta}^2) = -P - Mg\cos\theta \tag{3.5.3}$$

$$M(a\ddot{\theta}) = Q + Mg\sin\theta \tag{3.5.4}$$

$$\text{and} \quad MK^2\ddot{\theta} = Mga\sin\theta. \tag{3.5.5}$$

From (3.5.5) we have,

$$\ddot{\theta} = \frac{ga\sin\theta}{K^2} = \frac{2ga}{3a^2}\sin\theta = \frac{2g}{3a}\sin\theta,$$

$$\text{or,} \quad 2\dot{\theta}\ddot{\theta} = \frac{4g}{3a}\sin\theta\cdot\dot{\theta}. \tag{3.5.6}$$

Integrating,

$$\dot{\theta}^2 = -\frac{4g}{3a}\cos\theta + C.$$

Initially $\dot{\theta} = 0$, when $\theta = 0$. $\therefore\ C = \frac{4g}{3a}$.

$$\therefore \quad \dot{\theta}^2 = \frac{4g}{3a} - \frac{4g}{3a}\cos\theta = \frac{4g}{3a}(1-\cos\theta). \tag{3.5.7}$$

Using (3.5.7) in (3.5.3) we get,

$$\begin{aligned} P &= Ma\dot{\theta}^2 - Mg\cos\theta = M\left[\frac{4g}{3}(1-\cos\theta) - g\cos\theta\right] \\ &= \frac{Mg}{3}[4 - 4\cos\theta - 3\cos\theta] = \frac{Mg}{3}[4 - 7\cos\theta] \\ &= \frac{W}{3}[7\cos\theta - 4]. \\ \therefore \quad |P| &= \frac{W}{3}[7\cos\theta - 4]. \end{aligned}$$

Using (3.5.6) in (3.5.4) we get

$$\begin{aligned} Q &= Ma\ddot{\theta} - Mg\sin\theta = M\frac{2g}{3}\sin\theta - Mg\sin\theta \\ &= \frac{Mg}{3}[2\sin\theta - 3\sin\theta] = -\frac{W}{3}\sin\theta. \\ \therefore \quad |Q| &= \frac{W}{3}\sin\theta. \end{aligned}$$

► **Problem 3.5.2** *A heavy homogeneous cube, of weight W, can swing about an edge with is horizontal; it starts from rest being displaced from its unstable position of equilibrium; when the perpendicular from the centre of gravity upon the edge has turned through an angle θ, show that the components of the actions at the hinge along and right angle to this perpendicular are $\frac{W}{2}(3 - 5\cos\theta)$ and $\frac{W}{4}\sin\theta$.*

Solution: Let the cube of mass M and edge length $2a$, can turn about the edge OO'. Let $G(a, a, a)$ be the centre of gravity of the cube with respect to $O'x$, $O'y$ and $O'z$ as axis. Therefore, $OG = h = \sqrt{2}a =$ distance of G from the axis of rotation. Let P and Q be the component of reactions along and perpendicular to GO. Thus the m.i. of the cube about the axis of rotation OO'

$$= \frac{M}{3} \cdot 2a^2 + Mh^2 = \frac{2Ma^2}{3} + M \cdot 2a^2 = \frac{8Ma^2}{3}.$$

Thus the radius of gyration of the cube K is given by

$$K^2 = \frac{8a^2}{3}. \tag{3.5.8}$$

Now the equations of motion along and perpendicular to OG are

$$\begin{aligned} M(-h\dot{\theta}^2) &= -P - Mg\cos\theta & (3.5.9)\\ M(h\ddot{\theta}) &= Q + Mg\sin\theta & (3.5.10)\\ \text{and} \quad MK^2\ddot{\theta} &= Mgh\sin\theta. & (3.5.11)\end{aligned}$$

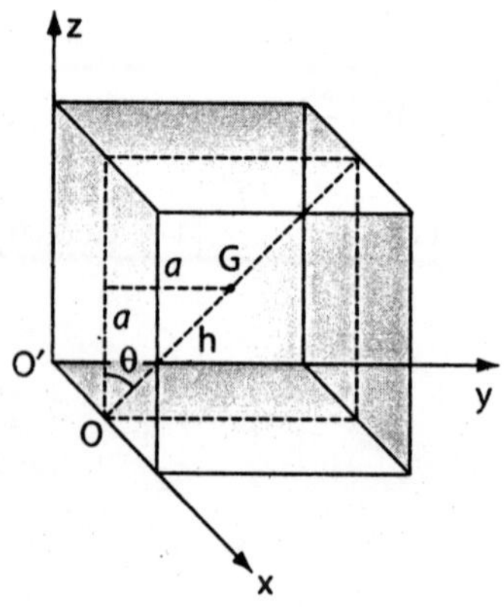

Fig 3.23

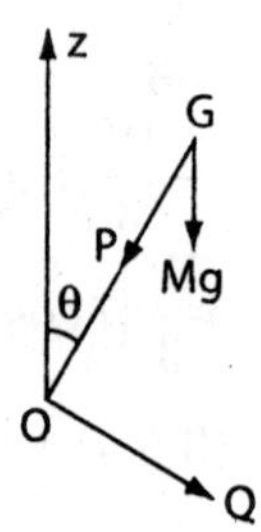

Fig 3.24

From (3.5.11), we get,

$$\ddot{\theta} = \frac{gh}{K^2}\sin\theta = \frac{g\cdot\sqrt{2}}{\frac{8a^2}{3}}\sin\theta = \frac{3\sqrt{2}\,g}{8a}\sin\theta \qquad (3.5.12)$$

$$\text{or,}\quad 2\dot{\theta}\ddot{\theta} = \frac{3\sqrt{2}}{4a}\,g\sin\theta\cdot\dot{\theta}.$$

Integrating

$$\dot{\theta}^2 = -\frac{3\sqrt{2}}{4a}\,g\cos\theta + C.$$

Initially $\dot{\theta} = 0$, when $\theta = 0$.

$$\therefore\quad C = \frac{3\sqrt{2}}{4a}\,g.$$

$$\therefore\quad \dot{\theta}^2 = \frac{3\sqrt{2}}{4a}\cdot g\cdot(1-\cos\theta). \qquad (3.5.13)$$

Using (3.5.13) in (3.5.9), we get

$$\begin{aligned} P &= Mh\dot{\theta}^2 - Mg\cos\theta = M\cdot\sqrt{2}a\cdot\frac{3\sqrt{2}}{4a}\,g(1-\cos\theta) - Mg\cos\theta \\ &= Mg\left[\frac{3}{2} - \frac{3}{2}\cos\theta - \cos\theta\right] = \frac{W}{2}[3-5\cos\theta] \quad [\because W = Mg]. \end{aligned}$$

Again, using (3.5.12) in (3.5.10) we get

$$\begin{aligned} Q &= -Mg\sin\theta + Mh\ddot{\theta} = -Mg\sin\theta + M\sqrt{2}a\cdot\frac{3\sqrt{2}\,g}{8a}\sin\theta \\ &= \frac{3Mg}{4}\cdot\sin\theta - Mg\sin\theta = -\frac{Mg}{4}\sin\theta = -\frac{W}{4}\sin\theta \\ \therefore\quad |Q| &= \frac{W}{4}\sin\theta. \end{aligned}$$

▶ **Problem 3.5.3** *A uniform rod of length 2a and weight W is turning about its end O and starts from the position in which it was vertically above O. When it has turned through an angle θ, show that the horizontal and vertical reactions about O are* $\frac{3W}{4}\sin\theta(2-3\cos\theta)$ *and* $\frac{W}{4}(1-3\cos\theta)^2$.

[CUH 1999; BUH 2004]

Solution: Let G be the c.g. of the rod OA of length $2a$ and of mass M, which can turn about O. Let X and Y be the horizontal and vertical reactions at O at any instant. Let at that instant the rod makes an angle θ with vertical.

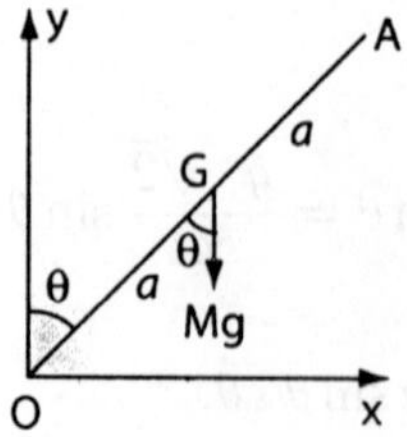

Fig 3.25

The equations of motions are

$$\begin{aligned} M(-a\dot{\theta}^2) &= X\sin\theta + Y\cos\theta - Mg\cos\theta & (3.5.14)\\ M(a\ddot{\theta}) &= X\cos\theta - Y\sin\theta + Mg\sin\theta & (3.5.15)\\ \text{and} \quad MK^2\ddot{\theta} &= Mga\sin\theta. & (3.5.16)\end{aligned}$$

The m.i. of the rod about O,

$$MK^2 = \frac{4}{3}a^2 \cdot M.$$

Thus

$$K^2 = \frac{4a^2}{3}.$$

Thus from (3.5.16), we get

$$\begin{aligned} \ddot{\theta} &= \frac{ga\sin\theta}{\frac{4a^2}{3}} = \frac{3g}{4a}\sin\theta, & (3.5.17)\\ \text{or,} \quad 2\dot{\theta}\ddot{\theta} &= \frac{3g}{2a}\sin\theta\cdot\dot{\theta}. \end{aligned}$$

Integrating

$$\dot{\theta}^2 = -\frac{3g}{2a}\cos\theta + C.$$

Initially $\dot{\theta} = 0$, when $\theta = 0$.

$$\therefore \quad C = \frac{3g}{2a}.$$

$$\therefore \quad \dot{\theta}^2 = \frac{3g}{2a} - \frac{3g}{2a}\cos\theta = \frac{3g}{2a}(1-\cos\theta). \qquad (3.5.18)$$

From (3.5.14) and (3.5.15) we get

$$\begin{aligned} X\sin\theta + Y\cos\theta &= Mg\cos\theta - Ma\dot{\theta}^2 & (3.5.19)\\ X\cos\theta - Y\sin\theta &= Ma\ddot{\theta} - Mg\sin\theta. & (3.5.20)\end{aligned}$$

Multiplying (3.5.19) by $\sin\theta$ and (3.5.20) by $\cos\theta$, then adding we get

$$\begin{aligned}
X &= Mg\sin\theta\cos\theta - Ma\dot{\theta}^2\sin\theta + Ma\ddot{\theta}\cos\theta - Mg\sin\theta\cos\theta \\
&= Ma\ddot{\theta}\cos\theta - Ma\dot{\theta}^2\sin\theta \\
&= M\frac{3g}{4}\sin\theta\cos\theta - M\frac{3g}{4}(1-\cos\theta)\sin\theta \\
&= \frac{3Mg}{4}\sin\theta\,[\cos\theta - 2 + 2\cos\theta] = \frac{3Mg}{4}\sin\theta\,(3\cos\theta - 2) \\
&= \frac{3W}{4}\sin\theta\,(3\cos\theta - 2), \\
\text{or,}\quad |X| &= \frac{3W}{4}\sin\theta\,(2 - 3\cos\theta).
\end{aligned}$$

Again multiplying (3.5.19) by $\cos\theta$ and (3.5.20) by $\sin\theta$, and then subtracting, we get

$$\begin{aligned}
Y &= (Mg\cos\theta - Ma\dot{\theta}^2)\cos\theta - (Ma\ddot{\theta} - Mg\sin\theta)\sin\theta \\
&= Mg - M(a\dot{\theta}^2\cos\theta + a\ddot{\theta}\sin\theta) \\
&= Mg - M\left[\frac{3g}{2}(1-\cos\theta)\cos\theta + \frac{3g}{4}\sin\theta\cdot\sin\theta\right] \\
&= Mg - Mg\left[\frac{3}{2}\cos\theta - \frac{3}{2}\cos^2\theta + \frac{3}{4}\sin^2\theta\right] \\
&= Mg\left[1 - \frac{3}{2}\cos\theta + \frac{3}{2}\cos^2\theta - \frac{3}{4}\sin^2\theta\right] \\
&= \frac{1}{4}Mg\,[4 - 6\cos\theta + 6\cos^2\theta - 3(1-\cos^2\theta)] \\
&= \frac{1}{4}Mg\,[1 - 6\cos\theta + 9\cos^2\theta] = \frac{W}{4}(1 - 3\cos\theta)^2.
\end{aligned}$$

Other Approach. Let P and Q be components of the reaction at O along and perpendicular to GO. Let G be the c.g. of the rod OA and at any instant, the rod makes an angle θ with the vertical. Therefore, $OG = GA = a$.

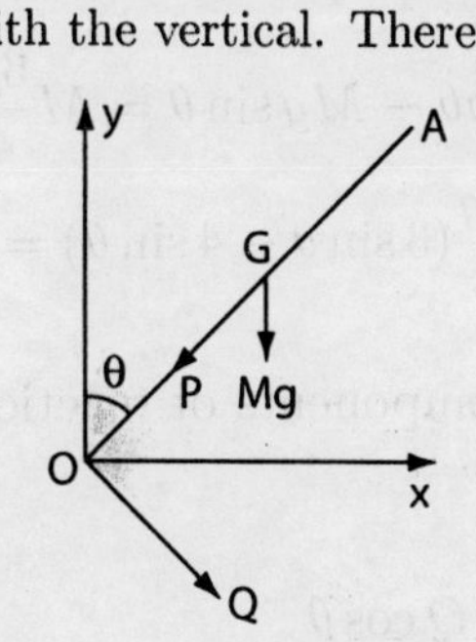

Fig 3.26

The equations of motion are

$$
\begin{aligned}
M(-a\dot{\theta}^2) &= -P - Mg\cos\theta && (3.5.21)\\
M(a\ddot{\theta}) &= Q + Mg\sin\theta && (3.5.22)\\
\text{and}\quad MK^2\ddot{\theta} &= Mga\sin\theta, && (3.5.23)
\end{aligned}
$$

From (3.5.23) we get

$$M\cdot\frac{4a^2}{3}\ddot{\theta} = Mga\sin\theta,\ \text{or},\ \ddot{\theta} = \frac{3g}{4a}\sin\theta,\ \text{or},\ 2\dot{\theta}\ddot{\theta} = \frac{3g}{2a}\sin\theta\cdot\dot{\theta}.$$

Integrating

$$\dot{\theta}^2 = -\frac{3g}{2a}\cos\theta + C.$$

Initially $\dot{\theta} = 0$, when $\theta = 0$. $\therefore\ C = \frac{3g}{2a}$.
Thus

$$\dot{\theta}^2 = \frac{3g}{2a} - \frac{3g}{2a}\cos\theta = \frac{3g}{2a}(1-\cos\theta).$$

$\therefore$ From (3.5.21), we get,

$$
\begin{aligned}
P &= Ma\dot{\theta}^2 - Mg\cos\theta = M\frac{3g}{2}(1-\cos\theta) - Mg\cos\theta\\
&= \frac{Mg}{2}[3 - 3\cos\theta - 2\cos\theta] = \frac{Mg}{2}[3 - 5\cos\theta].
\end{aligned}
$$

From (3.5.22) we get,

$$
\begin{aligned}
Q &= Ma\ddot{\theta} - Mg\sin\theta = M\frac{3g}{4}\sin\theta - Mg\sin\theta\\
&= \frac{Mg}{4}(3\sin\theta - 4\sin\theta) = -\frac{Mg}{4}\sin\theta.
\end{aligned}
$$

If X and Y be the components of reaction at O, horizontally and vertically, we have,

$$
\begin{aligned}
X &= -P\sin\theta + Q\cos\theta\\
&= -\frac{Mg}{4}\sin\theta\cos\theta - \frac{Mg}{2}[3 - 5\cos\theta]\sin\theta\\
&= -\frac{Mg}{4}\sin\theta[\cos\theta + 6 - 10\cos\theta] = -\frac{Mg}{4}\sin\theta[6 - 9\cos\theta]\\
&= -\frac{3W}{4}\sin\theta(2 - 3\cos\theta)
\end{aligned}
$$

$$|X| = \frac{3W}{4}\sin\theta(2-3\cos\theta)$$

$$\begin{aligned} Y &= -P\cos\theta - Q\sin\theta \\ &= -\left[\frac{Mg}{2}(3-5\cos\theta)\cos\theta - \frac{Mg}{4}\sin\theta\cdot\sin\theta\right] \\ &= -\frac{Mg}{4}\left[6\cos\theta - 10\cos^2\theta - \sin^2\theta\right] \\ &= -\frac{Mg}{4}\left[6\cos\theta - 10\cos^2\theta - 1 + \cos^2\theta\right] \\ &= -\frac{Mg}{4}\left[-1+6\cos\theta - 9\cos^2\theta\right] = \frac{Mg}{4}\left[9\cos^2\theta - 6\cos\theta + 1\right] \\ &= \frac{W}{4}(1-3\cos\theta)^2. \end{aligned}$$

► **Problem 3.5.4** *A right cone, of angle 2α, can turn freely about an axis passing through the centre of base and perpendicular to its axis, if the cone starts from rest with its axis horizontal, show that, when the axis is vertical the thrust on the fixed axis is to the weight of the cone as $1+\frac{1}{2}\cos^2\alpha$ to $1-\frac{1}{3}\cos^2\alpha$.*

[KUH 2005]

Solution: Let h be the height of the cone of mass M and G be the c.g. of the cone. Let DD' be the axis about which the cone rotates. Then $OG = \frac{h}{4}$. We first calculate the m.i. of the cone about the axis of rotation DD'. We consider a strip of thickness dz at a distance z from the vertex A, and parallel to the base BC. Let

$$m = \frac{\text{mass}}{\text{volume}} = \frac{M}{\frac{1}{3}\pi h^2\tan^2\alpha\cdot h}, \quad \text{or,} \quad m = \frac{3M}{\pi h^3\tan^2\alpha}.$$

The m.i. of the strip about an axis through the centre of the strip and parallel to DD'

$$= m\cdot\pi z^2\tan^2\alpha\, dz\cdot\frac{z^2\tan^2\alpha}{4}.$$

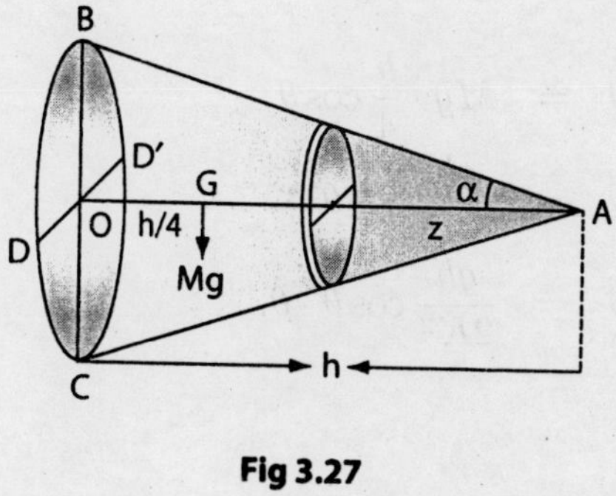

Fig 3.27

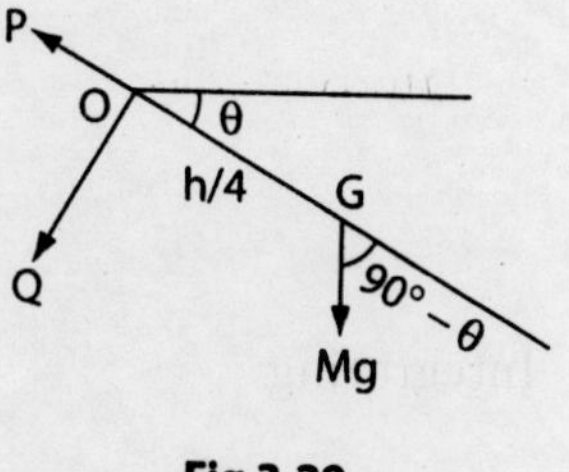

Fig 3.28

$\therefore$ The m.i. of the strip about DD'

$$
\begin{aligned}
&= m\cdot\pi z^2\tan^2\alpha\,dz\cdot\frac{z^2\tan^2\alpha}{4}+m\pi z^2\tan^2\alpha\,dz\cdot(h-z)^2\\
&= m\pi\tan^2\alpha\left[\frac{\tan^2\alpha}{4}z^4\,dz+z^2(h-z)^2\,dz\right].
\end{aligned}
$$

$\therefore$ m.i. of the cone about DD'

$$
\begin{aligned}
&= m\pi\tan^2\alpha\left[\frac{\tan^2\alpha}{4}\int_0^h z^4\,dz+h^2\int_0^h z^2\,dz-2h\int_0^h z^3\,dz+\int_0^h z^4\,dz\right]\\
&= \frac{3M}{\pi h^3\tan^2\alpha}\cdot\pi\tan^2\alpha\cdot\left[\frac{\tan^2\alpha}{4}\cdot\frac{h^5}{5}+\frac{h^5}{3}-2\frac{h^5}{4}+\frac{h^5}{5}\right]\\
&= \frac{3M}{h^3}\cdot h^5\left[\frac{\tan^2\alpha}{20}+\frac{1}{3}-\frac{1}{2}+\frac{1}{5}\right]=3Mh^2\left[\frac{3\tan^2\alpha+20-30+12}{60}\right]\\
&= \frac{Mh^2}{20}[3\tan^2\alpha+2].
\end{aligned}
$$

The radius of gyration of the cone about DD', K is given by

$$K^2=\frac{h^2}{20}\left[3\tan^2\alpha+2\right].$$

Let at any instant the axis of the cone OG makes an angle θ with horizontal and P and Q are the components of reaction along and perpendicular to GO. Thus the equations of motions are

$$M\left(-\frac{h}{4}\dot\theta^2\right) = -P+Mg\sin\theta \tag{3.5.24}$$

$$M\left(\frac{h}{4}\ddot\theta\right) = Q+Mg\cos\theta \tag{3.5.25}$$

and
$$MK^2\ddot\theta = Mg\cdot\frac{h}{4}\cos\theta \tag{3.5.26}$$

$$\text{or,}\quad \ddot\theta = \frac{gh}{4K^2}\cos\theta \tag{3.5.27}$$

$$\text{or,}\quad 2\dot\theta\ddot\theta = \frac{gh}{2K^2}\cos\theta\cdot\dot\theta.$$

Integrating

$$\dot\theta^2=\frac{gh}{2K^2}\sin\theta+C\quad\text{as }\dot\theta=0,\text{ when }\theta=0.\quad\therefore\quad C=0.$$

Thus

$$\dot\theta^2=\frac{gh}{2K^2}\sin\theta. \tag{3.5.28}$$

Using (3.5.28) in (3.5.24) we get,

$$\begin{aligned} P &= \frac{h}{4}\cdot M\cdot\dot{\theta}^2 + Mg\sin\theta = \frac{Mh}{4}\cdot\frac{gh}{2K^2}\sin\theta + Mg\sin\theta \\ &= Mg\sin\theta\left[1+\frac{h^2}{8K^2}\right] = Mg\cdot\sin\theta\left[1+\frac{h^2}{8\cdot\frac{h^2}{20}(3\tan^2\alpha+2)}\right] \\ &= Mg\sin\theta\left[1+\frac{5}{2(3\tan^2\alpha+2)}\right] = Mg\sin\theta\ \frac{6\tan^2\alpha+9}{6\tan^2\alpha+4} \\ &= Mg\sin\theta\cdot\frac{6\sin^2\alpha+9\cos^2\alpha}{6\sin^2\alpha+4\cos^2\alpha} = Mg\sin\theta\cdot\frac{6+3\cos^2\alpha}{6-2\cos^2\alpha} \\ &= Mg\sin\theta\cdot\frac{1+\frac{1}{2}\cos^2\alpha}{1-\frac{1}{3}\cos^2\alpha}. \end{aligned}$$

Now

$$\begin{aligned} P' &= P\big|_{\theta=\pi/2} = Mg\cdot\frac{1+\frac{1}{2}\cos^2\alpha}{1-\frac{1}{3}\cos^2\alpha}. \\ \therefore\quad \frac{P'}{Mg} &= \frac{P'}{W} = \frac{1+\frac{1}{2}\cos^2\alpha}{1-\frac{1}{3}\cos^2\alpha}. \end{aligned}$$

► **Problem 3.5.5** *A regular tetrahedron, of mass M, swings about one edge which horizontal. In initial position the perpendicular from the centre of mass upon the edge is horizontal. Show that when this line makes an angle θ with the vertical, the vertical component of thrust is*

$$\frac{Mg}{7}(2\sin^2\theta + 17\cos^2\theta).$$

Solution: Let a be the length of the edge of the regular tetrahedron of mass M and G be its c.g. Let D be the midpoint of BC and G' be the c.g. of the triangle ABC. Let the tetrahedron swing about the edge BC such that initially DG is horizontal. Let at any instant DG makes an angle θ with the vertical and P, Q are the components of thrust along and perpendicular to GD. Now we know that the moment of inertia of the tetrahedron of mass M about BC = m.i. of four particles each of mass $\frac{M}{20}$ placed at the vertices about BC + m.i. of a particle $\frac{4M}{5}$ placed at the c.g. G of the tetrahedron about BC.

Thus if K be the radius of gyration of the tetrahedron about BC we have,

$$MK^2 = \frac{M}{20}\cdot AD^2 + \frac{M}{20}\cdot OD^2 + \frac{4M}{5}\cdot GD^2.$$

Now

$$AD = OD = a\sin 60^\circ = \frac{\sqrt{3}\,a}{2}.$$

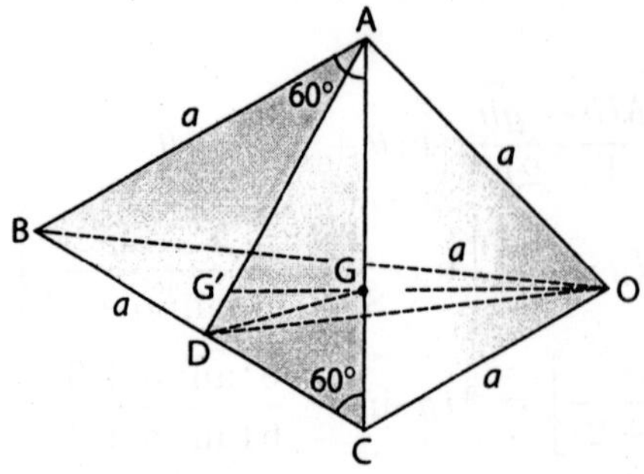

Fig 3.29

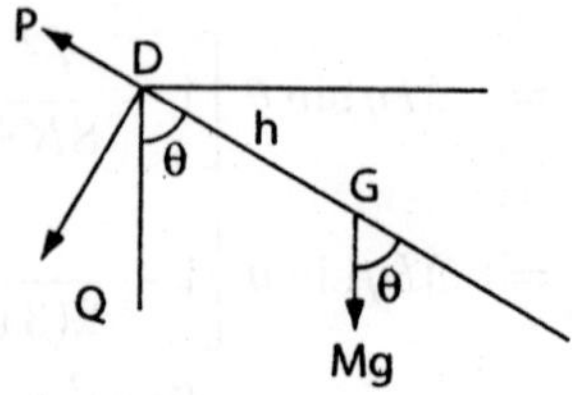

Fig 3.30

$$G'D = \frac{1}{3}AD = \frac{1}{3}\cdot\frac{\sqrt{3}\,a}{2} = \frac{a}{2\sqrt{3}}.$$

$$OG'^2 = OD^2 - G'D^2 = \left(\frac{\sqrt{3}\,a}{2}\right)^2 - \frac{a^2}{12} = \frac{3a^2}{4} - \frac{a^2}{12} = \frac{8a^2}{12} = \frac{2}{3}a^2.$$

$$\therefore\quad OG' = \sqrt{\frac{2}{3}}a,\quad GG' = \frac{1}{4}OG' = \frac{1}{4}\cdot\sqrt{\frac{2}{3}}a.$$

$$\therefore\quad GD^2 = GG'^2 + G'D^2 = \frac{2}{16.3}a^2 + \frac{a^2}{4.3} = \frac{a^2}{24} + \frac{a^2}{12} = \frac{a^2}{8}.$$

$$\therefore\quad GD = \frac{a}{2\sqrt{2}} = h. \qquad (3.5.29)$$

Thus

$$\begin{aligned} MK^2 &= \frac{M}{20}\cdot\frac{3}{4}a^2 + \frac{M}{20}\cdot\frac{3}{4}a^2 + \frac{4M}{5}\frac{a^2}{8} \\ &= \frac{3Ma^2}{40} + \frac{4Ma^2}{40} = \frac{7Ma^2}{40}. \end{aligned}$$

Thus

$$K^2 = \frac{7a^2}{40}. \qquad (3.5.30)$$

The equations of motions are

$$M(-h\dot{\theta}^2) = -P + Mg\cos\theta \qquad (3.5.31)$$

$$M(h\ddot{\theta}) = -Q - Mg\sin\theta \qquad (3.5.32)$$

$$\text{and}\quad MK^2\ddot{\theta} = -Mgh\sin\theta. \qquad (3.5.33)$$

Thus

$$\ddot{\theta} = -\frac{gh}{K^2}\sin\theta,$$

$$\text{or}\quad 2\dot{\theta}\ddot{\theta} = -\frac{2gh}{K^2}\sin\theta\cdot\dot{\theta}. \qquad (3.5.34)$$

Integrating

$$\dot{\theta}^2 = \frac{2gh}{K^2}\cos\theta + C.$$

Initially $\dot{\theta} = 0$, when $\theta = \pi/2$. $\therefore$ $C = 0$.
Thus

$$\dot{\theta}^2 = \frac{2gh}{K^2}\cos\theta. \tag{3.5.35}$$

Using (3.5.35) in (3.5.31) we get,

$$\begin{aligned} P &= Mh\dot{\theta}^2 + Mg\cos\theta = \frac{2Mgh^2}{K^2}\cos\theta + Mg\cos\theta \\ &= Mg\cos\theta\left[\frac{2h^2}{K^2} + 1\right] = Mg\cos\theta\left[2\cdot\frac{a^2}{8}\cdot\frac{40}{7a^2} + 1\right] \\ &= Mg\cos\theta\left(\frac{10}{7} + 1\right) \quad \text{[by (3.5.29) and (3.5.30)]} \\ &= \frac{17}{7}Mg\cos\theta. \end{aligned}$$

Using (3.5.34) in (3.5.32) we get,

$$\begin{aligned} Q &= -Mg\ddot{\theta} - Mh\sin\theta = \frac{Mh^2g}{K^2}\sin\theta - Mg\sin\theta \\ &= Mg\sin\theta\left[\frac{h^2}{K^2} - 1\right] = Mg\sin\theta\left[\frac{a^2}{8}\cdot\frac{40}{7a^2} - 1\right] \\ &= Mg\sin\theta\left[\frac{5}{7} - 1\right] \quad \text{[by (3.5.29) and (3.5.30)]} \\ &= -\frac{2}{7}Mg\sin\theta. \end{aligned}$$

Thus the vertical component of the thrust

$$\begin{aligned} &= P\cos\theta - Q\sin\theta = \frac{17}{7}Mg\cos^2\theta + \frac{2}{7}Mg\sin^2\theta \\ &= \frac{Mg}{7}[17\cos^2\theta + 2\sin^2\theta]. \end{aligned}$$

► **Problem 3.5.6** *A thin uniform rod has one end attached to a smooth hinge and is allowed to fall from a horizontal position, show that the horizontal strain on the hinge is greatest when the rod is inclined at an angle* 45° *to the vertical and that the vertical strain is then* $\frac{11}{8}$ *time the weight of the rod.*

[KU H 2001]

Solution: Let AB be the rod of length $2a$ and M is the mass. Let $G(\bar{x}, \bar{y})$ be c.g. of the rod with respect to Ax and Ay as axes. Let X and Y be the components of the strain along Ax and Ay at the hinge, at any instant when the rod makes an angle θ with the vertical.

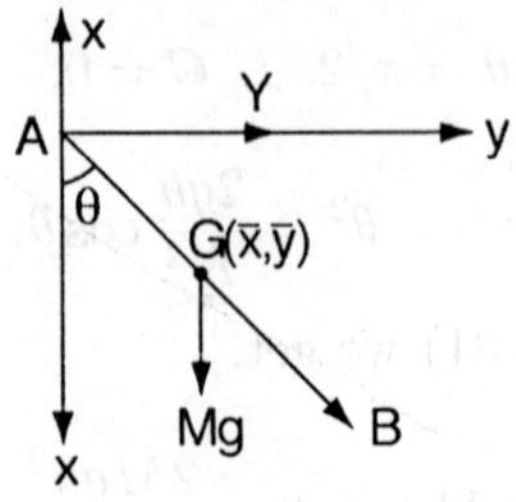

Fig 3.31

Therefore, the equations of motion are

$$M\ddot{\bar{x}} = -X + Mg \tag{3.5.36}$$
$$M\ddot{\bar{y}} = Y \tag{3.5.37}$$
$$\text{and} \quad MK^2\ddot{\theta} = -Mga\sin\theta. \tag{3.5.38}$$

From (3.5.38), we get

$$K^2\ddot{\theta} = -ga\sin\theta,$$
$$\text{or,} \quad \frac{4a^2}{3}\ddot{\theta} = -ga\sin\theta,$$
$$\text{or,} \quad \ddot{\theta} = -\frac{3g}{4a}\sin\theta, \tag{3.5.39}$$
$$\text{or,} \quad 2\dot{\theta}\ddot{\theta} = -\frac{3g}{4a}\sin\theta\dot{\theta}.$$

Integrating

$$\dot{\theta}^2 = \frac{3g}{2a}\cos\theta + C.$$

Initially $\dot{\theta} = 0$, when $\theta = \pi/2$. $\therefore$ $C = 0$.
Thus

$$\dot{\theta}^2 = \frac{3g}{2a}\cos\theta. \tag{3.5.40}$$

Again,

$$\left.\begin{aligned} \bar{x} = a\cos\theta; \quad \dot{\bar{x}} = -a\sin\theta\cdot\dot{\theta}, \quad \ddot{\bar{x}} = -a\cos\theta\cdot\dot{\theta}^2 - a\sin\theta\cdot\ddot{\theta} \\ \bar{y} = a\sin\theta; \quad \dot{\bar{y}} = a\cos\theta\dot{\theta}, \quad \ddot{\bar{y}} = -a\sin\theta\cdot\dot{\theta}^2 + a\cos\theta\cdot\ddot{\theta} \end{aligned}\right\}. \tag{3.5.41}$$

Using (3.5.41) in (3.5.36) and (3.5.37) we get,

$$\begin{aligned} X &= Mg - M\ddot{\bar{x}} = Mg + Ma\cos\theta\cdot\dot{\theta}^2 + Mg\sin\theta\cdot\ddot{\theta} \\ &= Mg + M\cos\theta\cdot\frac{3g}{2}\cos\theta - M\sin\theta\cdot\frac{3g}{4}\sin\theta \\ &= Mg\left[1+\frac{3}{2}\cos^2\theta - \frac{3}{4}\sin^2\theta\right] \qquad (3.5.42) \\ Y &= M\ddot{\bar{y}} = M(-a\sin\theta\cdot\dot{\theta}^2 + a\cos\theta\cdot\ddot{\theta}) \\ &= M\left[-\sin\theta\cdot\frac{3g}{2}\cos\theta - \cos\theta\cdot\frac{3g}{4}\sin\theta\right] \\ &= \frac{Mg}{8}[-12\sin\theta\cos\theta - 6\sin\theta\cos\theta] \\ &= -\frac{Mg}{8}\cdot 18\sin\theta\cos\theta = -\frac{9Mg}{8}\sin 2\theta. \qquad (3.5.43) \end{aligned}$$

Thus the horizontal component of strain will be maximum, i. e., $|Y|$ will be maximum when $\sin 2\theta = 1$, i. e., $2\theta = \pi/2$, or, $\theta = \frac{\pi}{4} = 45°$.

There

$$X\big|_{\theta=45°} = Mg\left[1+\frac{3}{4}-\frac{3}{8}\right] = \frac{Mg}{8}[8+6-3] = \frac{11}{8}Mg = \frac{11}{8}W,$$

where W is the weight of the rod.

▶ **Problem 3.5.7** *A uniform semi-circular arc, of mass m, and radius a fixed at its ends to two fixed points in the same vertical line and is rotates with constant angular velocity ω, show that the horizontal thrust on the upper end is $m\frac{g+a\omega^2}{\pi}$.*

Solution: Let O be the centre of the semi-circular arc APB and we take Ox and Oy as axis at O. Let X and X' are the thrust at A and B. Thus from symmetry the c.g. of the arc will be on the x-axis.

Let $\rho = \text{mass/arc} = \frac{m}{\pi a}$. We consider an element of length $a\,d\theta$ at P at a distance r from the axis of rotation AB.

Thus the distance of c.g. G from the axis of rotation OG

$$= \frac{\int_{-\pi/2}^{\pi/2} \rho a\,d\theta\cdot a\cos\theta}{\int_{-\pi/2}^{\pi/2}\rho a\,d\theta} = \frac{a^2\cdot|\sin\theta|_{-\pi/2}^{\pi/2}}{a|\theta|_{-\pi/2}^{\pi/2}} = \frac{2a}{\pi}$$

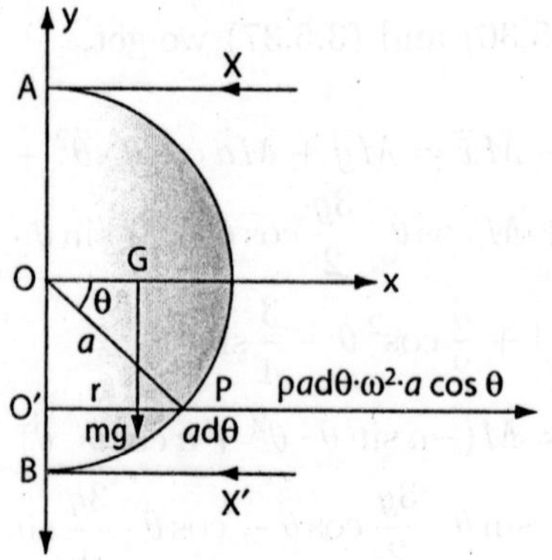

Fig 3.32

The effective force on the element $a\,d\theta$ at P

$$= \rho a\,d\theta\cdot\omega^2 r = \frac{m}{\pi a}\cdot a\,d\theta\omega^2 a\cos\theta$$

$$= \frac{m}{\pi a}a^2\omega^2\cos\theta\,d\theta = \frac{m}{\pi}a\omega^2\cos\theta\,d\theta.$$

Now taking moment about B, we get,

$$\begin{aligned}
X\cdot 2a &= mg\cdot OG + \int_{-\pi/2}^{\pi/2}\frac{m}{\pi}a\omega^2\cos\theta\,d\theta\cdot O'B\\
&= mg\cdot OG + \int_{-\pi/2}^{\pi/2}\frac{ma}{\pi}\omega^2\cos\theta\,d\theta(a-a\sin\theta)\\
&= mg\cdot\frac{2a}{\pi}+\frac{ma^2}{\pi}\omega^2\int_{-\pi/2}^{\pi/2}\left(\cos\theta-\frac{1}{2}\sin 2\theta\right)d\theta\\
&= \frac{2a\,mg}{\pi}+\frac{ma^2\omega^2}{\pi}\left[\sin\theta\Big|_{-\pi/2}^{\pi/2}+\frac{\cos 2\theta}{4}\Big|_{-\pi/2}^{\pi/2}\right]\\
&= \frac{2a\,mg}{\pi}+\frac{ma^2\omega^2}{\pi}\cdot 2\\
\text{or,}\quad X &= \frac{mg}{\pi}+\frac{ma\omega^2}{\pi}=\frac{m}{\pi}[g+a\omega^2].
\end{aligned}$$

3.6 Solved Problems

► **Problem 3.6.1** *A mass M resting on a rough inclined plane of inclination α is attached to one end of a flexible and inextensible cord, which runs parallel to a line of greatest slope and passes over a small pulley of mass m and radius a at the highest point of the plane and carries at the other end a mass M' which hangs freely. Find the acceleration of the masses, μ being the* *coefficient of friction between M and the plane and Q the constant torque supplied by friction on the axle of the pulley.*

Solution: Let the mass M moves down along the plane a distance x in time t.

$\therefore\ x = a\theta$, or, $\ddot{x} = a\ddot{\theta}$. Let T, T and T', T' are the tensions in the string. Let R be the normal reaction and μR be the force of friction.

Considering the motion of the mass M we get,

$$M\ddot{x} = Mg\sin\alpha - \mu R - T \tag{3.6.1}$$

$$R = Mg\cos\alpha. \tag{3.6.2}$$

$$\therefore\ M\ddot{x} = Mg\sin\alpha - Mg\mu\cos\alpha - T. \tag{3.6.3}$$

Considering the motion of the mass M', we get

$$M'\ddot{x} = T' - M'g. \tag{3.6.4}$$

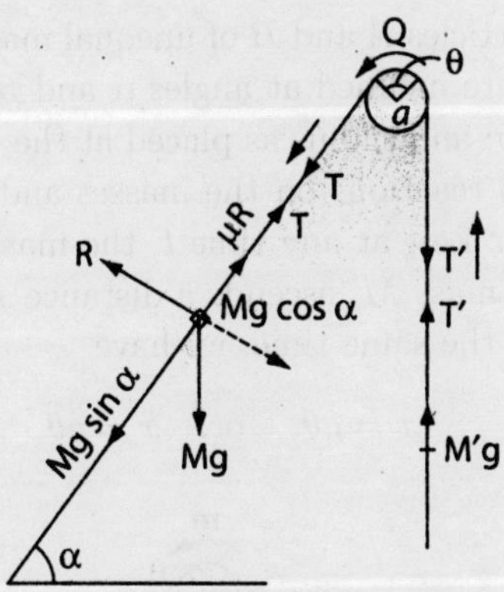

Fig 3.33

Let K is the radius of the gyration of the pulley about axis of rotation. Thus considering the motion of the pulley, we get

$$MK^2\ddot{\theta} + Q = Ta - T'a = a(T - T'). \tag{3.6.5}$$

Adding (3.6.3) and (3.6.4) we get

$$\begin{aligned} M\ddot{x} + M'\ddot{x} &= T' - T + Mg\sin\alpha - Mg\mu\cos\alpha - M'g \\ &= -\frac{MK^2\ddot{\theta} + Q}{a} + g[M(\sin\alpha - \mu\cos\alpha)] - M'g \\ &\qquad\qquad\qquad\qquad\qquad\qquad [\text{by } (3.6.5)] \\ &= -\frac{MK^2\ddot{x}}{a^2} - \frac{Q}{a} + g[M(\sin\alpha - \mu\cos\alpha)] - M'g, \end{aligned}$$

$$\text{or,}\quad \left(\frac{MK^2}{a^2} + M + M'\right)\ddot{x} = g[M(\sin\alpha - \mu\cos\alpha) - M'] - \frac{Q}{a}$$

$$\text{or,}\quad \ddot{x} = \frac{a^2 g[M(\sin\alpha - \mu\cos\alpha) - M'] - aQ}{MK^2 + a^2(M + M')}.$$

► **Problem 3.6.2** *Two unequal masses, M and M', rest on two rough planes inclined at angles α and β to the horizon; they are connected by a fine string passing over a small pulley, of mass m and radius a, which is placed at the common vertex of the two planes; show that the acceleration of either mass is*

$$g[M(\sin\alpha - \mu\cos\alpha) - M'(\sin\beta + \mu'\cos\beta)] \div \left[M + M' + m\frac{K^2}{a^2}\right],$$

where μ and μ' are the coefficient of frictions, K is the radius of gyration of the pulley about its axis and M is the mass which moves downwards.

[CUH 1982]

Solution: Let the particles A and B of unequal masses M and M' be placed on the planes which are inclined at angles α and β with the horizon. Let a small pulley of mass m and radius as placed at the common vertex O. Let R and R' be the normal reactions on the masses and T, T and T', T' are the tensions in the string. Let, at any time t, the mass M descends a distance x along OA, thus the mass M' ascends a distance x along BO. If the pulley rotates an angle θ, in the same time, we have

$$x = a\theta, \quad \text{or,} \quad \ddot{x} = a\ddot{\theta}. \tag{3.6.6}$$

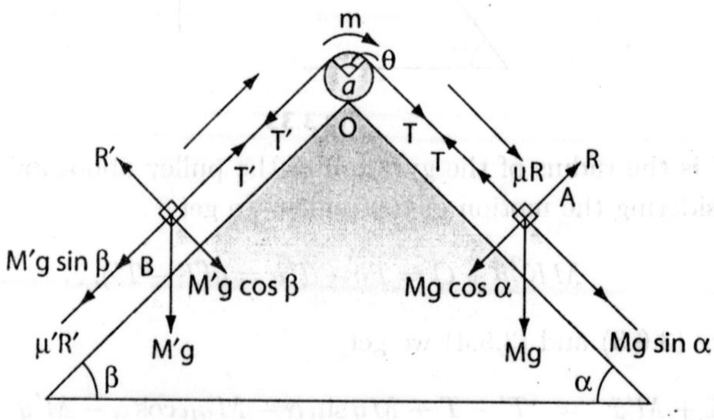

Fig 3.34

Considering the motion of the mass M, we get

$$\begin{aligned} M\ddot{x} &= -T + Mg\sin\alpha - \mu R. \\ R &= Mg\cos\alpha. \end{aligned}$$

$$\therefore \quad M\ddot{x} = Mg\sin\alpha - \mu Mg\cos\alpha - T. \tag{3.6.7}$$

Considering the motion of the mass M', we get

$$\begin{aligned} M'\ddot{x} &= T' - \mu' R' - M'g\sin\beta \\ R' &= M'g\cos\beta. \end{aligned}$$

$$\therefore \quad M'\ddot{x} = T' - \mu' M'g\cos\beta - M'g\sin\beta. \tag{3.6.8}$$

Adding (3.6.7) and (3.6.8) we get

$$(M + M')\ddot{x} = T' - T + g[M(\sin\alpha - \mu\cos\alpha) - M'(\sin\beta + \mu'\cos\beta)]. \tag{3.6.9}$$

Now considering the motion of the pulley, we get

$$mK^2\ddot{\theta} = T\cdot a - T'a = a(T - T'). \tag{3.6.10}$$

Using (3.6.10) in (3.6.9) we get

$$\begin{aligned} (M + M')\ddot{x} &= g\left[M(\sin\alpha - \mu\cos\alpha) - M'(\sin\beta + \mu'\cos\beta)\right] - \frac{mK^2\ddot{\theta}}{a} \\ &= g\left[M(\sin\alpha - \mu\cos\alpha) - M'(\sin\beta + \mu'\cos\beta)\right] - \frac{mK^2\ddot{x}}{a^2}, \\ &\qquad \text{[by (3.6.6)]} \end{aligned}$$

$$\text{or,} \quad \left(M + M' + \frac{mK^2}{a^2}\right)\ddot{x} = g\left[M(\sin\alpha - \mu\cos\alpha) - M'(\sin\beta + \mu'\cos\beta)\right].$$

$$\therefore \quad \ddot{x} = g[M(\sin\alpha - \mu\cos\alpha) - M'(\sin\beta + \mu'\cos\beta)] \div \left(M + M' + \frac{mK^2}{a^2}\right).$$

▶ **Problem 3.6.3** *A uniform vertical circular plate of radius a, is capable of revolving about a smooth horizontal axis through its centre; a rough perfectly flexible chain whose mass is equal to that of the plate and whose length is equal to its circumference, hangs over its rim in equilibrium; if one end be slightly displaced, show that the velocity of the chain when other end reaches the plate is* $\sqrt{\frac{\pi a g}{6}}$. **[CUH 1989]**

Solution: Let a be the radius of the circular plate of centre O. Thus the circumference of the plate is $2\pi a$. Therefore, the length of the chain $= 2\pi a$. Let M = mass of the plate = mass of the chain. Thus the length of the chain resting on the plate $= \pi a$ and πa length of the chain hangs on both sides. Let, in any time t, after displacement, the disc rotates an angle θ and the x-length of the chain is displaced from left to right.

$$\therefore \quad x = a\theta, \quad \dot{x} = a\dot{\theta}, \quad \ddot{x} = a\ddot{\theta}.$$

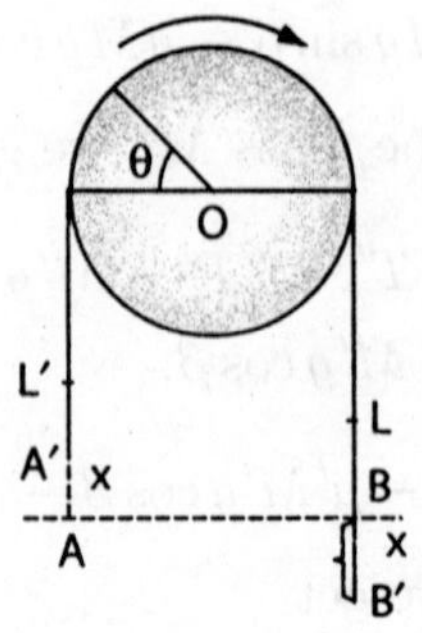

Fig 3.35

Thus at time t, the lengths of the chain on the right and left sides are respectively $\frac{\pi a}{2}+x$ and $\frac{\pi a}{2}-x$ and the depth of the midpoints of the chain on right and left sides below the centre of the disc L&$L'$are respectively $\frac{1}{2}\left(\frac{\pi a}{2}+x\right)$ and $\frac{1}{2}\left(\frac{\pi a}{2}-x\right)$.

Let K be the radius of gyration of the circular plate.

Therefore, by conservation of energy, K.E. + P.E. = constant, we have

$$\frac{1}{2}MK^2\dot{\theta}^2 \text{ (K.E. of disc)} + \frac{1}{2}M\dot{x}^2 \text{ (K.E. of the chain)}$$

$$-\frac{M}{2\pi a}\left(\frac{\pi a}{2}+x\right) g\cdot\frac{1}{2}\left(\frac{\pi a}{2}+x\right)-\frac{M}{2\pi a}\left(\frac{\pi a}{2}-x\right) g\cdot\frac{1}{2}\left(\frac{\pi a}{2}-x\right)$$

$$= \text{constant} = \text{Initial value}$$

$$= 0+0-\frac{M}{4}g.\frac{\pi a}{4}-\frac{M}{4}g.\frac{\pi a}{4} = -\frac{Mga\pi}{8},$$

$$\text{or,}\quad \frac{1}{2}MK^2\dot{\theta}^2+\frac{1}{2}M\dot{x}^2 = \frac{Mg}{4\pi a}\left[\left(\frac{\pi a}{2}+x\right)^2+\left(\frac{\pi a}{2}-x\right)^2\right]-\frac{Mga\pi}{8}$$

$$= \frac{Mg}{2\pi a}\left[\frac{\pi^2a^2}{4}+x^2\right]-\frac{Mga\pi}{8}$$

$$= \frac{Mg\pi a}{8}+\frac{Mgx^2}{2\pi a}-\frac{Mg\pi a}{8}=\frac{Mgx^2}{2\pi a},$$

$$\text{or,}\quad K^2\dot{\theta}^2+\dot{x}^2=\frac{gx^2}{\pi a},$$

$$\text{or,}\quad K^2\frac{\dot{x}^2}{a^2}+\dot{x}^2=\frac{gx^2}{\pi a},$$

$$\text{or,}\quad \dot{x}^2\left[\frac{K^2+a^2}{a^2}\right]=\frac{gx^2}{\pi a},$$

$$\text{or,}\quad \dot{x}^2=\frac{agx^2}{\pi(K^2+a^2)}=\frac{gax^2}{\left(\frac{a^2}{2}+a^2\right)\pi},$$

$$\text{or,}\quad \dot{x}^2=\frac{2gx^2}{3a\pi}.$$

When the end A' reaches the plate, $x = \frac{\pi a}{2}$.

$$\therefore \quad \dot{x}^2 \Big|_{x=\frac{\pi a}{2}} = \frac{2g \cdot \pi^2 a^2}{4.3\pi a} = \frac{g\pi a}{5}.$$

$$\therefore \quad \dot{x} \Big|_{x=\frac{\pi a}{2}} = \sqrt{\frac{\pi a g}{6}}.$$

► **Problem 3.6.4** *A uniform rod AB is freely movable on a rough inclined plane, whose inclination to the horizon is i and whose coefficient of friction is μ, about a smooth pin fixed through the end A; the bar is held in the horizontal position in the plane and allowed to fall from this position. If θ be the angle through which it falls from rest, show that $\frac{\sin\theta}{\theta} = \mu \cot i$.*

[CUH 1994, BUH 2004; VUH 1993]

Solution: Let AB be the rod of mass m and length $2a$. Let Ax and Ay are taken as axes at the fixed point A about which the rod rotates. Let the rod makes an angle θ with Ax at any time. We consider an element of the rod of thickness dz at P at a distance z from A. If m' be the mass of the element at P, then

$$m' = \frac{m}{2a} dz.$$

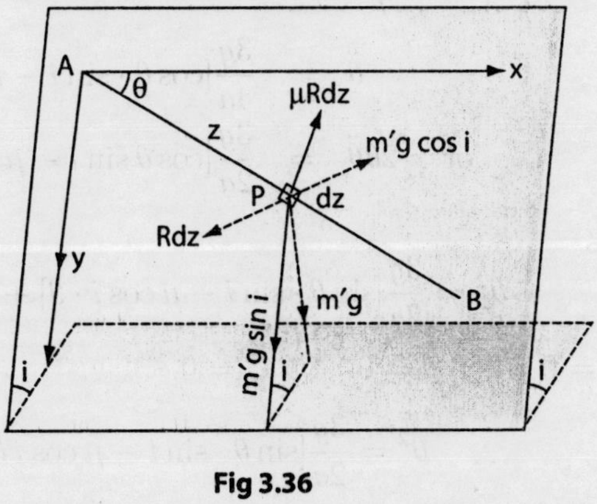

Fig 3.36

Let $R\,dz$ be the normal reaction on the element at P which is perpendicular to the plane and $\mu R\,dz$ be the frictional force perpendicular to the rod and along the plane.

Considering the equilibrium of the element dz at P, resolving perpendicular to the plane, we have

$$Rdz = m'g\cos i = \frac{m}{2a} dz \cdot g\cos i. \tag{3.6.11}$$

$$\therefore \quad R = \frac{m}{2a}\, g\cos i. \tag{3.6.12}$$

Now the algebraic sum of the moment of applied forces on the element at P about A, we get

$$\begin{aligned} & m'g\sin i\cdot z\cos\theta - \mu R\,dz\cdot z + R\,dz\cdot z - m'g\cos i\cdot z \\ &= m'g\sin i\cdot z\cos\theta - \mu R\,z\,dz \quad [\text{by } (3.6.11)] \\ &= \frac{m}{2a}\cdot g\sin i\cos\theta\cdot z\,dz - \mu\frac{m}{2a}\,g\cos i\,z\,dz. \end{aligned}$$

Considering the motion of the whole rod, taking moment about A, we get

$$\begin{aligned} mK^2\ddot{\theta} &= \int_0^{2a}\left[\frac{m}{2a}\cdot g\sin i\cos\theta\cdot z\,dz - \frac{mg}{2a}\mu\cos iz\,dz\right] \\ &= \frac{m}{2a}g\cos\theta\sin i\cdot\frac{4a^2}{2} - \frac{mg}{2a}\mu\cos i\cdot\frac{4a^2}{2} \\ &= mga[\cos\theta\cdot\sin i - \mu\cos i], \end{aligned}$$

where K is the radius of gyration of the rod about the end A.

$$\begin{aligned} m\cdot\left(\frac{a^2}{3}+a^2\right)\ddot{\theta} &= mga[\cos\theta\cdot\sin i - \mu\cos i]. \\ \ddot{\theta} &= \frac{3g}{4a}[\cos\theta\cdot\sin i - \mu\cos i] \\ \text{or,}\quad 2\dot{\theta}\ddot{\theta} &= \frac{3g}{2a}[\cos\theta\sin i - \mu\cos i]\dot{\theta}. \end{aligned}$$

Integrating

$$\dot{\theta}^2 = \frac{3g}{2a}[\sin\theta\cdot\sin i - \mu\cos i\cdot\theta] + C.$$

Initially $\dot{\theta} = 0$ when $\theta = 0$. $\therefore\ C = 0$.

$$\therefore \quad \dot{\theta}^2 = \frac{3g}{2a}[\sin\theta\cdot\sin i - \mu\cos i\,\theta].$$

As the rod falls till $\dot{\theta} = 0$, when $\dot{\theta} = 0$, we get

$$\sin\theta\cdot\sin i - \mu\cos i\cdot\theta = 0, \quad \text{or,} \quad \frac{\sin\theta}{\theta} = \mu\cot i.$$

► **Problem 3.6.5** *A fly wheel is movable in smooth bearing about a horizontal axis, and is set in motion by a body of mass P which hangs from end of a cord coil round the axle of radius r. The mass is bound to descend through h ft in t sec. Prove that the moment of inertia of the wheel about the axis is*

$$\left(\frac{gt^2}{2h} - 1\right)Pr^2.$$

Solution: Let $PK^2 = I$ be the moment of inertia of the wheel about its horizontal axis. Let it rotates by an angle θ about its axis, when the string descends a length x at any instant.

$$\therefore \quad x = r\theta, \quad \text{or,} \quad \ddot{x} = r\ddot{\theta}. \tag{3.6.13}$$

Let T be the tension in the string.

Considering the motion of the mass P, we get

$$P\ddot{x} = Pg - T. \tag{3.6.14}$$

Considering the motion of the wheel, we get

$$PK^2\ddot{\theta} = I\ddot{\theta} = T \cdot r, \quad \text{or,} \quad I.\frac{\ddot{x}}{r} = T \cdot r. \tag{3.6.15}$$

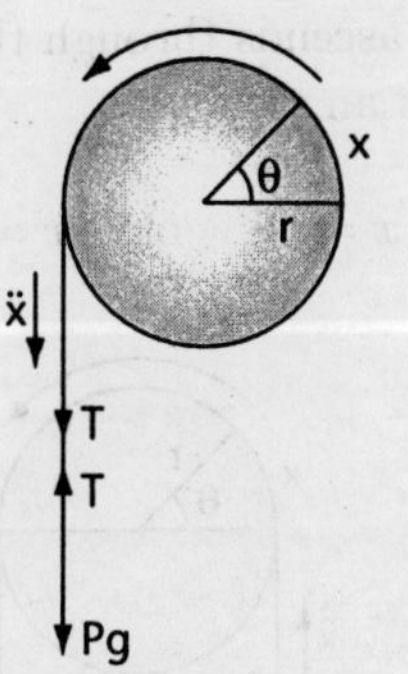

Fig 3.37

Thus from (3.6.14) and (3.6.15), we have.

$$P\ddot{x} = Pg - \frac{I\ddot{x}}{r^2}, \quad \text{or,} \quad (Pr^2 + I)\ddot{x} = Pgr^2, \quad \text{or,} \quad \ddot{x} = \frac{Pgr^2}{Pr^2 + I}.$$

Integrating we get

$$\dot{x} = \frac{Pg \cdot r^2}{Pr^2 + I} t + C; \quad \text{Initially } t = 0,\ \dot{x} = 0,\ C = 0.$$

$$\therefore \quad \dot{x} = \frac{Pgr^2}{I + Pr^2} \cdot t.$$

Integrating

$$x = \frac{Pgr^2}{I + Pr^2} . \frac{t^2}{2} + C_1, \quad \text{as when } t = 0,\ x = 0,\ C_1 = 0,$$

$$\text{or,} \quad x = \frac{Pgr^2}{I + Pr^2} . \frac{t^2}{2}.$$

If at time t, $x = h$, we have.

$$h = \frac{Pgr^2}{I + Pr^2} \cdot \frac{t^2}{2}, \quad \text{or,} \quad I + Pr^2 = \frac{Pgr^2}{2h} t^2,$$

$$\text{or,} \quad I = \frac{Pgr^2}{2h} t^2 - Pr^2 = \left(\frac{gt^2}{2h} - 1 \right) Pr^2.$$

► **Problem 3.6.6** *A string which passes round the rim of radius r mounted on horizontal axis through the centre carries masses m_1 and m_2 at its ends. If I is the moment of inertia of the wheel about its axis and $m_2 > m_1$, show that the acceleration of masses is $\frac{(m_2 - m_1)\, gr^2}{(m_1 + m_2) r^2 + I}$.* **[CUH 1992]**

Solution: Let the wheel is perfectly rough, so that the string cannot slide on the wheel. As $m_2 > m_1$, let at any time t, m_2 descends through a distance x and at the same time, m_1 ascends through the same distance x. Let during this time the wheel rotates an angle θ.

$$x = r\theta, \quad \text{or,} \quad \ddot{x} = r\ddot{\theta}. \tag{3.6.16}$$

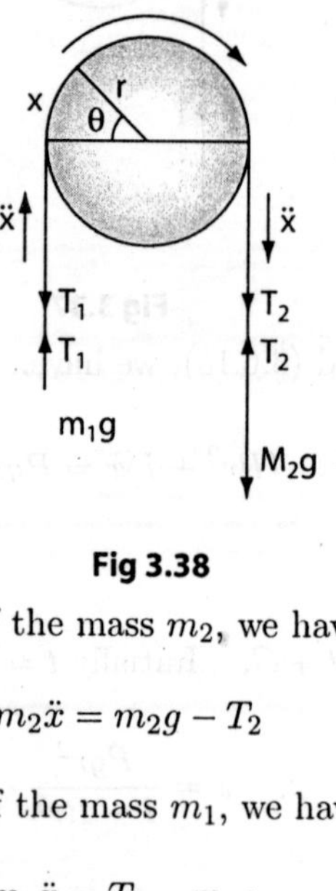

Fig 3.38

Considering the motion of the mass m_2, we have

$$m_2 \ddot{x} = m_2 g - T_2 \tag{3.6.17}$$

and considering the motion of the mass m_1, we have

$$m_1 \ddot{x} = T_1 - m_1 g. \tag{3.6.18}$$

Adding (3.6.17) and (3.6.18) we get

$$(m_1 + m_2)\ddot{x} = (m_2 - m_1)\, g + T_1 - T_2. \tag{3.6.19}$$

Considering the motion of the wheel, we have

$$I\ddot{\theta} = -T_1 r + T_2 r = (T_2 - T_1)r,$$

$$\text{or,}\quad T_1 - T_2 = -\frac{I\ddot{\theta}}{r} = -\frac{I\ddot{x}}{r^2} \quad \text{[by (3.6.16)]}. \tag{3.6.20}$$

Using (3.6.20) in (3.6.19), we get

$$(m_1 + m_2)\ddot{x} = (m_2 - m_1)\,g - \frac{I\ddot{x}}{r^2},$$

$$\text{or,}\quad \ddot{x}\left(m_1 + m_2 + \frac{I}{r^2}\right) = (m_2 - m_1)\,g,$$

$$\text{or,}\quad \ddot{x} = \frac{(m_2 - m_1)r^2 g}{(m_1 + m_2)r^2 + I}.$$

► **Problem 3.6.7** *The radius of gyration of a two-step pulley of mass M is K, the radii of the smaller and the bigger pulley are a and b. Two equal loads of masses P and P hang on the two sides at the ends of two vertical strings wound round the pulleys. Develop a formula for the downward acceleration of the following load fastened to the string passing round the bigger pulley.*

Solution: Let after time t, the mass P on the bigger pulley of radius b descends a distance x and the mass P on the smaller pulley of radius a ascends a distance y. Let also that the pulleys rotate by an angle θ.

$$\therefore\ x = b\theta \qquad \therefore\ \ddot{x} = b\ddot{\theta}$$

$$\text{and}\ \ y = a\theta \ \ \text{and}\ \ \ddot{y} = a\ddot{\theta}$$

$$\ddot{y} = \frac{a}{b}\ddot{x}. \tag{3.6.21}$$

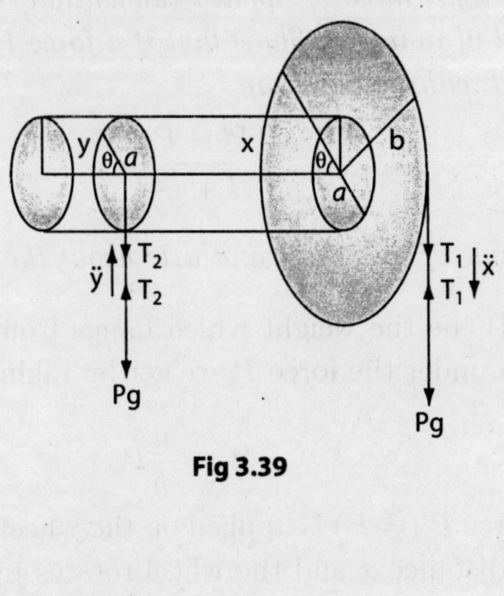

Fig 3.39

Now considering the motion of the masses, we get

$$P\ddot{x} = Pg - T_1 \tag{3.6.22}$$

$$\text{and} \quad P\ddot{y} = T_2 - Pg. \tag{3.6.23}$$

Considering the motion of the pulley, we get

$$MK^2\ddot{\theta} = T_1 \cdot b - T_2 \cdot a. \tag{3.6.24}$$

From (3.6.22) and (3.6.23) we get

$$\begin{aligned} T_1 b - T_2 a &= Pbg - Pb\ddot{x} - aP\ddot{y} - Pag \\ &= Pg(b-a) - P(b\ddot{x} + a\ddot{y}). \end{aligned}$$

From (3.6.24) we have

$$\begin{aligned} MK^2\ddot{\theta} &= Pg(b-a) - P(b\ddot{x} + a\ddot{y}) \\ &= Pg(b-a) - P\left[b\ddot{x} + \frac{a^2}{b}\ddot{x}\right] \\ &= Pg(b-a) - P\left(\frac{a^2+b^2}{b}\right)\ddot{x} \\ \text{or,} \quad MK^2 \cdot \frac{\ddot{x}}{b} &= Pg(b-a) - \frac{P(a^2+b^2)}{b}\ddot{x} \\ MK^2\ddot{x} &= Pbg(b-a) - P(a^2+b^2)\ddot{x}, \\ \text{or,} \quad \ddot{x} &= \frac{Pbg(b-a)}{MK^2 + P(a^2+b^2)}. \end{aligned}$$

► **Problem 3.6.8** *A weight hangs from an axle of radius b and it maintain equilibrium by a force P applied tangentially to the circumference of a concentric wheel of radius a. Show that if a force P′ be substituted for which the wheel ascend with acceleration*

$$\frac{(P'-P)ab}{I + \frac{Pab}{g}},$$

where I is the m.i. of the wheel and axle about the axis.

Solution: Let W be the weight which hangs from axle. When the system is in equilibrium under the force P, we get by taking moment about the axis $W \cdot b = P \cdot a$.

$$W = \frac{a}{b}P. \tag{3.6.25}$$

When the force $P'(\neq P)$ is applied on the wheel, suppose that the weight rises through a distance x and the wheel rotates by an angle θ.

$$x = b\theta, \quad \text{or,} \quad \ddot{x} = b\ddot{\theta}. \tag{3.6.26}$$

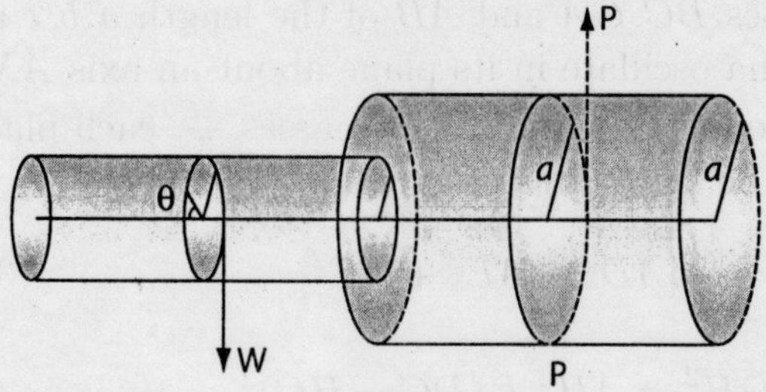

Fig 3.40

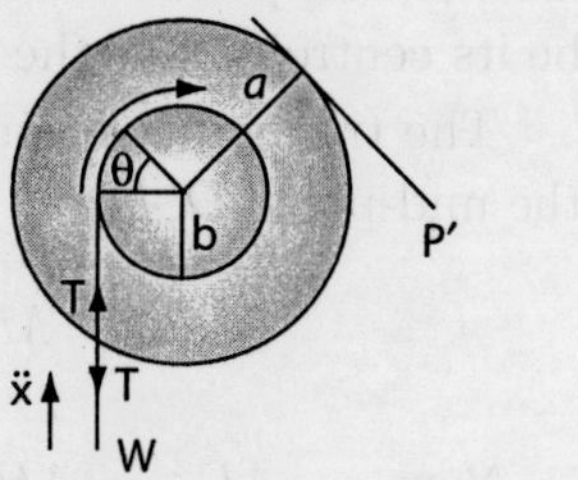

Fig 3.41

Let T be the tension in the string by which the weight is suspended. Consider the motion of the weight W, we have

$$\frac{W}{g}\ddot{x} = T - W. \tag{3.6.27}$$

Consider the motion of the wheel, we get

$$I\ddot{\theta} = P'a - T\cdot b. \tag{3.6.28}$$

Using (3.6.25) in (3.6.27) we get

$$\frac{Pa}{bg}\ddot{x} = T - \frac{Pa}{b}, \quad \text{or,} \quad T = \left(\frac{\ddot{x}}{g} + 1\right)\frac{Pa}{b}. \tag{3.6.29}$$

Again from (3.6.28) we have

$$I\frac{\ddot{x}}{b} = P'a - \left(\frac{\ddot{x}}{g} + 1\right)Pa, \quad \text{or,} \quad \ddot{x}\left[\frac{I}{b} + \frac{Pa}{g}\right] = (P' - P)a,$$

$$\text{or,} \quad \ddot{x} = \frac{(P' - P)ab}{I + \frac{Pab}{g}}.$$

► **Problem 3.6.9** *A uniform triangular lamina can oscillate in its own plane about an vertex. Prove that the length of the simple equivalent pendulum is* $\frac{3(b^2+c^2)-a^2}{4\sqrt{[2(b^2+c^2)-a^2]}}$, *where* a, b, c *are the lengths of the sides of the triangle* ABC *and the axis through* A *is horizontal.* **[KUH '01]**

Solution: Let M be the mass of the triangular lamina ABC, where D, E, F

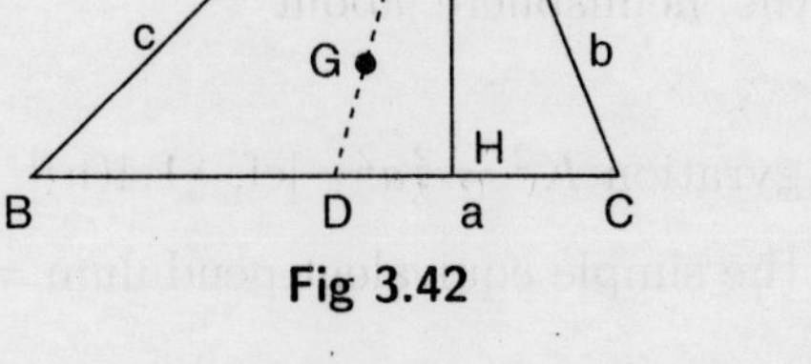

Fig 3.42

are the mid-points of the sides BC, CA and AB of the length a, b, c and G be its centroid. Let the lamina oscillate in its plane about an axis AX.

The triangular lamina is equimomental to the masses $\frac{M}{3}$, each placed at the mid-point D, E, F. If K be the radius of gyration. We have

$$MK^2 = \frac{M}{3}[AD^2 + AE^2 + AF^2] \tag{3.6.30}$$

Now
$$\begin{aligned} AD^2 &= AH^2 + DH^2 = AH^2 + (DC - HC)^2 \\ &= AH^2 + DC^2 - 2DC\cdot HC + HC^2 \\ &= AH^2 + HC^2 + DC^2 - 2\cdot DC\cdot AC\cos C \\ &= AC^2 + \frac{a^2}{4} - 2\cdot\frac{1}{2}BC\cdot AC\cos C \\ &= b^2 + \frac{a^2}{4} - 2\cdot\frac{1}{2}\cdot a\cdot b\cdot\frac{a^2+b^2-c^2}{2ab} \\ &= b^2 + \frac{a^2}{4} - \frac{a^2+b^2-c^2}{2} \\ &= \frac{4b^2 + a^2 - 2a^2 - 2b^2 + 2c^2}{4} = \frac{2b^2+2c^2-a^2}{4} \end{aligned} \tag{3.6.31}$$

and $AE = \frac{b}{2}$, $AD = \frac{c}{2}$.

$\therefore$ From (3.6.30),

$$MK^2 = \frac{M}{3}\left[\frac{2b^2+2c^2-a^2}{4} + \frac{b^2}{4} + \frac{c^2}{4}\right] = \frac{M}{12}\left[3b^2 + 3c^2 - a^2\right]$$

$$\therefore\ K^2 = \frac{1}{12}\left[3b^2 + 3c^2 - a^2\right]$$

The distance of c.g G from $AX = h = \frac{2}{3}AD = \frac{2}{3}\frac{\sqrt{2b^2+2c^2-a^2}}{2} = \frac{1}{3}\sqrt{2b^2+2c^2-a^2}$.
Thus the length of simple equivalent pendulum $= \frac{K^2}{h} = \frac{1}{4}\cdot\frac{3(b^2+c^2)-a^2}{\sqrt{2b^2+2c^2-a^2}}$.

► **Problem 3.6.10** *Show that the length of the simple equivalent pendulum of a solid hemisphere, whose diameter of the base is $2a$, and which oscillates about a diameter which is horizontal is $\frac{16}{15}a$.* **[KUH '06]**

Solution: Let O be the centre of the solid hemisphere and G be the c.g. of the hemisphere.

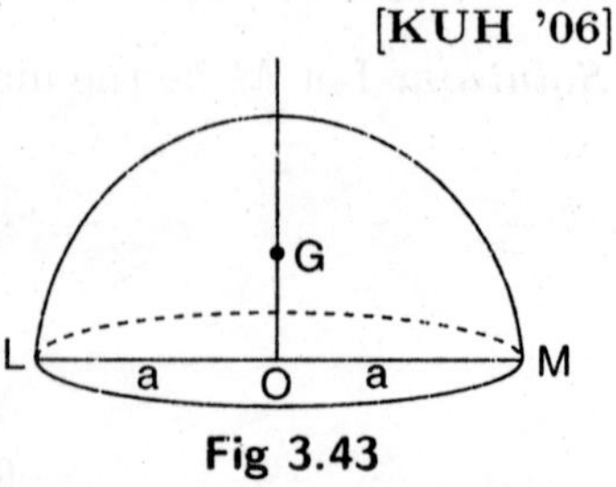

Fig 3.43

$$\therefore\ OG = \tfrac{3}{8}a = h$$

Thus the m.i. of the hemisphere about $LM = \frac{2}{5}Ma^2$.

$\therefore$ The radius of gyration $K^2 = \frac{2}{5}a^2$ [cf. §1.4(h)]

$\therefore$ The length of the simple equivalent pendulum $= \frac{K^2}{h} = \frac{\frac{2}{5}a^2}{\frac{3}{8}a} = \frac{16}{15}a$.

EXERCISES

1. A uniform rod AB of length $3a$ can turn freely in a vertical plane about a horizontal axis through the point O on the rod at a distance a from the centre. The rod starts from rest from its unstable equilibrium position. Show that its angular velocity when it comes to horizontal position is $\sqrt{\frac{8g}{7a}}$.

2. A homogeneous circular disc of radius a and weight W hangs in a vertical plane from a pin at its circumference. Show that the time for one complete small oscillation in its own vertical plane is
$$2\pi\sqrt{\frac{3a}{2g}}.$$

3. A solid homogeneous cone of height h and vertical angle 2α oscillates about a horizontal axis through the vertex. Show that, for small oscillations, the time period is
$$2\pi\sqrt{\frac{h}{5g}\,(4+\tan^2\alpha)}.$$

4. A triangular lamina ABC oscillates about the perpendicular from A on BC, the perpendicular being horizontal. Find the length of the simple equivalent pendulum when a, b, c are the lengths of the sides.

   $\left[Ans.\ \frac{a^4+3(b^2-c^2)^2}{8a(b^2-c^2)}\right]$

5. The door of a railway carriage stands open at right angles to the length of the strain when the later starts to move with an acceleration f, the door being supported to be smoothly hinged to the carriage and to be uniform and breadth $2a$, show that its angular velocity when it has turned through an angle θ is
$$\sqrt{\frac{3f}{2a}\sin\theta}.$$

Chapter 4

Motion in Two Dimension

4.1 Definition

When a rigid body is compelled to move in such a way that the path of every element of it remains parallel to a fixed plane, its motion is called a *two dimensional motion* and the plane is called the *plane of motion.*

If the rigid body be a lamina or symmetrical about the plane of motion then its position is completely determined if we know the coordinate of one of the point in the plane of motion, say its centre of gravity $(\bar{x}, \bar{y})$ and the angle θ that a fixed straight line in the lamina or in the plane of symmetry of the rigid body makes with a fixed straight line in the plane of motion. These quantities $\bar{x}$, $\bar{y}$, θ are called the *coordinates of the body.*

4.2 Equations of Motion

[KUH 2005, '06]

Let the plane of motion be taken as xy-plane and X, Y be the components of the external forces acting at an elementary mass m at $P(x, y)$ at any time t. By D'Alembert's Principle we get the following equations of motion.

$$\left.\begin{aligned} \sum m\ddot{x} = \sum X, \quad \sum m\ddot{y} = \sum Y \\ \text{and } \sum m\,(x\ddot{y} - y\ddot{x}) = \sum (xY - yX). \end{aligned}\right\}. \tag{4.2.1}$$

If $(\bar{x}, \bar{y})$ be the coordinates of centre of gravity G and (x', y') be the coordinates of P referred to parallel axes through $G(\bar{x}, \bar{y})$, then the equations of motion (4.2.1) an equivalent to

$$M\ddot{\bar{x}} = \sum X \tag{4.2.2}$$

$$M\ddot{\bar{y}} = \sum Y \tag{4.2.3}$$

and

$$\sum m\,(x'\ddot{y}' - y'\ddot{x}') = \sum\,(x'Y - y'X) = L, \tag{4.2.4}$$

where M is the mass of the body and L is the algebraic sum of the moments of the external forces about the centre of gravity G.

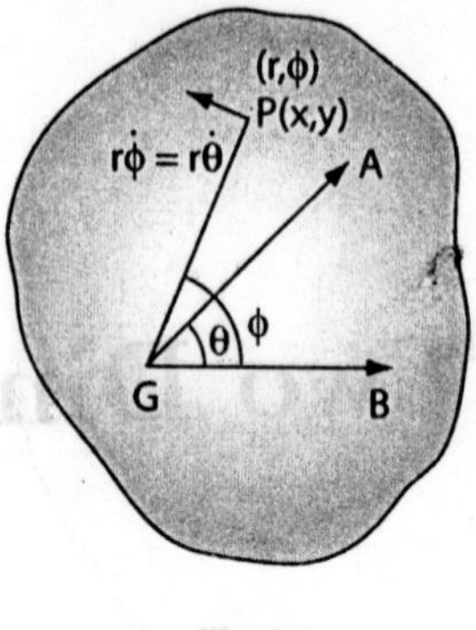

Fig 4.1

Let the system consists of a plane lamina or a rod. Let ϕ be the angle that the line P to G makes with a line GB fixed in space and θ be the angle made by GA, a fixed straight line in the body with GB.

Thus, $\phi - \theta =$ constant, i. e., $\dot{\theta} = \dot{\phi}$.

The equation (4.2.4) can be written as

$$\frac{d}{dt}\left[\sum m\,(x'\dot{y}' - y'\dot{x}')\right] = L. \tag{4.2.5}$$

Now

$$\sum m\,(x'\dot{y}' - y'\dot{x}')$$

is the algebraic sum of *moments of momentum* or *angular momentum* about centre of gravity G of all the elements of the body.

Let $GP = r$, the velocity of P relative to $G = r\dot{\phi} = r\dot{\theta}$ perpendicular GP in the sense of increasing θ. The momentum of the elementary particle m relative to c.g. $G = m.r\dot{\theta}$.

Thus the moment of momentum or angular momentum about c.g.

$$G = m \cdot r\dot{\theta} \cdot r = mr^2\dot{\theta}.$$

Thus the equation (4.2.5) becomes to

$$\frac{d}{dt}\left[\sum mr^2\dot{\theta}\right] = L, \quad \text{or,} \quad \frac{d}{dt}\left[\dot{\theta}\sum mr^2\right] = L$$

$$\frac{d}{dt}\left[MK^2\dot{\theta}\right] = L, \quad \text{or,} \quad MK^2\ddot{\theta} = L,$$

where $MK^2 =$ moment of inertia of the body about c.g. G of the body and K is the radius of gyration of the body about c.g. G. Thus the equation of motions are

$$M\ddot{\overline{x}} = \sum X, \quad M\ddot{\overline{y}} = \sum Y \quad \text{and} \quad MK^2\ddot{\theta} = L.$$

4.3 Kinetic Energy of a Rigid Body Moving in Two Dimensions

[CUH 2005; BUH 1997; VUH 2001, '03, '06; KUH 2005]

We take the plane of motion as xy-plane and Ox and Oy as fixed axes.

Let $(\bar{x}, \bar{y})$ be the coordinates of an element of mass m at P. Let $(\bar{x}, \bar{y})$ be the coordinates of centre of gravity G and (x', y') be the coordinates of P referred to parallel axes through centre of gravity G, i.e., Gx' and Gy'.

Let $M = \Sigma m$ is the mass of the rigid body.

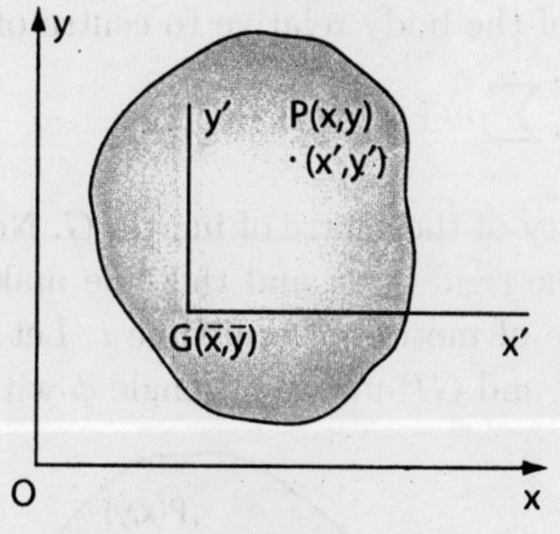

Fig 4.2

Therefore,

$$\left.\begin{aligned} x = \bar{x} + x', \quad y = \bar{y} + y' \\ \text{and} \quad \sum mx' = \sum my' = 0 \end{aligned}\right\}. \tag{4.3.1}$$

Thus Kinetic energy of the body

$$\begin{aligned}
\text{K.E.} &= \frac{1}{2}\sum m\left(\dot{x}^2 + \dot{y}^2\right) \\
&= \frac{1}{2}\sum m\left[\left\{\frac{d}{dt}(\bar{x} + x')\right\}^2 + \left\{\frac{d}{dt}(\bar{y} + y')\right\}^2\right] \\
&= \frac{1}{2}\sum m\left[\left\{\frac{d\bar{x}}{dt} + \frac{dx'}{dt}\right\}^2 + \left\{\frac{d\bar{y}}{dt} + \frac{dy'}{dt}\right\}^2\right] \\
&= \frac{1}{2}\sum m\left\{\left(\frac{d\bar{x}}{dt}\right)^2 + \left(\frac{d\bar{y}}{dt}\right)^2\right\} + \frac{1}{2}\sum m\left\{\left(\frac{dx'}{dt}\right)^2 + \left(\frac{dy'}{dt}\right)^2\right\} \\
&\quad + \sum m\frac{d\bar{x}}{dt}\cdot\frac{dx'}{dt} + \sum m\frac{d\bar{y}}{dt}\cdot\frac{dy'}{dt}
\end{aligned}$$

$$= \frac{1}{2}\left[\left(\frac{d\bar{x}}{dt}\right)^2 + \left(\frac{d\bar{y}}{dt}\right)^2\right]\sum m + \frac{1}{2}\sum m\left\{\left(\frac{dx'}{dt}\right)^2 + \left(\frac{dy'}{dt}\right)^2\right\}$$

$$+\frac{d\bar{x}}{dt}\frac{d}{dt}\left(\sum mx'\right) + \frac{d\bar{y}}{dt}\frac{d}{dt}\left(\sum my'\right)$$

$$= \frac{1}{2}M\left[\left(\frac{d\bar{x}}{dt}\right)^2 + \left(\frac{d\bar{y}}{dt}\right)^2\right] + \frac{1}{2}\sum m\left\{\left(\frac{dx'}{dt}\right)^2 + \left(\frac{dy'}{dt}\right)^2\right\}$$

[by (4.3.1)]

$$= \frac{1}{2}M\left[\dot{\bar{x}}^2 + \dot{\bar{y}}^2\right] + \frac{1}{2}\sum m\left(\dot{x}'^2 + \dot{y}'^2\right)$$

= The K.E. of a particle M placed at the centre of gravity G + K.E. of the body relative to centre of gravity G

$$= \frac{1}{2}Mv^2 + \frac{1}{2}\sum m\left(\dot{x}'^2 + \dot{y}'^2\right)$$

where v is the velocity of the centre of inertia G. Now let AB be a fixed line through c.g. G of the rigid body and this line makes an angle θ with fixed line GL in the plane of motion at any time t. Let $GP = r$ and GP makes an angle α with AB and GP makes an angle ϕ with GL.

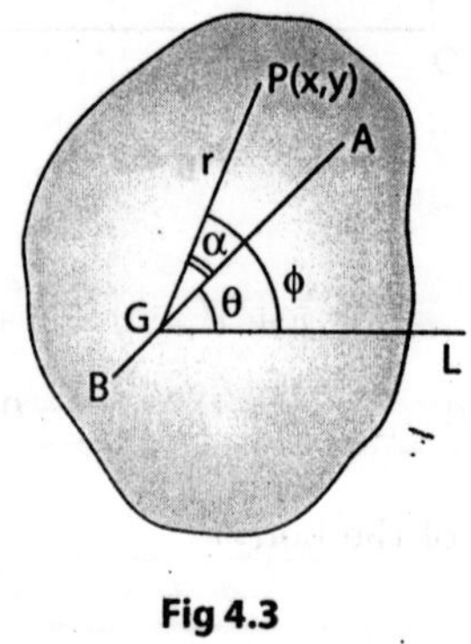

Fig 4.3

Thus, the velocity of P relative to G

$$= r\dot{\phi} = r\frac{d}{dt}(\theta + \alpha) = r\dot{\theta}. \tag{4.3.2}$$

Thus, K.E. relative to c.g. G

$$= \frac{1}{2}\sum m\left(r\dot{\theta}\right)^2 = \frac{1}{2}\dot{\theta}^2\sum mr^2 = \frac{1}{2}MK^2\dot{\theta}^2,$$

where K is the radius of gyration of the body about a line through c.g. G and perpendicular to the plane of motion.

Thus,

$$\begin{aligned} \text{K.E.} &= \frac{1}{2}M\left(\dot{\bar{x}}^2+\dot{\bar{y}}^2\right)+\frac{1}{2}MK^2\dot{\theta}^2=\frac{1}{2}M\left[\dot{\bar{x}}^2+\dot{\bar{y}}^2+K^2\dot{\theta}^2\right] \\ &= \frac{1}{2}Mv^2+\frac{1}{2}MK^2\dot{\theta}^2, \end{aligned}$$ **[KUH 2007]**

where v is the velocity of center of inertia.

4.4 Moment of Momentum About the Origin of a Rigid Body Moving in Two Dimension

[CUH 2003, '07; VUH 2004; KUH 2005]

We take the plane of the motion of the rigid body as xy-plane and Ox and Oy rectangular fixed axes at origin O. Let (x, y) be the coordinate of an element of mass m at P. Let $(\bar{x}, \bar{y})$ be the coordinates of c.g. G of the body and (x', y') be the coordinates of the element of mass m at P referred to parallel axes through c.g. G, i.e., Gx' and Gy'. Let $M = \Sigma m =$ mass of the rigid body.

Therefore

$$\left.\begin{aligned} x=\bar{x}+x', \quad y=\bar{y}+y' \\ \text{and } \sum mx'=\sum my'=0 \end{aligned}\right\}. \tag{4.4.1}$$

The expression of momentum or angular momentum about origin O

$$\begin{aligned}
&= \sum m\left(x\dot{y}-y\dot{x}\right)=\sum m\left(x\frac{dy}{dt}-y\frac{dx}{dt}\right) \\
&= \sum m\left\{\left(\bar{x}+x'\right)\frac{d}{dt}\left(\bar{y}+y'\right)-\left(\bar{y}+y'\right)\frac{d}{dt}\left(\bar{x}+x'\right)\right\} \\
&= \sum m\left(\bar{x}\frac{d\bar{y}}{dt}-\bar{y}\frac{d\bar{x}}{dt}\right)+\sum m\left(x'\frac{dy'}{dt}-y'\frac{dx'}{dt}\right) \\
&\quad +\sum m\left(\bar{x}\frac{dy'}{dt}-\bar{y}\frac{dx'}{dt}\right)+\sum m\left(x'\frac{d\bar{y}}{dt}-y'\frac{d\bar{x}}{dt}\right) \\
&= \sum m\left(\bar{x}\frac{d\bar{y}}{dt}-\bar{y}\frac{d\bar{x}}{dt}\right)+\sum m\left(x'\frac{dy'}{dt}-y'\frac{dx'}{dt}\right) \\
&\quad +\bar{x}\frac{d}{dt}\left(\sum my'\right)-\bar{y}\frac{d}{dt}\left(\sum mx'\right)+\frac{d\bar{y}}{dt}\sum mx'-\frac{d\bar{x}}{dt}\sum my' \\
&= M\left(\bar{x}\frac{d\bar{y}}{dt}-\bar{y}\frac{d\bar{x}}{dt}\right)+\sum m\left(x'\frac{dy'}{dt}-y'\frac{dx'}{dt}\right) \quad \text{[by (4.4.1)]}
\end{aligned} \tag{4.4.2}$$

$=$ moment of momentum of a particle of mass M placed at c.g. G about the origin + moment of momentum of the body relative to c.g. G.

Now if (r, θ) be the polar coordinates of the point P referred G as pole and Gx' as initial line, we get

$$x' = r\cos\theta; \quad \dot{x}' = -r\sin\theta\cdot\dot{\theta}$$

$$\text{and} \quad y' = r\sin\theta; \quad \dot{y}' = r\cos\theta\cdot\dot{\theta}.$$

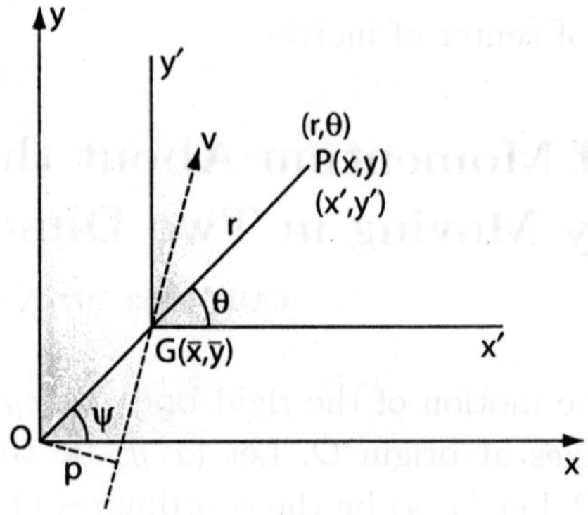

Fig 4.4

Thus

$$\sum m\left\{x'\frac{dy'}{dt} - y'\frac{dx'}{dt}\right\} = \sum m\left\{r\cos\theta(r\cos\theta\cdot\dot{\theta}) - r\sin\theta(-r\sin\theta\cdot\dot{\theta})\right\}$$

$$= \sum mr^2\dot{\theta} = \dot{\theta}\sum mr^2 = MK^2\dot{\theta},$$

where MK^2 = moment of inertia of the rigid body about a line through G and perpendicular to the plane of rotation of the body. Thus the moment of momentum or angular momentum about the origin O

$$= M\left(\bar{x}\frac{d\bar{y}}{dt} - \bar{y}\frac{d\bar{x}}{dt}\right) + MK^2\dot{\theta}. \tag{4.4.3}$$

If v be the velocity of c.g. G and p be the perpendicular distance from origin to the direction of v, then the moment of momentum of the mass M placed at c.g. $G = p\cdot Mv$. Thus the required moment of momentum

$$= Mv\cdot p + MK^2\dot{\theta}. \tag{4.4.4}$$

Again if (R, ψ) be the polar coordinates of c.g. G referred to origin O as pole and Ox as initial line, the velocity of c.g. G perpendicular to $OG = R\dot{\psi}$.

Then the moment of momentum of the mass M placed at c.g. G about $O = M\cdot R\dot{\psi}\cdot R = MR^2\dot{\psi}$.

Thus the moment of momentum or angular momentum of the body about the origin O

$$= MR^2\dot{\psi} + MK^2\dot{\theta}. \tag{4.4.5}$$

4.5 Kinetic Energy of a Rigid Body Moving in any Manner [VUH 2006]

Let the coordinates of an element of mass m at P be (x, y, z) and the coordinates of c.g. G be $(\bar{x}, \bar{y}, \bar{z})$ referred to a fixed set of axes Ox, Oy, Oz in space. Let (x', y', z') be the coordinates of the point P referred to a set of parallel axes through c.g. G, i.e., Gx', Gy' and Gz'. Let M be the mass of the rigid body.

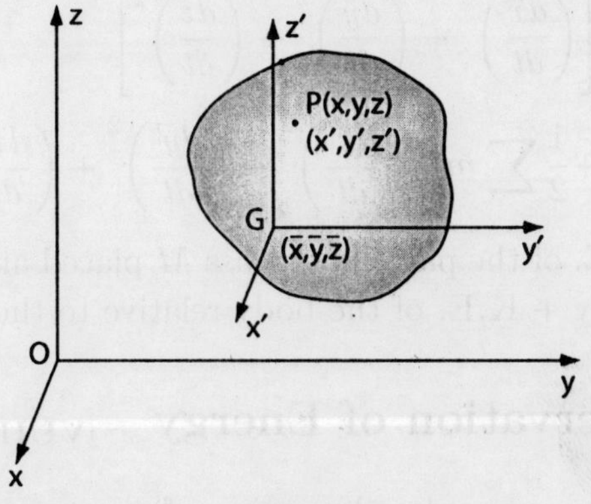

Fig 4.5

Thus

$$\left.\begin{array}{c} x = \bar{x} + x', \quad y = \bar{y} + y', \quad z = \bar{z} + z' \\ \text{and } \sum mx' = \sum my' = \sum mz' = 0 \end{array}\right\}. \tag{4.5.1}$$

Therefore the Kinetic energy of the rigid body is

$$\begin{aligned}
\text{K.E.} &= \frac{1}{2}\sum m\left(\dot{x}^2 + \dot{y}^2 + \dot{z}^2\right) \\
&= \frac{1}{2}\sum m\left\{\left(\frac{dx}{dt}\right)^2 + \left(\frac{dy}{dt}\right)^2 + \left(\frac{dz}{dt}\right)^2\right\} \\
&= \frac{1}{2}\sum m\left\{\left(\frac{d\bar{x}}{dt} + \frac{dx'}{dt}\right)^2 + \left(\frac{d\bar{y}}{dt} + \frac{dy'}{dt}\right)^2 + \left(\frac{d\bar{z}}{dt} + \frac{dz'}{dt}\right)^2\right\} \\
&\qquad \text{[by (4.5.1)]} \\
&= \frac{1}{2}\sum m\left\{\left(\frac{d\bar{x}}{dt}\right)^2 + \left(\frac{d\bar{y}}{dt}\right)^2 + \left(\frac{d\bar{z}}{dt}\right)^2\right\} \\
&\quad + \frac{1}{2}\sum m\left\{\left(\frac{dx'}{dt}\right)^2 + \left(\frac{dy'}{dt}\right)^2 + \left(\frac{dz'}{dt}\right)^2\right\} \\
&\quad + \sum m\left\{\frac{d\bar{x}}{dt}\cdot\frac{dx'}{dt} + \frac{d\bar{y}}{dt}\cdot\frac{dy'}{dt} + \frac{d\bar{z}}{dt}\cdot\frac{dz'}{dt}\right\}
\end{aligned}$$

$$
\begin{aligned}
&= \frac{1}{2}\left\{\left(\frac{d\bar{x}}{dt}\right)^2 + \left(\frac{d\bar{y}}{dt}\right)^2 + \left(\frac{d\bar{z}}{dt}\right)^2\right\}\sum m \\
&\quad + \frac{1}{2}\sum m\left\{\left(\frac{dx'}{dt}\right)^2 + \left(\frac{dy'}{dt}\right)^2 + \left(\frac{dz'}{dt}\right)^2\right\} \\
&\quad + \frac{d\bar{x}}{dt}\frac{d}{dt}\left(\sum mx'\right) + \frac{d\bar{y}}{dt}\frac{d}{dt}\left(\sum my'\right) + \frac{d\bar{z}}{dt}\frac{d}{dt}\left(\sum mz'\right) \\
&= \frac{1}{2}M\left[\left(\frac{d\bar{x}}{dt}\right)^2 + \left(\frac{d\bar{y}}{dt}\right)^2 + \left(\frac{d\bar{z}}{dt}\right)^2\right] \\
&\quad + \frac{1}{2}\sum m\left\{\left(\frac{dx'}{dt}\right)^2 + \left(\frac{dy'}{dt}\right)^2 + \left(\frac{dz'}{dt}\right)^2\right\} \quad \text{[by(4.5.1)]}
\end{aligned}
$$

= K.E. of the particle of mass M placed at the c.g. G of the body + K.E. of the body relative to the c.g. G.

4.6 Conservation of Energy [VUH 2003; KUH 2002, '04]

When a body moves under the action of a system of conservative system forces the sum of its kinetic energy and potential energy is constant throughout the motion.

If a system moves under the action of finite forces and if the geometrical equation of the system do not contain the time explicitly, the change in kinetic energy of the system in passing through from one configuration to the other is equal to the work done by the forces.

Let (X, Y, Z) are the components of the external force act at an element of mass m at $P(x, y, z)$ at time t. Then by D'Alembert's principle, the forces

$$X - m\frac{d^2x}{dt^2}, \quad Y - m\frac{d^2y}{dt^2}, \quad Z - m\frac{d^2z}{dt^2}$$

and similar forces acting on each particle of the system are in equilibrium.

Let δx, δy, δz are the virtual displacement of the element of particle of mass m at $P(x, y, z)$ consistent with geometrical conditions of the system at the time t. Then by principle of virtual work

$$\sum\left[\left(X - m\frac{d^2x}{dt^2}\right)\delta x + \left(Y - m\frac{d^2y}{dt^2}\right)\delta y + \left(Z - m\frac{d^2z}{dt^2}\right)\delta z\right] = 0 \quad (4.6.1)$$

If the geometrical relations of the system do not contain the time explicitly, then we may replaced δx, δy, δz by

$$\frac{dx}{dt}\delta t, \quad \frac{dy}{dt}\delta t, \quad \frac{dz}{dt}\delta t.$$

The equation (4.6.1) can be written as

$$\sum\left[m\left\{\left(\frac{d^2x}{dt^2}\cdot\frac{dx}{dt}+\frac{d^2y}{dt^2}\cdot\frac{dy}{dt}+\frac{d^2z}{dt^2}\cdot\frac{dz}{dt}\right)\right\}\delta t\right]$$

$$=\sum\left(X\frac{dx}{dt}+Y\frac{dy}{dt}+Z\frac{dz}{dt}\right)\delta t$$

$$=\sum(Xdx+Ydy+Zdz)$$

and integrating we get

$$\frac{1}{2}\sum\left[m\left\{\left(\frac{dx}{dt}\right)^2+\left(\frac{dy}{dt}\right)^2+\left(\frac{dz}{dt}\right)^2\right\}\right]_{t_1}^{t_2}$$

$$=\int_{t_1}^{t_2}\sum(X\,dx+Y\,dy+Z\,dz)\,. \tag{4.6.2}$$

The equation (4.6.2) shows that the change in kinetic energy of the system from time t_1 to time t_2 is equal to the work done by the external forces on the rigid body from the configuration of the body at time t_1 to the configuration at time t_2.

If the external forces are such that

$$\sum(Xdx+Ydy+Zdz)=-dV,$$

complete differential of some function V, where V is the function of the coordinates of all the particles of the body called the *potential energy.* the quantity

$$\sum(Xdx+Ydy+Zdz)$$

is independent to the path pursued from the initial to the final position of the body at time t_1 to t_2. Then the forces are called *conservation system* of forces. Then from (4.6.2) we get

K.E. of the body at time t_2- K.E. of the body at time t_1.

$$=\int_{t_1}^{t_2}-dV=V_{t_1}-V_{t_2}$$

Thus the K.E. of the body at time t_1+ P.E. of the body at time t_1

= K.E. of the body at time t_2+ P.E. of the body at time t_2.

Thus when a body moves under the action of conservation system of forces the sum of its K.E. and P.E. is constant throughout the motion.

4.7 Conservation of Linear Momentum Under Finite Forces

If a body moves under the action of some external forces whose sum of resolved parts along parallel to a certain line is zero throughout its motion, the momentum of the body parallel to that line will be constant throughout the motion.

Let x-axis be taken as the fixed line and X be the sum of resolved part of the external forces parallel to x-axis, acting on the particle of mass m at $P(x, y, z)$ at time t. Thus the equations of motion are

$$\sum m\frac{d^2x}{dt^2} = \Sigma X, \quad \text{or,} \quad \frac{d}{dt}\left(\sum m\frac{dx}{dt}\right) = \Sigma X.$$

Now if $\sum X = 0$, we have

$$\frac{d}{dt}\left(\sum m\frac{dx}{dt}\right) = 0, \quad \text{or,} \quad \sum m\frac{dx}{dt} = \text{constant}. \qquad (4.7.1)$$

Now if $\bar{x}$ be the x-coordinate of the centre of gravity G of the body of mass M, then

$$M\bar{x} = \sum mx, \quad \text{or,} \quad \sum m\frac{dx}{dt} = M\frac{d\bar{x}}{dt}.$$

$\therefore$ From (4.7.1) $M\frac{d\bar{x}}{dt}$ = constant, which shows that total momentum of the body parallel to any line remains constant throughout the motion as x-axis is arbitrary straight line.

4.8 Conservation of Moment of Momentum Under Finite Forces

[CUH 2003]

If a body moves under the action of some external forces whose sum of the moments about any given line be zero throughout the motion, then the moment of momentum or angular momentum of the body remains constant about the line throughout the motion.

Let $P(x, y, z)$ be the position of an element of mass m referred to the axes Ox, Oy, and Oz. Let (X, Y, Z) be the components of external forces along parallel to the coordinate axes act on mass m at P at time t. Let Ox be taken as the given line and $M = \Sigma m$.

Then by D'Alembert's Principle, we get, by taking moment about x-axis

$$\sum m\left(y\frac{d^2z}{dt^2} - z\frac{d^2y}{dt^2}\right) = \sum (yZ - zY)$$

if $\sum(yZ - zY) = 0$, we have

$$\sum m\left(y\frac{d^2z}{dt^2} - z\frac{d^2y}{dt^2}\right) = 0$$

$$\text{or,}\quad \frac{d}{dt}\left[\sum m\left(y\frac{dz}{dt} - z\frac{dy}{dt}\right)\right] = 0$$

or integrating,

$$\sum m\left(y\frac{dz}{dt} - z\frac{dy}{dt}\right) = \text{constant.}$$

As x-axis or given line is arbitrary the above results shows that the sum of moment of momentum or angular momentum of the body about any line is constant throughout the motion.

4.9 Some Important Results to be Remembered

- If R be the normal reaction, F be the frictional force and μ be the coefficient of friction, then

  (i) the body will roll if $\mu > F/R$;

  (ii) the body will slide if $\mu < F/R$;

  (iii) the body starts sliding or sliding begin if $\mu = F/R$.

- If R be the normal reaction on the body and the surface in contact, the body leaves the surface if R vanish and changes its sign.

- K.E. + P.E. = constant = initial value

  (i) K.E.$= \frac{1}{2}M\left(\dot{\bar{x}}^2 + \dot{\bar{y}}^2 + K^2\dot{\theta}^2\right)$, where $(\bar{x}, \bar{y})$ be the coordinate of centre of gravity and K = radius of gyration of the rod about centre of gravity.

  (ii) K.E. $= \frac{1}{2}MK^2\dot{\theta}^2$, where K = radius of gyration of the rod about one end point.

- $MK^2\ddot{\theta}$ = Algebraic sum of the moment of the applied forces about c.g. in the sense of θ increasing.

- The coefficient of friction $\mu = \tan\lambda$, when λ is the angle of friction.

4.10 Solved Problem on Rods

► **Problem 4.10.1** *A uniform rod is held in a vertical position with one end resting upon a perfectly rough table, and when released rotates about the end in contact with the table. Find the motion.*

[BUH 1998, '01; VUH 2004; KUH 2002]

Solution: Let AB be the rod of mass M and length $2a$ and the end A is in contact with the rough table. Let R be normal reaction and F be the frictional force at A respectively. Let $G(\bar{x}, \bar{y})$ be the centre of gravity of the rod through which the weight Mg acts vertically downwards and any time t the rod AB makes an angle θ with the vertical Ay.

$$\therefore \quad \bar{x} = a\sin\theta \quad \text{and} \quad \bar{y} = a\cos\theta.$$

$$\left.\begin{aligned} \therefore \quad \dot{\bar{x}} &= a\cos\theta\cdot\dot{\theta}\,; \quad \ddot{\bar{x}} = -a\sin\theta\cdot\dot{\theta}^2 + a\cos\theta\cdot\ddot{\theta} \\ \dot{\bar{y}} &= -a\sin\theta\cdot\dot{\theta}\,; \quad \ddot{\bar{y}} = -a\cos\theta\cdot\dot{\theta}^2 - a\sin\theta\cdot\ddot{\theta} \end{aligned}\right\}. \tag{4.10.1}$$

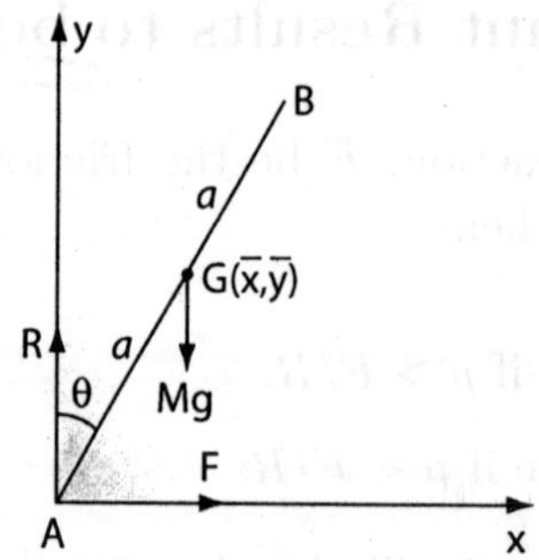

Fig 4.6

The equation of motion are

$$M\ddot{\bar{x}} = F \tag{4.10.2}$$

$$M\ddot{\bar{y}} = R - Mg \tag{4.10.3}$$

$$\text{and} \quad MK^2\ddot{\theta} = R\cdot a\sin\theta - Fa\cos\theta. \tag{4.10.4}$$

Using (4.10.1), we get from (4.10.2),

$$F = M\left(-a\sin\theta\cdot\dot{\theta}^2 + a\cos\theta\cdot\ddot{\theta}\right) \tag{4.10.5}$$

and from (4.10.3),

$$R - Mg = M\left(-a\cos\theta\cdot\dot{\theta}^2 - a\sin\theta\cdot\ddot{\theta}\right). \tag{4.10.6}$$

Substituting (4.10.5) and (4.10.6) in (4.10.4) we get

$$\begin{aligned} MK^2\ddot{\theta} &= a\sin\theta\left[Mg - Ma\left(\sin\theta\cdot\ddot{\theta} + \cos\theta\cdot\dot{\theta}^2\right)\right] \\ &\quad -Ma\cos\theta\left[a\cos\theta\cdot\ddot{\theta} - a\sin\theta\cdot\dot{\theta}^2\right] \\ &= Mag\sin\theta - Ma^2\ddot{\theta}, \end{aligned}$$

$$\text{or,}\quad \frac{a^2}{3}\ddot{\theta} = ag\sin\theta - a^2\ddot{\theta}, \quad \text{or,}\quad \frac{4a^2}{3}\ddot{\theta} = ag\sin\theta$$

$$\text{or,}\quad \ddot{\theta} = \frac{3g}{4a}\sin\theta, \tag{4.10.7}$$

$$\text{or,}\quad 2\dot{\theta}\ddot{\theta} = \frac{3g}{2a}\sin\theta\cdot\dot{\theta}.$$

Integrating

$$\dot{\theta}^2 = -\frac{3g}{2a}\cos\theta + C.$$

Initially $\dot{\theta} = 0$ when $\theta = 0$. $\therefore\ C = \frac{3g}{2a}$.

$$\therefore\quad \dot{\theta}^2 = -\frac{3g}{2a}\cos\theta + \frac{3g}{2a} = \frac{3g}{2a}(1-\cos\theta). \tag{4.10.8}$$

$\therefore$ From (4.10.5),

$$\begin{aligned} F &= M\left[-a\sin\theta\cdot\frac{3g}{2a}(1-\cos\theta) + a\cos\theta\cdot\frac{3g}{4a}\sin\theta\right] \\ &= M\cdot\frac{3g}{4}\left[\sin\theta\cos\theta - 2\sin\theta + 2\sin\theta\cos\theta\right] \\ &= \frac{3Mg}{4}\left[3\cos\theta - 2\right]\sin\theta. \end{aligned} \tag{4.10.9}$$

From (4.10.6) we get

$$\begin{aligned} R &= Mg + M\left[-a\cos\theta\cdot\frac{3g}{2a}(1-\cos\theta) - a\sin\theta\frac{3g}{4a}\sin\theta\right] \\ &= \frac{Mg}{4}\left[4 - 6\cos\theta(1-\cos\theta) - 3\sin^2\theta\right] \\ &= \frac{Mg}{4}\left[4 - 6\cos\theta + 6\cos^2\theta - 3 + 3\cos^2\theta\right] \\ &= \frac{Mg}{4}\left[1 - 6\cos\theta + 9\cos^2\theta\right] \\ &= \frac{Mg}{4}(1-3\cos\theta)^2. \end{aligned} \tag{4.10.10}$$

If the end leaves the plane, then R must vanish and changes its sign. If $1 - 3\cos\theta = 0$, i.e., $\cos\theta = 1/3$, R vanishes but does not changes its sign. Hence, A does not leave the plane.

When $\cos\theta = 1/3$, $R = 0$ then $\mu = F/R$ becomes infinite. Hence the rod slides provided the plane is not infinity rough.

From (4.10.9), we see that F changes its sign when $3\cos\theta - 2 < 0$, i.e., $\cos\theta < 2/3$ or $\theta < \cos^{-1}(2/3)$. Thus in this case the direction of the end A is reversed.

Thus the end of the rod begins to slip before $\cos\theta = 1/3$, and the end will slip backward or forward as the sliping occurs before and after the inclination of the rod is $\cos^{-1}(2/3)$.

► **Problem 4.10.2** *A uniform straight rod slides down in a vertical plane, its ends being in contact with two smooth planes, one horizontal and the other vertical. If it started from rest at an angle α with the horizontal. Find the motion.*

Solution: Let $AB(= 2a)$ be the rod and $G(\bar{x}, \bar{y})$ be the c.g. of the rod. Let the ends A and B resting on vertical and horizontal plane respectively. Let R and S be the normal reactions at A and B respectively. Let θ be angle made by the rod with horizontal at any time t. As $OL = LB$, then

$$\bar{x} = a\cos\theta \quad \text{and} \quad \bar{y} = a\sin\theta$$
$$\therefore\ \dot{\bar{x}} = -a\sin\theta\cdot\dot{\theta} \quad \text{and} \quad \dot{\bar{y}} = a\cos\theta\cdot\dot{\theta}$$

$$\left.\begin{aligned} \ddot{\bar{x}} &= -a\cos\theta\cdot\dot{\theta}^2 - a\sin\theta\cdot\ddot{\theta} \\ \ddot{\bar{y}} &= -a\sin\theta\cdot\dot{\theta}^2 + a\cos\theta\cdot\ddot{\theta} \end{aligned}\right\}. \qquad (4.10.11)$$

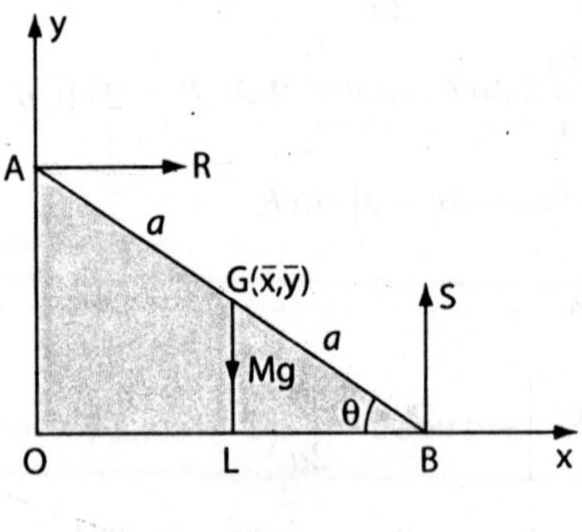

Fig 4.7

The equations of motion are

$$M\ddot{\bar{x}} = R \qquad (4.10.12)$$
$$M\ddot{\bar{y}} = S - Mg \qquad (4.10.13)$$
$$\text{and} \quad MK^2\ddot{\theta} = R\cdot a\sin\theta - S\cdot a\cos\theta \qquad (4.10.14)$$

Using (4.10.11), from (4.10.12) we get

$$R = M\left[-a\cos\theta\cdot\dot{\theta}^2 - a\sin\theta\cdot\ddot{\theta}\right] \tag{4.10.15}$$

and from (4.10.13) we get

$$S = Mg + M\left(-a\sin\theta\cdot\dot{\theta}^2 + a\cos\theta\cdot\ddot{\theta}\right). \tag{4.10.16}$$

Using (4.10.15) and (4.10.16) in (4.10.14) we get

$$\begin{aligned} MK^2\ddot{\theta} &= M\Big[a\sin\theta\cdot\left(-a\cos\theta\cdot\dot{\theta}^2 - a\sin\theta\cdot\ddot{\theta}\right) \\ &\qquad -a\cos\theta\left(g - a\sin\theta\cdot\dot{\theta}^2 + a\cos\theta\cdot\ddot{\theta}\right)\Big] \\ &= M\left[-ag\cos\theta - a^2\ddot{\theta}\right] \end{aligned}$$

$$\text{or,}\quad K^2\ddot{\theta} = -ag\cos\theta - a^2\ddot{\theta},\quad \text{or,}\quad \frac{a^2}{3}\ddot{\theta} = -ag\cos\theta - a^2\ddot{\theta},$$

$$\text{or,}\quad \frac{4a^2}{3}\ddot{\theta} = -ag\cos\theta,\quad \text{or,}\quad \ddot{\theta} = -\frac{3g}{4a}\cos\theta, \tag{4.10.17}$$

$$\text{or,}\quad 2\theta\ddot{\theta} = -\frac{3g}{2a}\cos\theta\cdot\dot{\theta}.$$

Integrating

$$\dot{\theta}^2 = -\frac{3g}{2a}\sin\theta + C_1.$$

Initially let $\dot{\theta} = 0$ when $\theta = \alpha$.

$$\therefore\quad C_1 = \frac{3g}{2a}\sin\alpha.$$

Thus

$$\dot{\theta}^2 = \frac{3g}{2a}(\sin\alpha - \sin\theta). \tag{4.10.18}$$

$\therefore$ From (4.10.15),

$$\begin{aligned} R &= M\left[a\sin\theta\cdot\frac{3g}{4a}\cos\theta - a\cos\theta\cdot\frac{3g}{2a}(\sin\alpha - \sin\theta)\right] \\ &= \frac{3Mg}{4}\left[3\sin\theta\cos\theta - 2\sin\alpha\cos\theta\right] \\ &= \frac{3Mg}{4}\left[3\sin\theta - 2\sin\alpha\right]\cos\theta \end{aligned} \tag{4.10.19}$$

and from (4.10.16),

$$\begin{aligned}
S &= M\left[g - a\sin\theta\cdot\frac{3g}{2a}(\sin\alpha-\sin\theta) + a\cos\theta\left(-\frac{3g}{4a}\cos\theta\right)\right] \\
&= \frac{Mg}{4}\left[4 - 6\sin\theta\sin\alpha + 6\sin^2\theta - 3\cos^2\theta\right] \\
&= \frac{Mg}{4}\left[4 - 6\sin\theta\sin\alpha + 6\sin^2\theta - 3 + 3\sin^2\theta\right] \\
&= \frac{Mg}{4}\left[1 - 6\sin\theta\sin\alpha + 9\sin^2\theta\right] \\
&= \frac{Mg}{4}\left[\cos^2\alpha + \sin^2\alpha - 6\sin\theta\sin\alpha + 9\sin^2\theta\right] \\
&= \frac{Mg}{4}\left[\cos^2\alpha + (3\sin\theta - \sin\alpha)^2\right]. \qquad (4.10.20)
\end{aligned}$$

From (4.10.19), when

$$\sin\theta = \frac{2}{3}\sin\alpha, \quad R = 0,$$

i. e., R vanishes and when θ decreases further, R changes its sign.

Therefore the end A leaves the vertical wall when

$$\sin\theta = \frac{2}{3}\sin\alpha.$$

Therefore the angular velocity at this instant, from (4.10.18) we get

$$\begin{aligned}
\dot{\theta}^2 &= \frac{3g}{2a}\left[\sin\alpha - \frac{2}{3}\sin\alpha\right] = \frac{g\sin\alpha}{2a} \\
\text{or,}\quad \dot{\theta}^2 &= \pm\sqrt{\frac{g\sin\alpha}{2a}}.
\end{aligned}$$

As θ decreases, the angular velocity at this instant $= -\sqrt{\frac{g\sin\alpha}{2a}}$. The horizontal velocity of c.g. G at that instant

$$\dot{x} = -a\sin\theta\cdot\dot{\theta} = \sqrt{\frac{g\sin\alpha}{2a}}\cdot\frac{2a}{3}\sin\alpha = \frac{1}{3}\sqrt{2ga\sin^3\alpha}.$$

For the second part of the motion, i. e., when the end A leaves the vertical wall, the angular velocity at that instant is $-\sqrt{\frac{g\sin\alpha}{2a}}$ and the horizontal velocity of c.g. G

$$= \frac{1}{3}\sqrt{2ga\sin^3\alpha}.$$

During the second part of motion, let ϕ be the angle made the rod with the horizon and S' be the normal reaction at B.

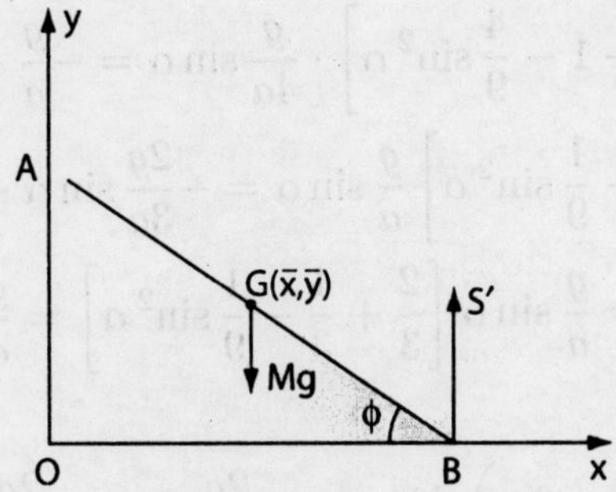

Fig 4.8

Thus the equations of motion are

$$M\ddot{\bar{x}} = 0 \tag{4.10.21}$$

$$M\ddot{\bar{y}} = S' - Mg \tag{4.10.22}$$

$$\text{and} \quad MK^2\ddot{\phi} = -S'a\cos\phi \tag{4.10.23}$$

when $\bar{x} = a\cos\phi$ and $\bar{y} = a\sin\phi$

$$\text{or,} \quad \dot{\bar{y}} = a\cos\phi\cdot\dot{\phi}, \quad \ddot{\bar{y}} = -a\sin\phi\cdot\dot{\phi}^2 + a\cos\phi\cdot\ddot{\phi}$$

$$\therefore \quad MK^2\ddot{\phi} = -a\cos\phi\,(M\ddot{\bar{y}} + Mg)$$

$$\text{or,} \quad K^2\ddot{\phi} = -a\cos\phi\left[-a\sin\phi\cdot\dot{\phi}^2 + a\cos\phi\cdot\ddot{\phi} + g\right]$$

$$\text{or,} \quad \frac{a^2}{3}\ddot{\phi} = a^2\sin\phi\cos\phi\dot{\phi}^2 - a^2\cos^2\phi\cdot\ddot{\phi} - ag\cos\phi$$

$$\text{or,} \quad \left(\frac{a^2}{3} + a^2\cos^2\phi\right)\ddot{\phi} - a^2\sin\phi\cos\phi\cdot\dot{\phi}^2 = -ag\cos\phi$$

$$\text{or,} \quad \left(\frac{1}{3} + \cos^2\phi\right)\ddot{\phi} - 2\sin\phi\cos\phi\left(\frac{\dot{\phi}^2}{2}\right) = -\frac{g}{a}\cos\phi$$

$$\text{or,} \quad \left(\frac{1}{3} + \cos^2\phi\right)\frac{d}{d\phi}\left(\frac{\dot{\phi}^2}{2}\right) - 2\sin\phi\cos\phi\left(\frac{\dot{\phi}^2}{2}\right) = -\frac{g}{a}\cos\phi$$

$$\text{or,} \quad \frac{d}{d\phi}\left[\left(\frac{1}{3} + \cos^2\phi\right)\frac{\dot{\phi}^2}{2}\right] = -\frac{g}{a}\cos\phi.$$

Integrating

$$\left(\frac{1}{3} + \cos^2\phi\right)\frac{\dot{\phi}^2}{2} = -\frac{g}{a}\sin\phi + C_2.$$

Initially

$$\dot{\phi}^2 = \frac{g}{2a}\sin\alpha, \text{ when } \sin\phi = \frac{2}{3}\sin\alpha.$$

$$\therefore \quad \left[\frac{1}{3}+1-\frac{4}{9}\sin^2\alpha\right]\cdot\frac{g}{4a}\sin\alpha = -\frac{g}{a}\cdot\frac{2}{3}\sin\alpha + C_2$$

$$\text{or,} \quad \left[\frac{1}{3}-\frac{1}{9}\sin^2\alpha\right]\frac{g}{a}\sin\alpha = -\frac{2g}{3a}\sin\alpha + C_2$$

$$\text{or,} \quad C_2 = \frac{g}{a}\sin\alpha\left[\frac{2}{3}+\frac{1}{3}-\frac{1}{9}\sin^2\alpha\right] = \frac{g}{a}\sin\alpha\left[1-\frac{1}{9}\sin^2\alpha\right].$$

$$\begin{aligned}\therefore \quad \left(\frac{1}{3}+\cos^2\phi\right)\dot{\phi}^2 &= -\frac{2g}{a}\sin\phi + \frac{2g}{a}\sin\alpha\left[1-\frac{1}{9}\sin^2\alpha\right]\\ &= \frac{2g}{a}\left[\sin\alpha\left(1-\frac{1}{9}\sin^2\alpha\right)-\sin\phi\right]\end{aligned}$$

when $\phi = 0$, i.e., when the rod reaches the horizontal plane. Let $\dot{\phi} = \Omega$, then

$$\begin{aligned}\Omega^2(4/3) &= \frac{2g}{a}\left[\sin\alpha\left(1-\frac{1}{9}\sin^2\alpha\right)\right]\\ \text{or,} \quad \Omega^2 &= \frac{3g}{2a}\left[\sin\alpha\left(1-\frac{1}{9}\sin^2\alpha\right)\right].\end{aligned}$$

Again from (4.10.21), we get by integration

$$\dot{\bar{x}} = \text{constant} = \text{initial value} = \frac{1}{3}\sqrt{2ag\sin^3\alpha}.$$

Hence, in the second part of motion, the horizontal velocity of the c.g. is constant throughout the motion.

▶ **Problem 4.10.3** *A uniform rod is held in a vertical position with one end resting upon a horizontal table, and when released, rotates about the end in contract with the table. Show that, when it is inclined at an angle of* 30° *to the horizontal, the force of friction that must be exerted to prevent sliding is approximately* 0.32 *of the weight.*

Solution: Let $AB(=2a)$ be the rod of mass M. $(\bar{x}, \bar{y})$ be the coordinates of the c.g. G, of the rod referred to the axes Ax and Ay.

Let R and F are the normal reaction and frictional force at A. Let any instant the rod makes an angle θ with the horizontal.

Thus the equations of motion are

$$M\ddot{\bar{x}} = F \tag{4.10.24}$$

$$M\ddot{\bar{y}} = R - Mg \tag{4.10.25}$$

$$\text{and} \quad MK^2\ddot{\theta} = F\cdot a\sin\theta - R\cdot a\cos\theta. \tag{4.10.26}$$

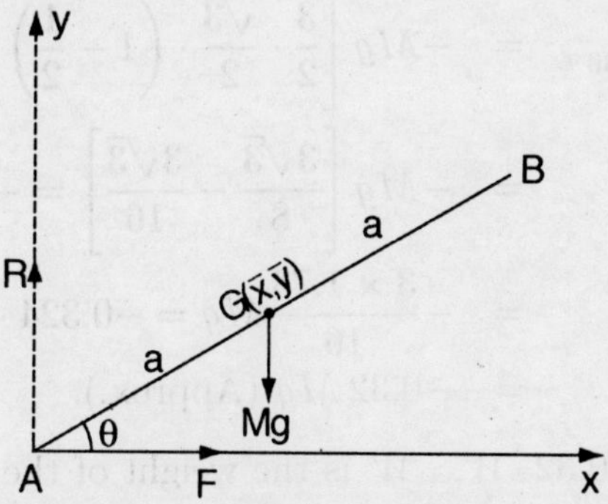

Fig 4.9

Again

$$\left.\begin{aligned}\bar{x} = a\cos\theta,\quad \dot{\bar{x}} = -a\sin\theta\cdot\dot{\theta}\,;\quad \ddot{\bar{x}} = -a\cos\theta\cdot\dot{\theta}^2 - a\sin\theta\cdot\ddot{\theta}\\ \bar{y} = a\sin\theta,\quad \dot{\bar{y}} = a\cos\theta\cdot\dot{\theta}\,;\quad \ddot{\bar{y}} = -a\sin\theta\cdot\dot{\theta}^2 + a\cos\theta\cdot\ddot{\theta}\end{aligned}\right\}. \quad (4.10.27)$$

Using (4.10.24) and (4.10.25) in (4.10.26) we get

$$\begin{aligned}MK^2\ddot{\theta} &= M\left[a\sin\theta\cdot\ddot{\bar{x}} - a\cos\theta\left(\ddot{\bar{y}}+g\right)\right]\\ \text{or,}\quad K^2\ddot{\theta} &= \Big[a\sin\theta\Big(-a\cos\theta\cdot\dot{\theta}^2 - a\sin\theta\cdot\ddot{\theta}\Big)\\ &\qquad -a\cos\theta\Big(-a\sin\theta\cdot\dot{\theta}^2 + a\cos\theta\cdot\ddot{\theta}+g\Big)\Big].\\ \frac{a^2}{3}\ddot{\theta} &= -a^2\ddot{\theta} - ga\cos\theta,\quad \text{or,}\quad \left(\frac{a^2}{3}+a^2\right)\ddot{\theta} = -ga\cos\theta,\\ \text{or,}\quad \frac{4a^2}{3}\ddot{\theta} &= -ga\cos\theta,\quad \text{or,}\quad \ddot{\theta} = -\frac{3g}{2a}\cos\theta, \qquad (4.10.28)\\ \text{or,}\quad 2\dot{\theta}\ddot{\theta} &= -\frac{3g}{2a}\cos\theta\cdot\dot{\theta}.\end{aligned}$$

Integrating

$$\dot{\theta}^2 = -\frac{3g}{2a}\sin\theta + C_1.$$

Initially $\dot{\theta} = 0$ when $\theta = \pi/2$. $\quad\therefore\ C_1 = \frac{3g}{2a}$.
Thus

$$\dot{\theta}^2 = \frac{3g}{2a}(1-\sin\theta). \qquad (4.10.29)$$

Now from (4.10.24), we get

$$\begin{aligned}F &= M\left(-a\cos\theta\cdot\dot{\theta}^2 - a\sin\theta\cdot\ddot{\theta}\right)\\ &= -M\left[\cos\theta\cdot\frac{3g}{2}(1-\sin\theta) - \sin\theta\cdot\frac{3g}{4}\cos\theta\right] \quad [\text{by (4.10.28) and (4.10.29)}]\\ &= -Mg\left[\frac{3}{2}\cos\theta(1-\sin\theta) - \frac{3}{4}\sin\theta\cos\theta\right].\end{aligned}$$

$$\therefore \quad F\big|_{\theta=30^\circ} = -Mg\left[\frac{3}{2}\cdot\frac{\sqrt{3}}{2}\cdot\left(1-\frac{1}{2}\right)-\frac{3}{4}\cdot\frac{1}{2}\cdot\frac{\sqrt{3}}{2}\right]$$

$$= -Mg\left[\frac{3\sqrt{3}}{8}-\frac{3\sqrt{3}}{16}\right] = -\frac{3\sqrt{3}}{16}Mg$$

$$= -\frac{3\times 1.73}{16}Mg = -0.324\cdot Mg$$

$$= -0.32\,Mg \text{ (Approx.)}.$$

$\therefore \quad F\big|_{\theta=30^\circ} = 0.32\cdot W,\quad W$ is the weight of the rod.

▶ **Problem 4.10.4** *A uniform rod is placed with one end in contact with a horizontal table, and then at an angle α to the horizon and is allowed to fall. When it becomes horizontal, show that its angular velocity is $\sqrt{\frac{3g}{2a}}\sin\alpha$, where the plane be perfectly smooth or perfectly rough. Show also that the end of the rod will not leave the plane in either case.*

[BUH 1999; VUH 2003]

Solution:

Case I. *When the table is perfectly smooth.*

Let at any time t, the rod $AB(=2a)$ of mass M makes an angle θ with the horizontal table. Let $(\bar{x},\bar{y})$ be the coordinates of the c.g. G, of the rod referred to axes Ax and Ay. Let R is the normal reaction at A.

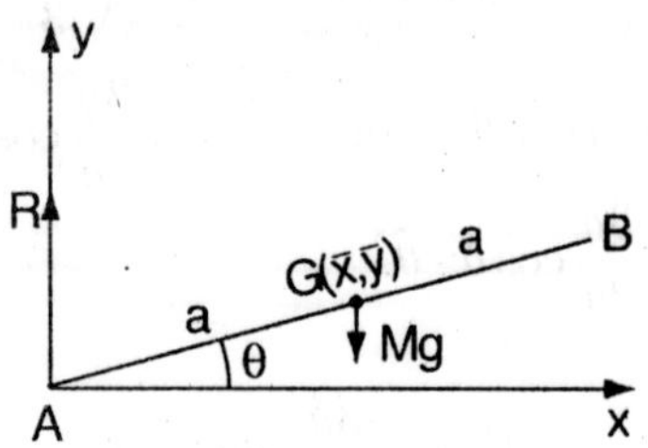

Fig 4.10

The equations of motion are

$$M\ddot{\bar{x}} = 0 \tag{4.10.30}$$

$$M\ddot{\bar{y}} = R - Mg \tag{4.10.31}$$

$$\text{and}\quad MK^2\ddot{\theta} = -R\cdot a\cos\theta \tag{4.10.32}$$

$$= -M\,(\ddot{\bar{y}}+g)\,a\cos\theta \quad \text{[by (4.10.31)]}$$

$$\text{or,}\quad \frac{a^2}{3}\ddot{\theta} = -\,(\ddot{\bar{y}}+g)\,a\cos\theta,$$

$$\text{or,}\quad \frac{a\ddot{\theta}}{3} = -\,(\ddot{\bar{y}}+g)\,a\cos\theta. \tag{4.10.33}$$

Now

$$\bar{y} = a\sin\theta; \quad \dot{\bar{y}} = a\cos\theta\cdot\dot{\theta}; \quad \ddot{\bar{y}} = -a\sin\theta\cdot\dot{\theta}^2 + a\cos\theta\cdot\ddot{\theta}.$$

Substituting in (4.10.33), we get

$$\begin{aligned}
\frac{a\ddot{\theta}}{3} &= -\left[-a\sin\theta\cdot\dot{\theta}^2 + a\cos\theta\cdot\ddot{\theta} + g\right]\cos\theta,\\
&= a\sin\theta\cos\theta\cdot\dot{\theta}^2 - a\cos^2\theta\cdot\ddot{\theta} - g\cos\theta,\\
\text{or,}\quad \frac{\ddot{\theta}}{3} &= \sin\theta\cos\theta\cdot\dot{\theta}^2 - \cos^2\theta\cdot\ddot{\theta} - \frac{g}{a}\cos\theta,\\
\text{or,}\quad \left(\frac{1}{3}+\cos^2\theta\right)\ddot{\theta} &- \sin\theta\cos\theta\cdot\dot{\theta}^2 = -\frac{g}{a}\cos\theta, \qquad (4.10.34)\\
\text{or,}\quad \left(\frac{1}{3}+\cos^2\theta\right)\frac{d}{d\theta}\left(\frac{\dot{\theta}^2}{2}\right) &- 2\sin\theta\cos\theta\cdot\left(\frac{\dot{\theta}^2}{2}\right) = -\frac{g}{a}\cos\theta,\\
\text{or,}\quad \frac{d}{d\theta}\left[\left(\frac{1}{3}+\cos^2\theta\right)\frac{\dot{\theta}^2}{2}\right] &= -\frac{g}{a}\cos\theta.
\end{aligned}$$

Integrating

$$\left(\frac{1}{3}+\cos^2\theta\right)\frac{\dot{\theta}^2}{2} = -\frac{g}{a}\sin\theta + C.$$

Initially $\dot{\theta} = 0$, when $\theta = \alpha$. $\quad\therefore\ C = \frac{g}{a}\sin\alpha$.

$$\therefore\quad \left(\frac{1}{3}+\cos^2\theta\right)\frac{\dot{\theta}^2}{2} = \frac{g}{a}(\sin\alpha - \sin\theta).$$

Thus

$$\dot{\theta}^2 = \frac{2g}{a}\frac{(\sin\alpha - \sin\theta)}{\left(\frac{1}{3}+\cos^2\theta\right)}. \qquad (4.10.35)$$

Therefore

$$\dot{\theta}^2\big|_{\theta=0} = \frac{2g}{a}\cdot\frac{\sin\alpha}{\frac{1}{3}+1} = \frac{3g\sin\alpha}{2a}.$$

Thus

$$\dot{\theta}\big|_{\theta=0} = \sqrt{\frac{3g\sin\alpha}{2a}}.$$

Using (4.10.35) in (4.10.34) we get,

$$\begin{aligned}
\left(\frac{1}{3}+\cos^2\theta\right)\ddot{\theta} &= \sin\theta\cos\theta\cdot\dot{\theta}^2-\frac{g}{a}\cos\theta\\
&= -\frac{g}{a}\cos\theta+\sin\theta\cos\theta\cdot\frac{2g}{a}\frac{(\sin\alpha-\sin\theta)}{\left(\frac{1}{3}+\cos^2\theta\right)}\\
&= \frac{g}{a}\left[\frac{2\sin\theta\cos\theta(\sin\alpha-\sin\theta)-\cos\theta\left(\frac{1}{3}+\cos^2\theta\right)}{\left(\frac{1}{3}+\cos^2\theta\right)}\right]\\
\text{or,}\quad \ddot{\theta} &= \frac{g}{a}\left[\frac{2\sin\theta\cos\theta(\sin\alpha-\sin\theta)-\cos\theta\left(\frac{1}{3}+\cos^2\theta\right)}{\left(\frac{1}{3}+\cos^2\theta\right)^2}\right].
\end{aligned} \tag{4.10.36}$$

Now from (4.10.31), we get

$$\begin{aligned}
R &= M(\ddot{y}+g)=M\left[-a\sin\theta\cdot\dot{\theta}^2+a\cos\theta\cdot\ddot{\theta}+g\right]\\
&= M\Bigg[-\sin\theta\cdot\frac{2g(\sin\alpha-\sin\theta)}{\left(\frac{1}{3}+\cos^2\theta\right)}\\
&\qquad+\cos\theta\cdot g\left\{\frac{2\sin\theta\cos\theta(\sin\alpha-\sin\theta)-\cos\theta\left(\frac{1}{3}+\cos^2\theta\right)}{\left(\frac{1}{3}+\cos^2\theta\right)^2}\right\}+g\Bigg]\\
&= \frac{Mg}{\left(\frac{1}{3}+\cos^2\theta\right)^2}\Bigg[-2\sin\theta(\sin\alpha-\sin\theta)\left(\frac{1}{3}+\cos^2\theta\right)\\
&\qquad+2\sin\theta\cos^2\theta(\sin\alpha-\sin\theta)\\
&\qquad-\cos^2\theta\left(\frac{1}{3}+\cos^2\theta\right)+\left(\cos^2\theta+\frac{1}{3}\right)^2\Bigg]\\
&= \frac{Mg}{\left(\frac{1}{3}+\cos^2\theta\right)^2}\Bigg[-2\sin\theta\Bigg(\frac{1}{3}\sin\alpha+\sin\alpha\cos^2\theta\\
&\qquad-\frac{1}{3}\sin\theta-\sin\theta\cos^2\theta+2\sin\theta\cos^2\theta\sin\alpha\\
&\qquad-2\sin^2\theta\cos^2\theta-\frac{1}{3}\cos^2\theta-\cos^4\theta+\cos^4\theta+\frac{2}{3}\cos^2\theta+\frac{1}{9}\Bigg)\Bigg]\\
&= \frac{Mg}{\left(\frac{1}{3}+\cos^2\theta\right)^2}\Bigg[-\frac{2}{3}\sin\alpha\sin\theta-2\sin\theta\cos^2\theta\sin\alpha+\frac{2}{3}\sin^2\theta\\
&\qquad+2\sin^2\theta\cos^2\theta+2\sin\theta\cos^2\theta\sin\alpha-2\sin^2\theta\cos^2\theta\\
&\qquad-\frac{1}{3}\cos^2\theta+\frac{2}{3}\cos^2\theta+\frac{1}{9}\Bigg]\\
&= \frac{Mg}{9\left(\frac{1}{3}+\cos^2\theta\right)^2}\left[-6\sin\theta\sin\alpha+6\sin^2\theta-3\cos^2\theta+6\cos^2\theta+1\right]
\end{aligned}$$

$$= \frac{Mg}{9\left(\frac{1}{3}+\cos^2\theta\right)^2}\left[-6\sin\theta\sin\alpha - 3\cos^2\theta + 7\right]$$

$$= \frac{Mg}{9\left(\frac{1}{3}+\cos^2\theta\right)^2}\left[-6\sin\theta\sin\alpha - 3 + 3\sin^2\theta + 7\right]$$

$$= \frac{Mg}{9\left(\frac{1}{3}+\cos^2\theta\right)^2}\left[3\left(\sin^2\theta - 2\sin\theta\sin\alpha + \sin^2\alpha\right) + 4 - 3\sin^2\alpha\right]$$

$$= \frac{Mg}{9\left(\frac{1}{3}+\cos^2\theta\right)^2}\left[3\left(\sin\theta - \sin\alpha\right)^2 + 1 + 3\cos^2\alpha\right] > 0.$$

As the normal reaction R always > 0, then the rod will not leave the plane.

Case II. *When the plane be perfectly rough.*

As the plane be perfectly rough, the end in contact with the plane cannot slide. Let R' and F be the normal reaction and the frictional force at the point A, in contact with the plane. Let $G\,(\bar{x}, \bar{y})$ be the centre of gravity of the rod referred to Ax and Ay as axes at any time. And let the rod makes an angle θ with the horizontal plane.

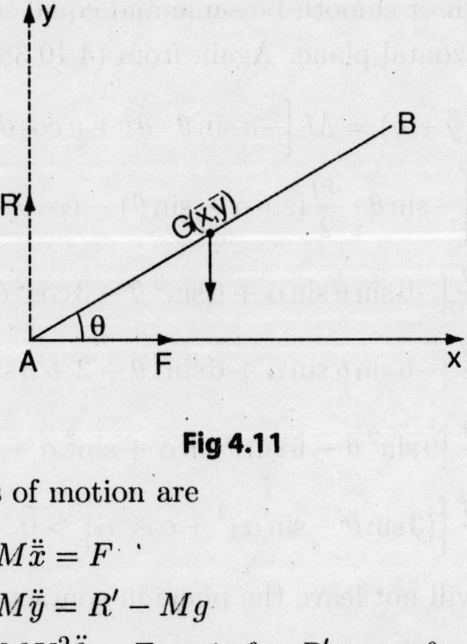

Fig 4.11

Thus the equations of motion are

$$M\ddot{\bar{x}} = F \tag{4.10.37}$$

$$M\ddot{\bar{y}} = R' - Mg \tag{4.10.38}$$

$$\text{and} \quad MK^2\ddot{\theta} = F\cdot a\sin\theta - R'\cdot a\cos\theta \tag{4.10.39}$$

$$\text{or,} \quad MK^2\ddot{\theta} = M\left[\ddot{\bar{x}}\cdot a\sin\theta - (\ddot{\bar{y}} + g)\,a\cos\theta\right]$$

[by (4.10.37) and (4.10.38)]

Now

$$\bar{x} = a\cos\theta;\quad \dot{\bar{x}} = -a\sin\theta\cdot\dot{\theta};\quad \ddot{\bar{x}} = -a\cos\theta\cdot\dot{\theta}^2 - a\sin\theta\cdot\ddot{\theta}$$

$$\bar{y} = a\sin\theta;\quad \dot{\bar{y}} = a\cos\theta\cdot\dot{\theta};\quad \ddot{\bar{y}} = -a\sin\theta\cdot\dot{\theta}^2 + a\cos\theta\cdot\ddot{\theta}.$$

$$\therefore \quad K^2\ddot{\theta} = \left[\left(-a\cos\theta\,;\dot{\theta}^2 - a\sin\theta\cdot\ddot{\theta}\right)a\sin\theta\right.$$
$$\left.-\left(-a\sin\theta\cdot\dot{\theta}^2 + a\cos\theta\cdot\ddot{\theta} + g\right)a\cos\theta\right],$$

$$\text{or,}\quad \frac{a^2}{3}\ddot{\theta} = -a^2\ddot{\theta} - ga\cos\theta, \quad \text{or,}\quad \left(\frac{a^2}{3} + a^2\right)\ddot{\theta} = -ga\cos\theta,$$

$$\text{or,}\quad \ddot{\theta} = -\frac{3g}{4a}\cos\theta, \quad \text{or,}\quad 2\dot{\theta}\ddot{\theta} = -\frac{3g}{2a}\cos\theta\cdot\dot{\theta}. \tag{4.10.40}$$

Integrating, we get

$$\dot{\theta}^2 = -\frac{3g}{2a}\sin\theta + C_2$$

Initially $\dot{\theta} = 0$ when $\theta = \alpha$. $\quad\therefore\ C_2 = \frac{3g}{2a}\sin\alpha$.

$$\therefore \quad \dot{\theta}^2 = \frac{3g}{2a}(\sin\alpha - \sin\theta). \tag{4.10.41}$$

$$\therefore \quad \dot{\theta}^2\Big|_{\theta=0} = \frac{3g}{2a}\sin\alpha, \quad \text{or,}\quad \dot{\theta}\Big|_{\theta=0} = \sqrt{\frac{3g}{2a}\sin\alpha}.$$

Thus the angular velocity of the rod, for both the cases, i.e., when the plane perfectly rough or smooth be same and equal to $\sqrt{\frac{3g}{2a}\sin\alpha}$, when the rod becomes to horizontal plane. Again from (4.10.38)

$$\begin{aligned}
R' &= M(\ddot{y} + g) = M\left[-a\sin\theta\cdot\dot{\theta}^2 + a\cos\theta\cdot\ddot{\theta} + g\right] \\
&= M\left[-\sin\theta\cdot\frac{3g}{2}(\sin\alpha - \sin\theta) - \cos\theta\cdot\frac{3g}{4}\cos\theta + g\right] \\
&= \frac{Mg}{4}\left[-6\sin\theta\sin\alpha + 6\sin^2\theta - 3\cos^2\theta + 4\right] \\
&= \frac{Mg}{4}\left[-6\sin\theta\sin\alpha + 6\sin^2\theta - 3 + 3\sin^2\theta + 4\right] \\
&= \frac{Mg}{4}\left[9\sin^2\theta - 6\sin\theta\sin\alpha + \sin^2\alpha + 1 - \sin^2\alpha\right] \\
&= \frac{Mg}{4}\left[(3\sin\theta - \sin\alpha)^2 + \cos^2\alpha\right] > 0.
\end{aligned}$$

Hence, the rod will not leave the plane in contact in either of the cases.

▶ **Problem 4.10.5** *A rough uniform rod of length 2a, is placed on a rough table at right angle to its edge; if its centre of gravity be initially at a distance b beyond the edge, show that the rod will begin to slide when it has turned through an angle*

$$\tan^{-1}\frac{\mu a^2}{a^2 + 9b^2},$$

where μ is the coefficient of friction.

Solution: We take the x and y axes as shown in the Fig. 4.12. Let $AB(=2a)$ be the rod. Let R and F are the normal reaction and frictional force at the point of contact O. Let $(\bar{x}, \bar{y})$ be the coordinates of the c.g. G, of the rod referred to Ox and Oy as axes.

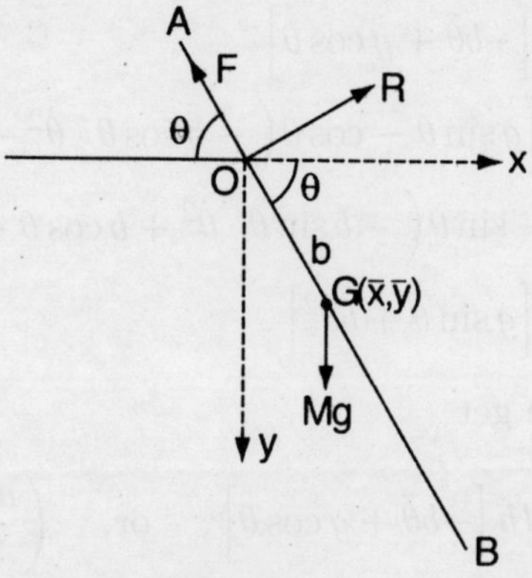

Fig 4.12

Therefore the equations of motion are

$$M\ddot{\bar{x}} = R\sin\theta - F\cos\theta \qquad (4.10.42)$$

$$M\ddot{\bar{y}} = -F\sin\theta - R\cos\theta + Mg \qquad (4.10.43)$$

$$\text{and} \quad MK^2\ddot{\theta} = R \cdot b \qquad (4.10.44)$$

And

$$\bar{x} = b\cos\theta; \qquad \bar{y} = b\sin\theta,$$

$$\text{or,} \quad \dot{\bar{x}} = -b\sin\theta\cdot\dot{\theta}\,; \quad \dot{\bar{y}} = b\cos\theta\cdot\dot{\theta}.$$

$$\left.\begin{aligned} \ddot{\bar{x}} &= -b\cos\theta\cdot\dot{\theta}^2 - b\sin\theta\cdot\ddot{\theta}\,; \\ \ddot{\bar{y}} &= -b\sin\theta\cdot\dot{\theta}^2 + b\cos\theta\cdot\ddot{\theta}. \end{aligned}\right\}. \qquad (4.10.45)$$

Now multiplying (4.10.42) by $\sin\theta$ and (4.10.43) by $\cos\theta$ and then subtracting we get

$$R - Mg\cos\theta = M\ddot{\bar{x}}\sin\theta - M\ddot{\bar{y}}\cos\theta,$$

$$\text{or,} \quad R = M\left[\ddot{\bar{x}}\sin\theta - \ddot{\bar{y}}\cos\theta + g\cos\theta\right]. \qquad (4.10.46)$$

Again multiplying (4.10.42) by $\cos\theta$ and (4.10.43) by $\sin\theta$ and then adding we get

$$\begin{aligned} F &= Mg\sin\theta - M\ddot{\bar{x}}\cos\theta - M\ddot{\bar{y}}\sin\theta \\ &= M\left[g\sin\theta - \ddot{\bar{x}}\cos\theta - \ddot{\bar{y}}\sin\theta\right]. \end{aligned} \qquad (4.10.47)$$

Thus

$$\begin{aligned} R &= M\Big[\sin\theta\Big(-b\cos\theta\cdot\dot\theta^2 - b\sin\theta\cdot\ddot\theta\Big) \\ &\qquad -\cos\theta\Big(-b\sin\theta\cdot\dot\theta^2 + b\cos\theta\cdot\ddot\theta\Big) + g\cos\theta\Big] \\ &= M\left[-b\ddot\theta + g\cos\theta\right] \\ F &= M\Big[g\sin\theta - \cos\theta\Big(-b\cos\theta\cdot\dot\theta^2 - b\sin\theta\cdot\ddot\theta\Big) \\ &\qquad -\sin\theta\Big(-b\sin\theta\cdot\dot\theta^2 + b\cos\theta\cdot\ddot\theta\Big)\Big] \\ &= M\left[g\sin\theta + b\dot\theta^2\right]. \end{aligned}$$

From (4.10.44), we get

$$MK^2\ddot\theta = Mb\left[-b\ddot\theta + g\cos\theta\right], \quad \text{or,} \quad \left(\frac{a^2}{3} + b^2\right)\ddot\theta = bg\cos\theta,$$

$$\text{or,} \quad \ddot\theta = \frac{3bg}{a^2+3b^2}\cos\theta, \quad \text{or,} \quad 2\dot\theta\ddot\theta = \frac{6bg}{a^2+3b^2}\cos\theta\cdot\dot\theta. \tag{4.10.48}$$

Integrating

$$\dot\theta^2 = \frac{6bg}{a^2+3b^2}\sin\theta + C_1.$$

Initially $\dot\theta = 0$ when $\theta = 0$. $\quad\therefore\ C_1 = 0$.
Thus

$$\dot\theta^2 = \frac{6bg}{a^2+3b^2}\sin\theta. \tag{4.10.49}$$

Therefore

$$\begin{aligned} R &= M\left[-b\cdot\frac{3bg}{a^2+3b^2}\cos\theta + g\cos\theta\right] = Mg\left[1 - \frac{3b^2}{a^2+3b^2}\right]\cos\theta \\ &= Mg\cdot\frac{a^2}{a^2+3b^2}\cos\theta \end{aligned} \tag{4.10.50}$$

and

$$\begin{aligned} F &= M\left[g\sin\theta + b\cdot\frac{6gb}{a^2+3b^2}\sin\theta\right] = Mg\left[1 + \frac{6b^2}{a^2+3b^2}\right]\sin\theta \\ &= Mg\cdot\frac{a^2+9b^2}{a^2+3b^2}\sin\theta. \end{aligned} \tag{4.10.51}$$

The rod will begin to slide when the coefficient of friction

$$\mu = \frac{F}{R} = \frac{a^2+9b^2}{a^2}\tan\theta, \quad \text{or,} \quad \tan\theta = \frac{\mu a^2}{a^2+9b^2},$$

$$\text{or,} \quad \theta = \tan^{-1}\left(\frac{\mu a^2}{a^2+9b^2}\right).$$

► **Problem 4.10.6** *A uniform rod is held one end in contact with a rough horizontal table at an inclination α to the horizon and then released. If μ be the coefficient of friction, show that it will commence to slide if*

$$\mu < \frac{3\sin\alpha\cos\alpha}{1+3\sin^2\alpha}, \quad or, \quad \mu < \frac{3\tan\alpha}{1+4\tan^2\alpha}.$$

[CUH 1994, 96, 99, 2002, 2007; VUH 2002, 2005]

Solution: Let $AB(=2a)$ be the rod of mass M and G be its centric of gravity. Let $(\bar{x}, \bar{y})$ be the coordinates of c.g. referred to Ax and Ay as axes. Let R and F are the normal reaction and frictional force at the end A. Let θ be the inclination of the rod at any time t.

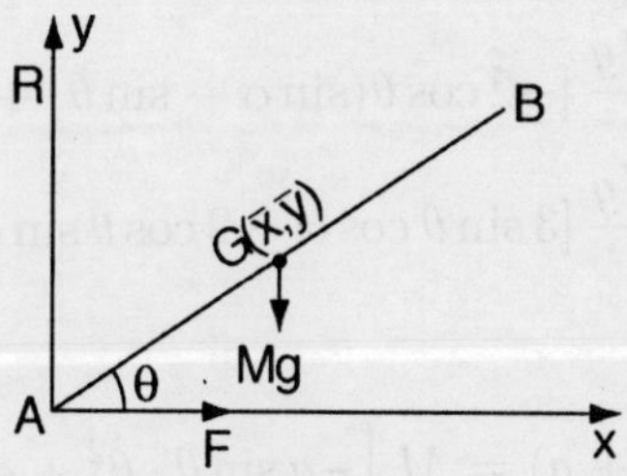

Fig 4.13

The equations of motion are

$$M\ddot{\bar{x}} = F \tag{4.10.52}$$

$$M\ddot{\bar{y}} = R - Mg \tag{4.10.53}$$

$$\text{and} \quad MK^2\ddot{\theta} = F\cdot a\sin\theta - R\cdot a\cos\theta \tag{4.10.54}$$

where

$$\left.\begin{aligned} &\bar{x} = a\cos\theta; \quad \dot{\bar{x}} = -a\sin\theta\cdot\dot{\theta}; \quad \ddot{\bar{x}} = -a\cos\theta\cdot\dot{\theta}^2 - a\sin\theta\cdot\ddot{\theta} \\ &\bar{y} = a\sin\theta; \quad \dot{\bar{y}} = a\cos\theta\cdot\dot{\theta}; \quad \ddot{\bar{y}} = -a\sin\theta\cdot\dot{\theta}^2 + a\cos\theta\cdot\ddot{\theta} \end{aligned}\right\}. \tag{4.10.55}$$

From (4.10.54), using (4.10.52) and (4.10.53) we get,

$$MK^2\ddot{\theta} = M\left[a\sin\theta\cdot\ddot{\bar{x}} - (\ddot{\bar{y}}+g)\,a\cos\theta\right],$$

$$\text{or,} \quad K^2\ddot{\theta} = \Big[a\sin\theta\cdot\Big(-a\cos\theta\cdot\dot{\theta}^2 - a\sin\theta\cdot\ddot{\theta}\Big)$$

$$-a\cos\theta\Big(-a\sin\theta\cdot\dot{\theta}^2 + a\cos\theta\cdot\ddot{\theta}\Big) - ag\cos\theta\Big]$$

$$= -a^2\ddot{\theta} - ag\cos\theta,$$

$$\text{or,} \quad \frac{a^2}{3}\ddot{\theta} + a^2\ddot{\theta} = -ag\cos\theta, \quad \text{or,} \quad \frac{4a^2}{3}\ddot{\theta} = -ag\cos\theta$$

$$\text{or,} \quad \ddot{\theta} = -\frac{3g}{4a}\cos\theta, \quad \text{or,} \quad 2\dot{\theta}\ddot{\theta} = -\frac{3g}{2a}\cos\theta\cdot\dot{\theta}. \tag{4.10.56}$$

Integrating

$$\dot{\theta}^2 = -\frac{3g}{2a}\sin\theta + C_1.$$

Initially $\dot{\theta} = 0$, when $\theta = \alpha$. $\therefore\ C_1 = \frac{3g}{2a}\sin\alpha$.
Thus

$$\dot{\theta}^2 = \frac{3g}{2a}(\sin\alpha - \sin\theta). \tag{4.10.57}$$

From (4.10.52)

$$\begin{aligned}
\therefore\ F &= M\ddot{x} = M\left(-a\cos\theta\cdot\dot{\theta}^2 - a\sin\theta\cdot\ddot{\theta}\right)\\
&= M\left[-\cos\theta\cdot\frac{3g}{2}(\sin\alpha - \sin\theta) + \sin\theta\cdot\frac{3g}{4}\cos\theta\right]\\
&= \frac{3Mg}{4}\left[-2\cos\theta(\sin\alpha - \sin\theta) + \sin\theta\cos\theta\right]\\
&= \frac{3Mg}{4}\left[3\sin\theta\cos\theta - 2\cos\theta\sin\alpha\right].
\end{aligned}$$

From (4.10.53)

$$\begin{aligned}
\therefore\ F &= M(\ddot{y} + g) = M\left[-a\sin\theta\cdot\dot{\theta}^2 + a\cos\theta\cdot\ddot{\theta} + g\right]\\
&= M\left[-\sin\theta\cdot\frac{3g}{2}(\sin\alpha - \sin\theta) - \cos\theta\cdot\frac{3g}{4}\cos\theta + g\right]\\
&= \frac{Mg}{4}\left[-6\sin\theta(\sin\alpha - \sin\theta) - 3\cos^2\theta + 4\right]\\
&= \frac{Mg}{4}\left[6\sin^2\theta - 6\sin\theta\sin\alpha - 3 + 3\sin^2\theta + 4\right]\\
&= \frac{Mg}{4}\left[9\sin^2\theta - 6\sin\theta\sin\alpha + 1\right].
\end{aligned}$$

When $\theta = \alpha$,

$$F = \frac{3Mg}{4}\sin\alpha\cos\alpha \quad \text{and} \quad R = \frac{Mg}{4}\left(1 + 3\sin^2\alpha\right).$$

Thus for sliding

$$\begin{aligned}
\mu < \left.\frac{F}{R}\right|_{\theta=\alpha}, \quad \text{or,} \quad \mu &< \frac{\frac{3Mg}{4}\sin\alpha\cos\alpha}{\frac{Mg}{4}(1 + 3\sin^2\alpha)} = \frac{3\sin\alpha\cos\alpha}{1 + 3\sin^2\alpha}\\
&= \frac{3\tan\alpha}{\sec^2\alpha + 3\tan^2\alpha} = \frac{3\tan\alpha}{1 + 4\tan^2\alpha}.
\end{aligned}$$

► **Problem 4.10.7** *A uniform rod of length 2a is placed on a rough table at right angle to its edge. If its centre of gravity be initially at a distance a/3 beyond the edge, show that the rod will begin to slide when it has turned through an angle* $\tan^{-1}(\mu/2)$, *μ being the coefficient of friction.*

[CUH 1997]

Solution: Let $AB(=2a)$ be the rod of mass M and G be the c.g. of the rod. We take the x and y axes as shown in the Fig 4.14. Let $(\bar{x}, \bar{y})$ be the coordinates of c.g. G referred to axes Ox and Oy. Let the rod makes an angle θ with the horizon at any instant. Let R and F are the normal reaction and frictional force at the point of contact O.

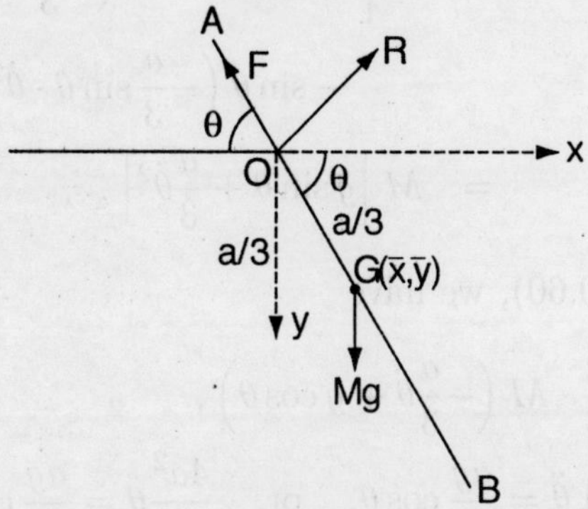

Fig 4.14

The equations of motion are

$$M\ddot{\bar{x}} = R\sin\theta - F\cos\theta \tag{4.10.58}$$

$$M\ddot{\bar{y}} = Mg - R\cos\theta - F\sin\theta \tag{4.10.59}$$

$$MK^2\ddot{\theta} = R\cdot\frac{a}{3} \tag{4.10.60}$$

and

$$\left.\begin{aligned} \bar{x} &= \tfrac{a}{3}\cos\theta; \quad \dot{\bar{x}} = -\tfrac{a}{3}\sin\theta\cdot\dot{\theta}\,; \quad \ddot{\bar{x}} = -\tfrac{a}{3}\cos\theta\cdot\dot{\theta} - \tfrac{a}{3}\sin\theta\cdot\ddot{\theta} \\ \bar{y} &= \tfrac{a}{3}\sin\theta; \quad \dot{\bar{y}} = \tfrac{a}{3}\cos\theta\cdot\dot{\theta}\,; \quad \ddot{\bar{y}} = -\tfrac{a}{3}\sin\theta\cdot\dot{\theta}^2 + \tfrac{a}{3}\cos\theta\cdot\ddot{\theta} \end{aligned}\right\}. \tag{4.10.61}$$

Multiplying (4.10.58) by $\sin\theta$ and (4.10.59) by $\cos\theta$, then subtracting we get,

$$\begin{aligned} R &= M\ddot{\bar{x}}\sin\theta - M\ddot{\bar{y}}\cos\theta + Mg\cos\theta \\ &= M\Bigg[\sin\theta\left(-\frac{a}{3}\cos\theta\cdot\dot{\theta}^2 - \frac{a}{3}\sin\theta\cdot\ddot{\theta}\right) \\ &\qquad -\cos\theta\left(-\frac{a}{3}\sin\theta\cdot\dot{\theta}^2 + \frac{a}{3}\cos\theta\cdot\ddot{\theta}\right) + g\cos\theta\Bigg] \\ &= M\left[-\frac{a}{3}\ddot{\theta} + g\cos\theta\right]. \end{aligned} \tag{4.10.62}$$

Multiplying (4.10.58) by $\cos\theta$ and (4.10.59) by $\sin\theta$ and then adding we get

$$\begin{aligned} M\ddot{x}\cos\theta + M\ddot{y}\sin\theta &= -F + Mg\sin\theta \\ \text{or,}\quad F &= M\left[g\sin\theta - \ddot{x}\cos\theta - \ddot{y}\sin\theta\right] \\ &= M\left[g\sin\theta - \cos\theta\left(-\frac{a}{3}\cos\theta\cdot\dot{\theta}^2 - \frac{a}{3}\sin\theta\cdot\ddot{\theta}\right)\right. \\ &\qquad \left. -\sin\theta\left(-\frac{a}{3}\sin\theta\cdot\dot{\theta}^2 + \frac{a}{3}\cos\theta\cdot\ddot{\theta}\right)\right] \\ &= M\left[g\sin\theta + \frac{a}{3}\dot{\theta}^2\right]. \end{aligned}$$

Thus from (4.10.60), we have

$$MK^2\ddot{\theta} = \frac{a}{3}\cdot M\left(-\frac{a}{3}\ddot{\theta} + g\cos\theta\right),$$

$$\text{or,}\quad \left(\frac{a^2}{3} + \frac{a^2}{9}\right)\ddot{\theta} = \frac{ag}{3}\cos\theta,\quad \text{or,}\quad \frac{4a^2}{9}\ddot{\theta} = \frac{ag}{3}\cos\theta,\quad \left[\because\ K^2 = \frac{a^2}{3}\right]$$

$$\text{or,}\quad \ddot{\theta} = \frac{3g}{4a}\cos\theta,$$

$$\text{or,}\quad 2\dot{\theta}\ddot{\theta} = \frac{3g}{2a}\cos\theta\cdot\dot{\theta}. \tag{4.10.63}$$

Integrating

$$\dot{\theta}^2 = \frac{3g}{2a}\sin\theta + C_1.$$

Initially $\dot{\theta} = 0$ when $\theta = 0$. $\therefore\ C_1 = 0$.
Thus

$$\dot{\theta}^2 = \frac{3g}{2a}\sin\theta. \tag{4.10.64}$$

Therefore

$$R = M\left[-\frac{a}{3}\cdot\frac{3g}{4a}\cos\theta + g\cos\theta\right] = \frac{3Mg}{4}\cos\theta$$

and

$$F = M\left[g\sin\theta + \frac{a}{3}\cdot\frac{3g}{2a}\sin\theta\right] = \frac{Mg}{2}\left[3\sin\theta\right] = \frac{3Mg}{2}\sin\theta.$$

The sliding will begin, when

$$\mu = \frac{F}{R} = \frac{\frac{3Mg}{2}\sin\theta}{\frac{3Mg}{4}\cos\theta} = 2\tan\theta,\quad \text{or,}\quad \theta = \tan^{-1}(\mu/2).$$

▶ **Problem 4.10.8** *A uniform rod of length 2a is placed with one end in contact with a smooth horizontal table and is then allowed to fall; if α be its* *initial inclination to the vertical, show that its angular velocity when it is inclined at an angle θ is*

$$\left\{\frac{6g}{a}\cdot\frac{\cos\alpha-\cos\theta}{1+3\sin^2\theta}\right\}^{1/2}.$$

Find also the reaction of the table.

[CUH 2000, B.H.U. 2003; VUH 2005]

Solution: Let $AB(=2a)$ be the rod of mass M and G be its centre of gravity. Let $(\bar{x},\bar{y})$ be the coordinates of c.g. G of the rod, referred to Ax and Ay as axes. Let R be the reaction at the end A and the rod makes an angle θ with the vertical at any time t.

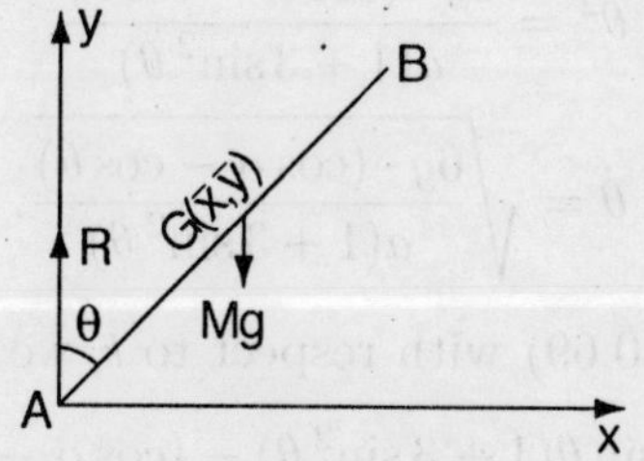

Fig 4.15

The equations of motion are

$$M\ddot{\bar{x}}=0 \tag{4.10.65}$$
$$M\ddot{\bar{y}}=R-Mg \tag{4.10.66}$$
$$\text{and}\quad MK^2\ddot{\theta}=R\cdot a\sin\theta, \tag{4.10.67}$$

where

$$\bar{x}=a\sin\theta;\quad \dot{\bar{x}}=a\cos\theta\cdot\dot{\theta};\quad \ddot{\bar{x}}=-a\sin\theta\cdot\dot{\theta}^2+a\cos\theta\cdot\ddot{\theta}$$
$$\bar{y}=a\cos\theta;\quad \dot{\bar{y}}=-a\sin\theta\cdot\dot{\theta};\quad \ddot{\bar{y}}=-a\cos\theta\cdot\dot{\theta}^2-a\sin\theta\cdot\ddot{\theta}.$$

From (4.10.66), we get

$$R=M\left[\ddot{\bar{y}}+g\right]=M\left[-a\cos\theta\cdot\dot{\theta}^2-a\sin\theta\,\ddot{\theta}+g\right]. \tag{4.10.68}$$

From (4.10.67) we get

$$MK^2\ddot{\theta}=M\left[a\sin\theta\left(-a\cos\theta\cdot\dot{\theta}^2-a\sin\theta\cdot\ddot{\theta}+g\right)\right],$$
$$\text{or,}\quad K^2\ddot{\theta}=-a^2\sin\theta\cos\theta\cdot\dot{\theta}^2-a^2\sin^2\theta\cdot\ddot{\theta}+ag\sin\theta,$$
$$\text{or,}\quad \frac{a^2}{3}\ddot{\theta}+a^2\sin^2\theta\cdot\ddot{\theta}+a^2\sin\theta\cos\theta\cdot\dot{\theta}^2=ag\sin\theta,$$

$$\text{or,}\quad \left(\frac{1}{3}+\sin^2\theta\right)\ddot{\theta}+2\sin\theta\cos\theta\cdot\frac{\dot{\theta}^2}{2}=\frac{g}{a}\sin\theta,$$

$$\text{or,}\quad \frac{d}{d\theta}\left[\left(\frac{1}{3}+\sin^2\theta\right)\frac{\dot{\theta}^2}{2}\right]=\frac{g}{a}\sin C.$$

Integrating

$$\left(\frac{1}{3}+\sin^2\theta\right)\frac{\dot{\theta}^2}{2}=-\frac{g}{a}\cos\theta+C_1.$$

Initially $\dot{\theta}=0$ when $\theta=\alpha$. $\therefore\ C=\frac{g}{a}\cos\alpha$.

$$\therefore\quad \left(\frac{1}{3}+\sin^2\theta\right)\frac{\dot{\theta}^2}{2}=\frac{g}{a}(\cos\alpha-\cos\theta),$$

$$\text{or,}\quad \dot{\theta}^2=\frac{6g\cdot(\cos\alpha-\cos\theta)}{a(1+3\sin^2\theta)}, \tag{4.10.69}$$

$$\text{or,}\quad \dot{\theta}=\sqrt{\frac{6g\cdot(\cos\alpha-\cos\theta)}{a(1+3\sin^2\theta)}}.$$

Differentiating (4.10.69) with respect to t, we get

$$2\dot{\theta}\ddot{\theta}=\frac{6g}{a}\left[\frac{\sin\theta(1+3\sin^2\theta)-(\cos\alpha-\cos\theta)6\sin\theta\cos\theta}{(1+3\sin^2\theta)^2}\right]\dot{\theta},$$

$$\text{or,}\quad \ddot{\theta}=\frac{3g}{a}\left[\frac{1+3\sin^2\theta-6\cos\theta(\cos\alpha-\cos\theta)}{(1+3\sin^2\theta)^2}\right]\sin\theta. \tag{4.10.70}$$

Substituting (4.10.69) and (4.10.70) in (4.10.68) we will get the normal reaction R at A.

▶ **Problem 4.10.9** *The lower end of an uniform rod, inclined at an angle α to the horizon, is placed on a smooth table. A horizontal force is applied to its lower end of such a magnitude that the rod rotates in a vertical plane with constant angular velocity ω. Show that when the rod inclined at an angle θ to the horizon the magnitude of the force is $Mg\cot\theta - Ma\omega^2\cos\theta$; where M is the mass of the rod.*

Solution: Let $AB(=2a)$ be the uniform rod of mass M and G be its c.g. Let $(\bar{x},\bar{y})$ be the coordinates of the c.g. G of the rod referred to the axes Ax and Ay at the end A in contact with the table. Let ϕ be the angle made by the rod with the table at any instant. Let R be the normal reaction and F be the applied force at A. The equations of motion are

$$M\ddot{\bar{x}}=F \tag{4.10.71}$$

$$M\ddot{\bar{y}}=R-Mg \tag{4.10.72}$$

$$\text{and}\quad MK^2\ddot{\phi}=F\cdot a\sin\phi-R\cdot a\cos\phi. \tag{4.10.73}$$

As $\dot{\phi} =$ constant $= \omega$ (say). $\quad \therefore \ \ddot{\phi} = 0.$

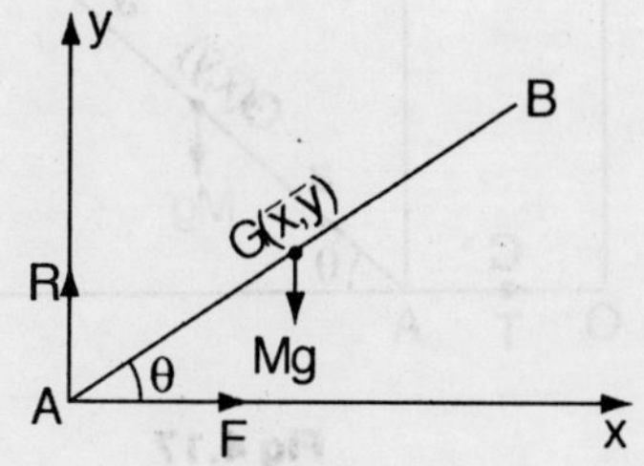

Fig 4.16

Thus from (4.10.73), we get

$$R = F \tan \phi. \tag{4.10.74}$$

Now

$$\bar{x} = a \cos \phi, \qquad \bar{y} = a \sin \phi,$$

$$\dot{\bar{y}} = a \cos \phi \cdot \dot{\phi}; \quad \ddot{\bar{y}} = -a \sin \phi \cdot \dot{\phi}^2 + a \cos \phi \cdot \ddot{\phi} = -a \sin \phi \cdot \omega^2$$

[as $\dot{\phi} = \omega$ and $\ddot{\phi} = 0$]

Using (4.10.74) we get from (4.10.72)

$$\begin{aligned} R &= F \tan \phi = M\left(\ddot{\bar{y}} + g\right) = M\left(-a \sin \phi \cdot \omega^2\right) + Mg, \\ \text{or,} \quad F &= Mg \cot \phi - Ma\omega^2 \cos \phi. \end{aligned}$$

$$\therefore \quad F\Big|_{\phi=\theta} = Mg \cot \theta - Ma\omega^2 \cos \theta.$$

▶ **Problem 4.10.10** *A uniform rod, of length 2a, is held with one end on a smooth horizontal plane, this end being attached by a light inextensible string to a point in the plane, the string is tight and in the same vertical plane as the rod and makes with it an acute angle α. If the rod be now allowed to fall under the action of gravity, find its inclination to the horizon when the string ceases to tight, and show that its angular velocity Ω just before it becomes horizontal is given by* $6a\Omega^2 = g \sin \alpha (8 + \cos^2 \alpha)$.

Solution: Let $AB(= 2a)$ be the rod of mass M and G be its c.g. Let $(\bar{x}, \bar{y})$ be the coordinates of c.g. G, referred to Ox and Oy as axis. Let OA be string of length c and T be the tension in the string. Let θ be the inclination of the rod with the plane at any time t and R be the normal reaction at A.

$$\therefore \quad \bar{x} = c + a \cos \theta.$$

$$\therefore \quad \dot{\bar{x}} = -a \sin \theta \cdot \dot{\theta}, \quad \ddot{\bar{x}} = -a \cos \theta \cdot \dot{\theta}^2 - a \sin \theta \cdot \ddot{\theta}$$

$$\text{and} \quad \bar{y} = a \sin \theta, \qquad \dot{\bar{y}} = a \cos \theta \cdot \dot{\theta},$$

$$\ddot{\bar{y}} = -a \sin \theta \cdot \dot{\theta}^2 + a \cos \theta \cdot \ddot{\theta}.$$

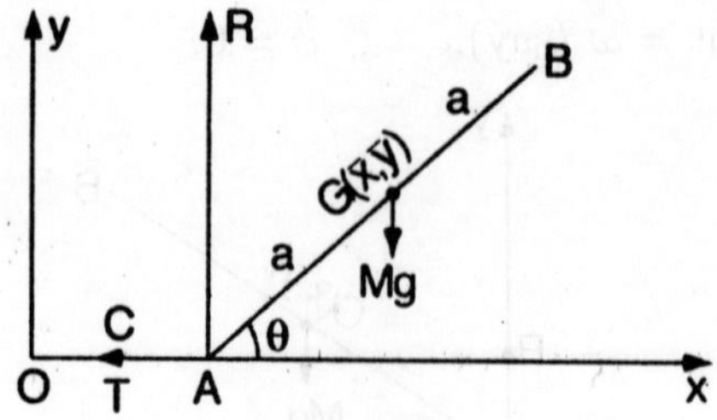

Fig 4.17

The equations of motion are

$$M\ddot{x} = -T \tag{4.10.75}$$

$$M\ddot{y} = R - Mg \tag{4.10.76}$$

$$\text{and} \quad MK^2\ddot{\theta} = -T \cdot a\sin\theta - R \cdot a\cos\theta. \tag{4.10.77}$$

Using (4.10.75) and (4.10.76) in (4.10.77) we get,

$$MK^2\ddot{\theta} = a\sin\theta \cdot M\ddot{x} - a\cos\theta\,(M\ddot{y} + Mg)\,,$$

$$\text{or,} \quad K^2\ddot{\theta} = a\sin\theta\big(-a\cos\theta \cdot \dot{\theta}^2 - a\sin\theta \cdot \ddot{\theta}\big)$$

$$-a\cos\theta\left(-a\sin\theta \cdot \dot{\theta}^2 + a\cos\theta \cdot \ddot{\theta}\right) - ag\cos\theta,$$

$$\text{or,} \quad \frac{a^2}{3}\ddot{\theta} = -a^2\ddot{\theta} - ag\cos\theta \quad \left[\text{as } K^2 = \frac{a^2}{3}\right]$$

$$\left(\frac{a^2}{3} + a^2\right)\ddot{\theta} = -ag\cos\theta, \quad \text{or,} \quad \ddot{\theta} = -\frac{3g}{4a}\cos\theta, \tag{4.10.78}$$

$$\text{or,} \quad 2\dot{\theta}\ddot{\theta} = -\frac{3g}{2a}\cos\theta \cdot \dot{\theta}.$$

Integrating

$$\dot{\theta}^2 = -\frac{3g}{2a}\sin\theta + C.$$

Initially $\dot{\theta} = 0$ when $\theta = \alpha$. $\therefore\ C = \frac{3g}{2a}\sin\alpha$.

$$\therefore \quad \dot{\theta}^2 = \frac{3g}{2a}(\sin\alpha - \sin\theta). \tag{4.10.79}$$

Thus from (4.10.75),

$$\begin{aligned} -T &= M\ddot{x} = M\left(-a\cos\theta \cdot \dot{\theta}^2 - a\sin\theta \cdot \ddot{\theta}\right) \\ &= M\left[-a\cos\theta \cdot \frac{3g}{2a}(\sin\alpha - \sin\theta) + a\sin\theta \cdot \frac{3g}{4a}\cos\theta\right] \\ &= \frac{3Mg}{4}\left[-2\sin\alpha\cos\theta + 2\sin\theta\cos\theta + \sin\theta\cos\theta\right] \\ &= \frac{3Mg}{4}\left[3\sin\theta - 2\sin\alpha\right]\cos\theta. \end{aligned}$$

When $T = 0$, i.e., when the string cases to tight,

$$\sin\theta = \frac{2}{3}\sin\alpha \quad \text{as } \cos\theta \neq 0 \tag{4.10.80}$$

and at this moment the angular velocity is

$$\dot{\theta}^2 = \frac{3g}{2a}\cdot\frac{\sin\alpha}{3} = \frac{g\sin\alpha}{2a} \quad \text{[by (4.10.79)]}. \tag{4.10.81}$$

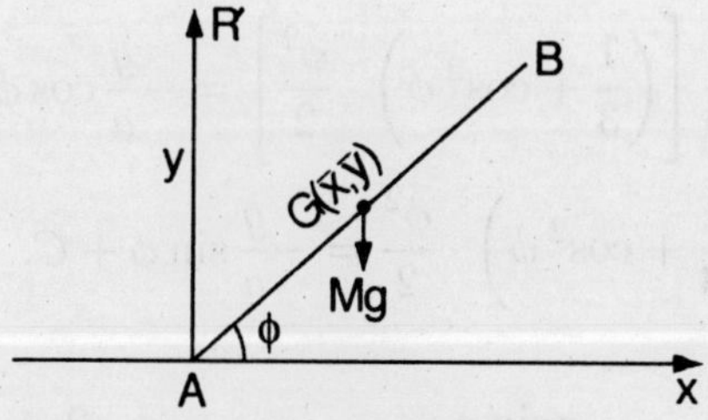

Fig 4.18

After rotating an angle

$$\theta = \sin^{-1}\left(\frac{2}{3}\sin\alpha\right),$$

the rod rotates under the action of gravity only, with angular velocity $\frac{g\sin\alpha}{2a}$ and initial angle

$$\sin^{-1}\left(\frac{2}{3}\sin\alpha\right).$$

Let ϕ be the angle made by the rod at any instant in the second part of the motion and R' be the normal reaction at A and $(\bar{x}, \bar{y})$ be the coordinates of c.g. G.

Now the equations of motion are

$$M\ddot{\bar{x}} = 0 \tag{4.10.82}$$

$$M\ddot{\bar{y}} = R' - Mg \tag{4.10.83}$$

$$\text{and} \quad MK^2\ddot{\phi} = -R'\cdot a\cos\phi, \tag{4.10.84}$$

where $\bar{x} = a\cos\phi$, $\bar{y} = a\sin\phi$; $\dot{\bar{y}} = a\cos\phi\cdot\dot{\phi}$, $\ddot{\bar{y}} = -a\sin\phi\cdot\dot{\phi}^2 + a\cos\phi\cdot\ddot{\phi}$.

Thus from (4.10.83) and (4.10.84) we get

$$MK^2\ddot{\phi} = -Ma\cos\phi\,(\ddot{y}+g)$$

$$\text{or,}\quad K^2\ddot{\phi} = -a\cos\phi\left[-a\sin\phi\cdot\dot{\phi}^2 + a\cos\phi\cdot\ddot{\phi}+g\right]$$

$$\text{or,}\quad \frac{a^2}{3}\ddot{\phi} = a^2\sin\phi\cos\phi\cdot\dot{\phi}^2 - a^2\cos^2\phi\ddot{\phi} - ag\cos\phi$$

$$\text{or,}\quad \left(\frac{1}{3}+\cos^2\phi\right)\ddot{\phi} - \sin\phi\cos\phi\cdot\dot{\phi}^2 = -\frac{g}{a}\cos\phi$$

$$\text{or,}\quad \left(\frac{1}{3}+\cos^2\phi\right)\ddot{\phi} - 2\sin\phi\cos\phi\cdot\left(\frac{\dot{\phi}^2}{2}\right) = -\frac{g}{a}\cos\phi$$

$$\text{or,}\quad \frac{d}{d\phi}\left[\left(\frac{1}{3}+\cos^2\phi\right)\cdot\frac{\dot{\phi}^2}{2}\right] = -\frac{g}{a}\cos\phi$$

$$\text{or,}\quad \left(\frac{1}{3}+\cos^2\phi\right)\cdot\frac{\dot{\phi}^2}{2} = -\frac{g}{a}\sin\phi + C.$$

Initially

$$\dot{\phi} = \frac{g\sin\alpha}{2a}\quad\text{and}\quad \sin\phi = \frac{2}{3}\sin\alpha.$$

$$\therefore\quad \frac{g\sin\alpha}{4a}\left[\frac{1}{3}+1-\frac{4}{9}\sin^2\alpha\right] = -\frac{g}{a}\cdot\frac{2}{3}\sin\alpha + C.$$

$$\begin{aligned}\text{or,}\quad C &= \frac{g\sin\alpha}{4a}\left[\frac{3+9-4\sin^2\alpha}{9}\right] + \frac{2g}{3a}\sin\alpha\\ &= \frac{g\sin\alpha}{9a}(3-\sin^2\alpha) + \frac{2g}{3a}\sin\alpha\\ &= \frac{g}{9a}\left(3-\sin^2\alpha+6\right)\sin\alpha\\ &= \frac{g}{9a}\left[9-\sin^2\alpha\right]\sin\alpha\\ &= \frac{g}{9a}\left(8+\cos^2\alpha\right)\sin\alpha.\end{aligned}$$

$$\therefore\quad \left(\frac{1}{3}+\cos^2\phi\right)\frac{\dot{\phi}^2}{2} + \frac{g}{a}\sin\phi = \frac{g}{9a}\left(8+\cos^2\alpha\right)\sin\alpha.$$

As $\dot{\phi} = \Omega$, when $\phi = 0$, we have

$$\left(\frac{1}{3}+1\right)\frac{\Omega^2}{2} + 0 = \frac{g}{9a}\left(8+\cos^2\alpha\right)\sin\alpha,$$

$$\text{or,}\quad \frac{2}{3}\Omega^2 = \frac{g}{9a}\left(8+\cos^2\alpha\right)\sin\alpha,$$

$$\text{or,}\quad 6a\Omega^2 = g\sin\alpha\left(8+\cos^2\alpha\right).$$

▶ **Problem 4.10.11** *A straight uniform rod, of mass m, is placed at right angle to a smooth plane of inclination α to the horizon with one end in contact with it; the rod is released. Show that, when its inclination to the plane is ϕ, the reaction of the plane will be*

$$mg\frac{3\,(1-\sin\phi)^2+1}{(3\cos^2\phi+1)^2}\cos\alpha.$$

Solution: Let $AB(=2a)$ be the rod of mass m. Let θ be inclination of the rod with the plane inclined at an angle α to the horizon at any time t. Let R be the normal reaction at the end A in contact with the plane. Let $(\bar{x},\bar{y})$ be the coordinates of the c.g. G of the rod, referred to axes Ax and Ay.

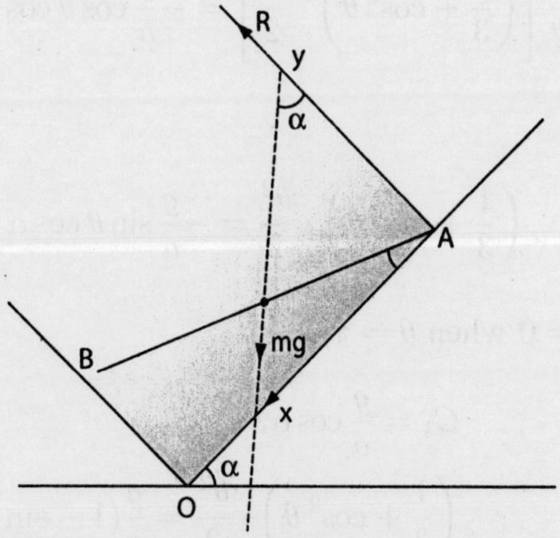

Fig 4.19

The equations of motion are

$$M\ddot{x} = -mg\sin\alpha \tag{4.10.85}$$

$$M\ddot{y} = R - Mg\cos\alpha \tag{4.10.86}$$

$$\text{and}\quad mK^2\ddot{\theta} = -R\cdot a\cos\theta. \tag{4.10.87}$$

Now

$$\bar{y} = a\sin\theta,\quad \dot{\bar{y}} = a\cos\theta\cdot\dot{\theta},\quad \ddot{\bar{y}} = -a\sin\theta\cdot\dot{\theta}^2 + a\cos\theta\cdot\ddot{\theta}.$$

From (4.10.86),

$$R = m\left\{-a\sin\theta\cdot\dot{\theta}^2 + a\cos\theta\cdot\ddot{\theta} + g\cos\alpha\right\}.$$

From (4.10.87), we get

$$mK^2\ddot{\theta} = -ma\cos\theta\left[-a\sin\theta\cdot\dot{\theta}^2 + a\cos\theta\cdot\ddot{\theta} + g\cos\alpha\right],$$

$$\text{or,}\quad \frac{a^2}{3}\ddot{\theta} = a^2\sin\theta\cos\theta\cdot\dot{\theta}^2 - a^2\cos^2\theta\cdot\ddot{\theta} - ag\cos\alpha\cos\theta,$$

$$\text{or,}\quad \left(\frac{1}{3}+\cos^2\theta\right)\ddot{\theta} - \sin\theta\cos\theta\cdot\dot{\theta}^2 = -\frac{g}{a}\cos\alpha\cos\theta,$$

$$\text{or,}\quad \left(\frac{1}{3}+\cos^2\theta\right)\ddot{\theta} - 2\sin\theta\cos\theta\cdot\left(\frac{\dot{\theta}^2}{2}\right) = -\frac{g}{a}\cos\theta\cos\alpha,$$

$$\text{or,}\quad \frac{d}{d\theta}\left[\left(\frac{1}{3}+\cos^2\theta\right)\cdot\frac{\dot{\theta}^2}{2}\right] = -\frac{g}{a}\cos\theta\cos\alpha.$$

Integrating,

$$\left(\frac{1}{3}+\cos^2\theta\right)\cdot\frac{\dot{\theta}^2}{2} = -\frac{g}{a}\sin\theta\cos\alpha + C_1.$$

Initially, $\dot{\theta} = 0$ when $\theta = \pi/2$.

$$\therefore\quad C_1 = \frac{g}{a}\cos\alpha.$$

$$\therefore\quad \left(\frac{1}{3}+\cos^2\theta\right)\cdot\frac{\dot{\theta}^2}{2} = \frac{g}{a}(1-\sin\theta)\cos\alpha,$$

$$\text{or,}\quad \dot{\theta}^2 = \frac{6g}{a}\frac{(1-\sin\theta)\cos\alpha}{(1+3\cos^2\theta)}. \tag{4.10.88}$$

Differentiating with respect to t we get

$$\begin{aligned}
2\ddot{\theta} &= \frac{6g}{a}\left[\frac{-\cos\theta(1+3\cos^2\theta)+(1-\sin\theta)6\sin\theta\cos\theta}{(1+3\cos^2\theta)^2}\right]\cos\alpha.\\
\therefore\quad \ddot{\theta} &= \frac{3g}{a}\cos\theta\cos\alpha\left[\frac{-1-3\cos^2\theta+6\sin\theta-6\sin^2\theta}{(1+3\cos^2\theta)^2}\right]\\
&= \frac{3g}{a}\cos\theta\cos\alpha\left[\frac{-4+3\sin^2\theta+6\sin\theta-6\sin^2\theta}{(1+3\cos^2\theta)^2}\right]\\
&= \frac{3g}{a}\cos\theta\cos\alpha\left[\frac{-3-1+6\sin\theta-3\sin^2\theta}{(1+3\cos^2\theta)^2}\right]\\
&= \frac{3g}{a}\cos\theta\cos\alpha\left[\frac{-1-3(1-\sin\theta)^2}{(1+3\cos^2\theta)^2}\right]\\
&= -\frac{3g}{a}\cos\theta\cos\alpha\left[\frac{1+3(1-\sin\theta)^2}{(1+3\cos^2\theta)^2}\right].
\end{aligned}$$

Therefore from (4.10.87)

$$\begin{aligned} R &= -\frac{mK^2\ddot{\theta}}{a\cos\theta} = \frac{ma}{3\cos\theta}\cdot\frac{3g}{a}\cos\theta\cos\alpha\left[\frac{1+3(1-\sin\theta)^2}{(1+3\cos^2\theta)^2}\right] \\ &= mg\cos\alpha\left[\frac{1+3(1-\sin\theta)^2}{(1+3\cos^2\theta)^2}\right] \end{aligned}$$

when $\theta = \phi$, we have

$$R = mg\frac{1+3(1-\sin\phi)^2}{(1+3\cos^2\phi)^2}\cdot\cos\alpha.$$

▶ **Problem 4.10.12** *A uniform rod is projected in any manner in a vertical plane. Assuming that there is no resisting force to the motion. Prove that its centre of gravity describe a parabolic path while the rod rotates with uniform angular velocity about the centre of gravity.* **[CUH 1994]**

Solution: Let the rod $AB(=2a)$ is projected with initial velocity V at an angle α with the horizon. Let $(\bar{x},\bar{y})$ be the coordinates of c.g. G of the rod, referred to the axes Ox and Oy at any time t and in this position the rod makes an angle θ with the horizon. Let M be the mass of the rod.

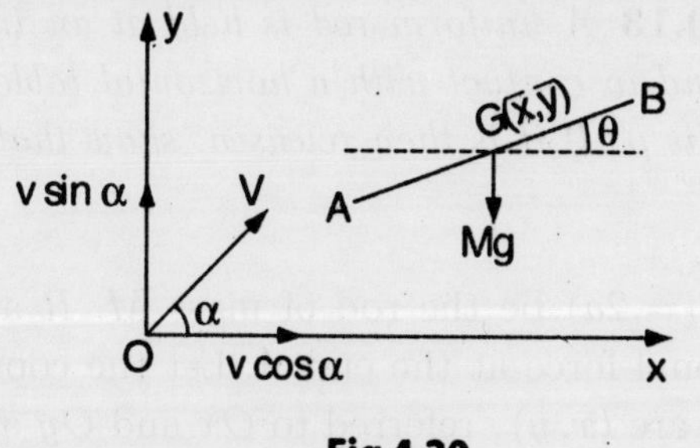

Fig 4.20

Now the equation of motion are

$$M\ddot{\bar{x}} = 0 \quad (4.10.89)$$

$$M\ddot{\bar{y}} = -Mg \quad (4.10.90)$$

$$\text{and} \quad MK^2\ddot{\theta} = 0. \quad (4.10.91)$$

From (4.10.91), we get $\ddot{\theta} = 0$, or, $\dot{\theta} = \text{constant} = \Omega$ (say).

From (4.10.89), $\ddot{\bar{x}} = 0$, or, $\dot{\bar{x}} = \text{constant} = \text{initial value} = V\cos\alpha$, or, $\bar{x} = V\cos\alpha\cdot t + C_1$.

Initially $\bar{x} = 0$ when $t = 0$ $\therefore$ $C_1 = 0$

$$\therefore \quad \bar{x} = V\cos\alpha\cdot t \quad (4.10.92)$$

From (4.10.90), $\ddot{\bar{y}} = -g$ or $\dot{\bar{y}} = -gt + C_2$.

Initially $\dot{y} = V \sin \alpha$, when $t = 0$. $\therefore\ C_2 = V \sin \alpha$.

$$\therefore \quad \dot{\bar{y}} = V \sin \alpha - gt \quad \text{or,} \quad \bar{y} = V \sin \alpha \cdot t - \tfrac{1}{2} g t^2 + C_3$$

when $\bar{y} = 0$, $t = 0$, then $C_3 = 0$

$$\therefore \quad \bar{y} = V \sin \alpha \cdot t - \tfrac{1}{2} g t^2. \tag{4.10.93}$$

Eliminating t, from (4.10.92) and (4.10.93) we get

$$\begin{aligned} \bar{y} &= V \sin \alpha \cdot \frac{\bar{x}}{V \cos \alpha} - \frac{1}{2} g \frac{\bar{x}^2}{V^2 \cos^2 \alpha} \\ &= \tan \alpha \cdot \bar{x} - \frac{1}{2} g \frac{\bar{x}^2}{V^2 \cos^2 \alpha}. \end{aligned}$$

Thus, the locus of path of c.g. G is

$$y = x \tan \alpha - \frac{1}{2} \frac{g x^2}{V^2 \cos^2 \alpha},$$

which is obviously a parabola.

▶ **Problem 4.10.13** *A uniform rod is held at an inclination 45° to the vertical with one end in contact with a horizontal table whose coefficient of friction of friction is μ. If it is then released, show that it will commence to slide if $\mu < 3/5$.* **[B.U.H 1999]**

Solution: Let $AB(= 2a)$ be the rod of mass M. R and F be the normal reaction and frictional force at the end A. Let the coordinates of c.g. G of the rod of mass M are $(\bar{x}, \bar{y})$, referred to Ox and Oy as axes.

$$\therefore \quad \bar{x} = a \sin \theta; \quad \dot{\bar{x}} = a \cos \theta \cdot \dot{\theta}; \quad \ddot{\bar{x}} = -a \sin \theta \cdot \dot{\theta}^2 + a \cos \theta \cdot \ddot{\theta}$$

$$\bar{y} = a \cos \theta; \quad \dot{\bar{y}} = -a \sin \theta \cdot \dot{\theta}; \quad \ddot{\bar{y}} = -a \cos \theta \cdot \dot{\theta}^2 - a \sin \theta \cdot \ddot{\theta}.$$

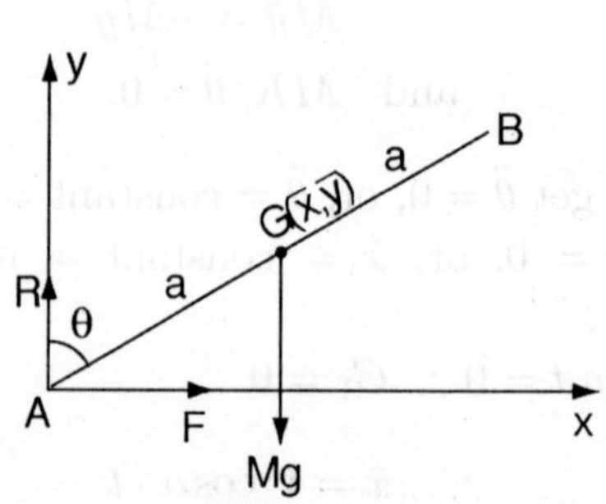

Fig 4.21

The equations of motion are

$$M\ddot{x} = F \tag{4.10.94}$$

$$M\ddot{y} = R - Mg \tag{4.10.95}$$

$$\text{and} \quad MK^2\ddot{\theta} = R \cdot a\sin\theta - Fa\cos\theta. \tag{4.10.96}$$

Substituting (4.10.94), (4.10.95) in (4.10.96) we get

$$MK^2\ddot{\theta} = Ma\sin\theta \cdot (g + \ddot{y}) - Ma\cos\theta\,\ddot{x}$$

$$\text{or,} \quad \frac{a^2}{3}\ddot{\theta} = a\sin\theta\left[g - a\cos\theta \cdot \dot{\theta}^2 - a\sin\theta \cdot \ddot{\theta}\right]$$

$$-a\cos\theta\left[-a\sin\theta \cdot \dot{\theta}^2 + a\cos\theta \cdot \ddot{\theta}\right]$$

$$= ag\sin\theta - a^2\ddot{\theta},$$

$$\text{or,} \quad a^2\left(1 + \frac{1}{3}\right)\ddot{\theta} = ag\sin\theta, \quad \text{or,} \quad a\ddot{\theta} = \frac{3g\sin\theta}{4}, \tag{4.10.97}$$

$$\text{or,} \quad 2a\dot{\theta}\ddot{\theta} = \frac{3g}{2}\sin\theta \cdot \dot{\theta},$$

Integrating

$$a\dot{\theta}^2 = -\frac{3g}{2}\cos\theta + C.$$

$$\therefore \quad C = 0 + \frac{3g}{2\sqrt{2}} \qquad \left[\text{as } \dot{\theta} = 0 \text{ when } \theta = 45^\circ\right].$$

Thus

$$a\dot{\theta}^2 = \frac{3g}{2}\left[\frac{1}{\sqrt{2}} - \cos\theta\right]. \tag{4.10.98}$$

Thus from (4.10.94),

$$F = M\left(-a\sin\theta \cdot \dot{\theta}^2 + a\cos\theta \cdot \ddot{\theta}\right)$$

$$= M\left[-\sin\theta\left(\frac{3g}{2\sqrt{2}} - \frac{3g}{2}\cos\theta\right) - \cos\theta \cdot \frac{3g}{4} \cdot \sin\theta\right].$$

$$\therefore \quad F\big|_{\theta=45^\circ} = M\left[-\frac{1}{\sqrt{2}}\left(\frac{3g}{2\sqrt{2}} - \frac{3g}{2\sqrt{2}}\right) - \frac{3g}{4} \cdot \frac{1}{\sqrt{2}} \cdot \frac{1}{\sqrt{2}}\right] = -\frac{3Mg}{8}$$

From (4.10.95),

$$R = M(\ddot{y} + g) = M\left[-a\cos\theta \cdot \dot{\theta}^2 - a\sin\theta \cdot \ddot{\theta} + g\right]$$

$$= M\left[-\cos\theta \cdot \frac{3g}{2}\left(\frac{1}{\sqrt{2}} - \cos\theta\right) - \sin\theta\left(\frac{3g}{4}\sin\theta\right) + g\right].$$

$$\therefore \quad R\big|_{\theta=45^\circ} = -Mg\left(1 - \frac{3}{8}\right) = -\frac{5Mg}{8}.$$

For sliding

$$\mu < \left|\frac{F}{R}\right| = \frac{3}{5}.$$

4.11 Application of Conservation of Energy and Momentum

► **Problem 4.11.1** *A uniform straight rod, of length $2a$, has two small rings at its ends which can respectively slide on thin smooth horizontal and vertical wire Ox and Oy. The rod starts at an angle α to the horizon with an angular velocity*

$$\sqrt{\frac{3g}{2a}(1-\sin\alpha)},$$

and moves downwards. Show that it will strike the horizontal wire at the end of time

$$2\sqrt{\frac{a}{3g}}\left[\log\left\{\cot\left(\frac{\pi}{8}-\frac{\alpha}{4}\right)\tan\frac{\pi}{8}\right\}\right].$$ **[BUH 2005]**

Solution: Let $AB(=2a)$ be the rod of mass M and G be its c.g. Let $(\bar{x},\bar{y})$ be the coordinates of G, referred to axes Ox and Oy. Let R_1 and R_2 be the normal reactions at A and B. Let θ be the inclination of the rod with horizontal at any time t.

$$\therefore \quad \bar{x} = a\cos\theta \quad \text{and} \quad \bar{y} = a\sin\theta$$

$$\left.\begin{aligned} \therefore\ \dot{\bar{x}} &= -a\sin\theta\cdot\dot{\theta}; \quad \ddot{\bar{x}} = -a\cos\theta\cdot\dot{\theta}^2 - a\sin\theta\cdot\ddot{\theta} \\ \dot{\bar{y}} &= a\cos\theta\cdot\dot{\theta}; \quad \ddot{\bar{y}} = -a\sin\theta\cdot\dot{\theta}^2 + a\cos\theta\cdot\ddot{\theta} \end{aligned}\right\}. \tag{4.11.1}$$

The equations of motion are

$$M\ddot{\bar{x}} = R_2 \tag{4.11.2}$$

$$M\ddot{\bar{y}} = R_1 - Mg. \tag{4.11.3}$$

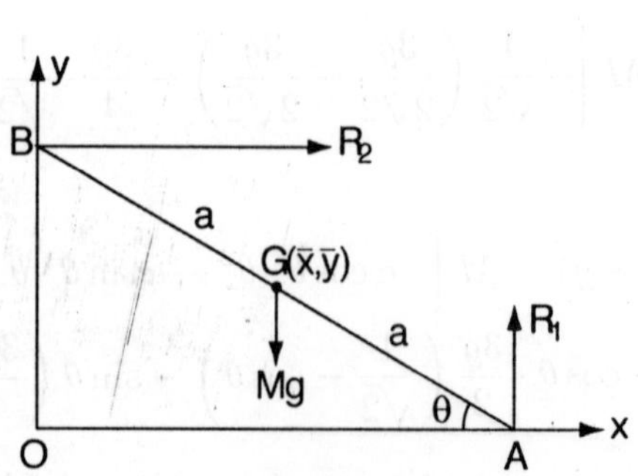

Fig 4.22

By principle of energy, K.E. + P.E. = constant = initial value, we have

$$\frac{1}{2}M\left[\dot{\bar{x}}^2+\dot{\bar{y}}^2+K^2\dot{\theta}^2\right]+Mg\cdot a\sin\theta=\text{constant}$$

$$\text{or,}\quad \frac{1}{2}M\left[a^2\dot{\theta}^2+\frac{a^2}{3}\cdot\dot{\theta}^2\right]+Mga\sin\theta=\text{constant}\quad[\text{by }(4.11.1)]$$

$$\text{or,}\quad \frac{2Ma^2}{3}\cdot\dot{\theta}^2+Mga\sin\theta=\text{constant}=\text{initial value}$$

$$=\frac{2Ma^2}{3}\cdot\frac{3g}{2a}(1-\sin\alpha)+Mga\sin\alpha$$

$$=Mga(1-\sin\alpha)+Mga\sin\alpha=Mga$$

$$\text{or,}\quad \frac{2a^2}{3}\cdot\dot{\theta}^2+ga\sin\theta=ga$$

$$\text{or,}\quad \dot{\theta}^2=\frac{3ga(1-\sin\theta)}{2a^2}=\frac{3g(1-\sin\theta)}{2a}=\frac{3g}{2a}\left(\cos\frac{\theta}{2}-\sin\frac{\theta}{2}\right)^2$$

$$\text{or,}\quad \dot{\theta}=-\sqrt{\frac{3g}{2a}}\left(\cos\frac{\theta}{2}-\sin\frac{\theta}{2}\right),$$

since the rod move in the sense of θ decreasing, i. e., $\cos\frac{\theta}{2}>\sin\frac{\theta}{2}$

$$\therefore\quad \dot{\theta}=\sqrt{\frac{3g}{2a}}\left(\sin\frac{\theta}{2}-\cos\frac{\theta}{2}\right)\quad\text{or,}\quad dt=\sqrt{\frac{2a}{3g}}\frac{1}{(\sin\frac{\theta}{2}-\cos\frac{\theta}{2})}d\theta.$$

If t_1 be the time taken by the rod to strike the horizontal wire, we have

$$\int_0^{t_1}dt=\sqrt{\frac{2a}{3g}}\int_\alpha^0\frac{d\theta}{(\sin\frac{\theta}{2}-\cos\frac{\theta}{2})}=\sqrt{\frac{a}{3g}}\int_\alpha^0\frac{d\theta}{\frac{1}{\sqrt{2}}\sin\frac{\theta}{2}-\frac{1}{\sqrt{2}}\cos\frac{\theta}{2}},$$

$$\text{or,}\quad t_1=\sqrt{\frac{a}{3g}}\int_0^\alpha\frac{d\theta}{\frac{1}{\sqrt{2}}\cos\frac{\theta}{2}-\frac{1}{\sqrt{2}}\sin\frac{\theta}{2}}=\sqrt{\frac{a}{3g}}\int_0^\alpha\sec\left(\frac{\pi}{4}+\frac{\theta}{2}\right)d\theta$$

$$=\sqrt{\frac{a}{3g}}\cdot2\left|\log\tan\left(\frac{\pi}{4}+\frac{\pi}{8}+\frac{\theta}{4}\right)\right|_0^\alpha$$

$$=2\sqrt{\frac{a}{3g}}\left[\log\tan\left(\frac{\pi}{4}+\frac{\pi}{8}+\frac{\alpha}{4}\right)-\log\tan\left(\frac{\pi}{4}+\frac{\pi}{8}\right)\right]$$

$$=2\sqrt{\frac{a}{3g}}\left[\log\tan\left(\frac{3\pi}{8}+\frac{\alpha}{4}\right)-\log\tan\frac{3\pi}{8}\right]$$

$$=2\sqrt{\frac{a}{3g}}\left[\log\tan\left(\frac{\pi}{2}-\frac{\pi}{8}+\frac{\alpha}{4}\right)-\log\tan\left(\frac{\pi}{2}-\frac{\pi}{8}\right)\right]$$

$$=2\sqrt{\frac{a}{3g}}\left[\log\cot\left(\frac{\pi}{8}-\frac{\alpha}{4}\right)-\log\cot\frac{\pi}{8}\right]$$

$$=2\sqrt{\frac{a}{3g}}\log\left\{\cot\left(\frac{\pi}{8}-\frac{\alpha}{4}\right)\cdot\tan\frac{\pi}{8}\right\}.$$

▶ **Problem 4.11.2** *A uniform rod, of length 2a, is freely jointed at one end to a small ring, whose mass is equal to that of the rod. The ring is free to slide on a smooth horizontal wire and initially it is at rest and the rod is vertical and below the ring and rotates with angular velocity* $\sqrt{\frac{3g}{a}}$ *in a vertical plane passing through the wire. When the rod is inclined at an angle* θ *to the vertical, show that its angular velocity is*

$$\sqrt{\frac{3g}{a} \cdot \frac{1 + 4\cos\theta}{8 - \cos^2\theta}}$$

and find the velocity of ring then.

Solution: Let $AB(= 2a)$ be the rod of mass m and $(\bar{x}, \bar{y})$ be the coordinates of the c.g. G of the rod, referred to the axes Ox and Oy. Let at any time t, the rod makes an angle θ with the horizontal. R be the normal reaction at the ring A. Let the coordinates of the ring be $A(x, 0)$. Thus

$$\bar{x} = x + a\cos\theta \quad \text{and} \quad \bar{y} = a\sin\theta$$

$$\text{or,}\quad \dot{\bar{x}} = \dot{x} - a\sin\theta\cdot\dot{\theta}; \quad \ddot{\bar{x}} = \ddot{x} - a\cos\theta\cdot\dot{\theta}^2 - a\sin\theta\cdot\ddot{\theta}$$

$$\dot{\bar{y}} = a\cos\theta\cdot\dot{\theta}; \qquad \ddot{\bar{y}} = -a\sin\theta\cdot\dot{\theta}^2 + a\cos\theta\cdot\ddot{\theta}$$

Thus the equations of motion of the system are

$$m\ddot{x} + m\ddot{\bar{x}} = 0,$$

$$\text{or,}\quad \ddot{x} + \ddot{x} - a\cos\theta\cdot\dot{\theta}^2 - a\sin\theta\cdot\ddot{\theta} = 0,$$

$$\text{or,}\quad 2\ddot{x} - a\cos\theta\cdot\dot{\theta}^2 - a\sin\theta\cdot\ddot{\theta} = 0. \tag{4.11.4}$$

$$\text{or,}\quad m\ddot{\bar{y}} = 2mg - R,$$

$$\text{or,}\quad m(-a\sin\theta\cdot\dot{\theta}^2 + a\cos\theta\cdot\ddot{\theta}) = 2mg - R. \tag{4.11.5}$$

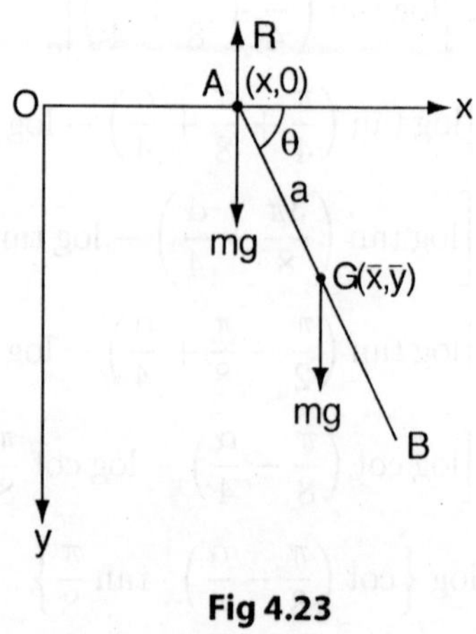

Fig 4.23

Now

$$
\begin{aligned}
\text{K.E. of the system} &= \text{K.E. of the ring} + \text{K.E. of the rod} \\
&= \frac{1}{2}m\dot{x}^2 + \frac{1}{2}m\left[\dot{\bar{x}}^2 + \dot{\bar{y}}^2 + K^2\dot{\theta}^2\right] \\
&= \frac{1}{2}m\left[\dot{x}^2 + \dot{\bar{x}}^2 + \dot{\bar{y}}^2 + K^2\dot{\theta}^2\right] \\
&= \frac{1}{2}m\left[\dot{x}^2 + \left(\dot{x} - a\sin\theta\cdot\dot{\theta}\right)^2 + (a\cos\theta\cdot\dot{\theta})^2 + K^2\dot{\theta}^2\right] \\
&= \frac{1}{2}m\left[2\dot{x}^2 - 2a\dot{x}\sin\theta\cdot\dot{\theta} + a^2\dot{\theta}^2 + K^2\dot{\theta}^2\right] \\
&= \frac{1}{2}m\left[2\dot{x}^2 - 2a\sin\theta\cdot\dot{\theta}\dot{x} + a^2\dot{\theta}^2 + \frac{a^2}{3}\dot{\theta}^2\right] \\
&= \frac{1}{2}m\left[2\dot{x}^2 - 2a\sin\theta\cdot\dot{\theta}\dot{x} + \frac{4a^2}{3}\dot{\theta}^2\right] \\
&= m\left[\dot{x}^2 - a\sin\theta\cdot\dot{\theta}\dot{x} + \frac{2}{3}a^2\dot{\theta}^2\right] \\
&= m\left[\left(\dot{x} - \frac{a\sin\theta}{2}\cdot\dot{\theta}\right)^2 - \frac{a^2\sin^2\theta}{4}\cdot\dot{\theta}^2 + \frac{2}{3}a^2\dot{\theta}^2\right]. \qquad (4.11.6)
\end{aligned}
$$

From (4.11.4), we have

$$
\begin{aligned}
2\ddot{x} - a\cos\theta\cdot\dot{\theta}^2 - a\sin\theta\cdot\ddot{\theta} &= 0, \\
\text{or,}\quad 2\ddot{x} + \frac{d^2}{dt^2}(a\cos\theta) &= 0.
\end{aligned}
$$

Integrating

$$
\begin{aligned}
2\dot{x} + \frac{d}{dt}(a\cos\theta) &= C, \\
\text{or,}\quad 2\dot{x} - a\sin\theta\cdot\dot{\theta} &= C.
\end{aligned}
$$

Initially $\dot{x} = 0$; $\dot{\theta} = \sqrt{\frac{3g}{a}}$ and $\theta = \pi/2$. $\quad\therefore\ C = -a\sqrt{\frac{3g}{a}}$.

$$
\therefore\quad 2\dot{x} - a\sin\theta\cdot\dot{\theta} = -a\sqrt{\frac{3g}{a}}
$$

$$
\text{or,}\quad \dot{x} - \frac{a\sin\theta\cdot\dot{\theta}}{2} = -\frac{a}{2}\sqrt{\frac{3g}{a}}.
$$

Thus, from (4.11.6), we get

$$
\begin{aligned}
\text{K.E. of the system} &= m\left[\left(-\frac{a}{2}\sqrt{\frac{3g}{a}}\right)^2 - \frac{a^2\sin^2\theta}{4}\cdot\dot{\theta}^2 + \frac{2}{3}a^2\dot{\theta}^2\right] \\
&= \frac{3mga}{4} + \frac{ma^2}{12}\left(8 - 3\sin^2\theta\right)\dot{\theta}^2.
\end{aligned}
$$

P.E. $= -mga\sin\theta$.

Thus by principle of energy, K.E. + P.E. = constant = initial value, we have

$$\frac{3mga}{4} + \frac{ma^2}{12}(8 - 3\sin^2\theta)\cdot\dot{\theta}^2 - mga\sin\theta$$

$$= \text{initial value} = \frac{3mga}{4} + \frac{5}{12}ma^2\cdot\frac{3g}{a} - mga,$$

or, $$\frac{a^2}{12}(8 - 3\sin^2\theta)\cdot\dot{\theta}^2 = ga\sin\theta + \frac{1}{4}ga = \frac{1}{4}ga(1 + 4\sin\theta),$$

or, $$\dot{\theta}^2 = \frac{3g}{a}\cdot\frac{1 + 4\sin\theta}{8 - 3\sin^2\theta}.$$

If θ is the angle made by the rod with vertical, we replace θ by $\frac{\pi}{2} - \theta$.

$$\therefore\quad \dot{\theta}^2 = \frac{3g}{a}\cdot\frac{1 + 4\cos\theta}{8 - 3\cos^2\theta},\quad \text{or,}\quad \dot{\theta} = \sqrt{\frac{3g}{a}\cdot\frac{1 + 4\cos\theta}{8 - 3\cos^2\theta}}.$$

► **Problem 4.11.3** *Two like rods, AB and BC, each of length 2a, are freely jointed at B; AB can turn round the end A and C can move freely on a vertical straight line through A. Initially the rods are held in a horizontal line, C being in coincidence with A, and they are then released. Show that when the rods are inclined at an angle θ to the horizontal the angular velocity is*

$$\sqrt{\frac{3g}{a}\cdot\frac{\sin\theta}{1 + 3\cos^2\theta}}.$$

Solution: Let AB and BC are the two rods each of weight m and each makes an angle θ with the horizontal at any time t.

Let G and G' be the c.g.'s of the rod AB and BC respectively. Let $(\bar{x}, \bar{y})$ and $(\bar{x}, \bar{y}')$ are the coordinates of G and G' referred to axes Ax and Ay.

Thus

$$\bar{x} = a\cos\theta,\qquad \bar{y} = a\sin\theta,\qquad \bar{y}' = 3a\sin\theta$$

$$\therefore\quad \dot{\bar{x}} = -a\sin\theta\cdot\dot{\theta},\quad \dot{\bar{y}} = a\cos\theta\cdot\dot{\theta},\quad \dot{\bar{y}}' = 3a\cos\theta\cdot\dot{\theta}$$

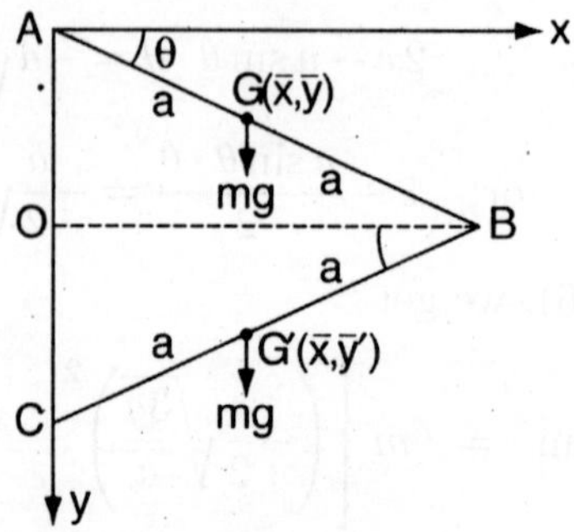

Fig 4.24

Kinetic energy of the system

$$\begin{aligned}
&= \text{K.E. of the rod } AB + \text{K.E. of the rod } BC\\
&= \frac{1}{2}m\left[\dot{x}^2+\dot{y}^2+K^2\dot{\theta}^2\right]+\frac{1}{2}m\left[\dot{x}^2+\dot{y}'^2+K^2\dot{\theta}^2\right]\\
&= \frac{1}{2}m\left[a^2\dot{\theta}^2+2K^2\dot{\theta}^2+a^2\sin^2\theta\cdot\dot{\theta}^2+9a^2\cos^2\theta\cdot\dot{\theta}^2\right]\\
&= \frac{1}{2}m\left[a^2\dot{\theta}^2+\frac{2a^2}{3}\dot{\theta}^2+a^2\left(1-\cos^2\theta\right)\cdot\dot{\theta}^2+9a^2\cos^2\theta\cdot\dot{\theta}^2\right]\\
&= \frac{1}{2}ma^2\left[8\cos^2\theta+\left(2+\frac{2}{3}\right)\right]\dot{\theta}^2=\frac{1}{2}ma^2\frac{24\cos^2\theta+8}{3}\cdot\dot{\theta}^2\\
&= \frac{4ma^2}{3}\left[1+3\cos^2\theta\right]\cdot\dot{\theta}^2.
\end{aligned}$$

$$\begin{aligned}
\text{P.E. of the system} &= \text{P.E. of rod } AB + \text{P.E. of the rod } BC\\
&= -mg\sin\theta-3mg\sin\theta=-4mg\sin\theta.
\end{aligned}$$

From the principle of energy K.E. + P.E. = constant = initial value, we have

$$\frac{4ma^2}{3}\left[1+3\cos^2\theta\right]\dot{\theta}^2-4mga\sin\theta=0 \;\left[\text{as } \dot{\theta}=0 \text{ when } \theta=0\right]$$

$$\text{or,}\quad \dot{\theta}^2=\frac{3g}{a}\cdot\frac{\sin\theta}{1+3\cos^2\theta},\quad\text{or,}\quad \dot{\theta}=\sqrt{\frac{3g}{a}\cdot\frac{\sin\theta}{1+3\cos^2\theta}}.$$

▶ **Problem 4.11.4** *A uniform rod of mass M moves so that the ends have parallel velocity U and V normal to the rod and no velocity along the rod. Show that the kinetic energy is*

$$\tfrac{1}{6}M\left(U^2+V^2+UV\right).$$

Solution: Let AB be the rod of length $2a$ which makes an angle θ with Ox. Let G be the c.g of the rod. Let u be the only velocity of c.g. perpendicular to the rod and there is no velocity along the rod. Now the velocity of B = velocity is c.g $G+$ velocity of B relative to G.

$$V=u-a\dot{\theta}. \tag{4.11.7}$$

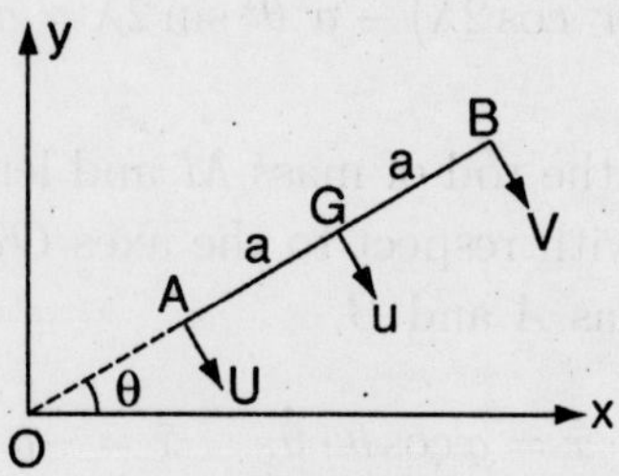

Fig 4.25

Again velocity of A = velocity of c.g. $G+$ velocity of A relative to c.g. G

$$U = u + a\dot{\theta}. \tag{4.11.8}$$

Now adding we get

$$U + V = 2u, \quad \text{or,} \quad u = \frac{U+V}{2}$$

and subtracting

$$U - V = 2a\dot{\theta}, \quad \text{or,} \quad a\dot{\theta} = \frac{U-V}{2}.$$

Now the Kinetic energy of the rod

$$\begin{aligned}
&= \frac{1}{2}M\left(u^2 + 0^2 + K^2\dot{\theta}^2\right) \\
&= \frac{1}{2}M\left[\left(\frac{U+V}{2}\right)^2 + 0^2 + \frac{a^2}{3}\cdot\left(\frac{U-V}{2a}\right)^2\right] \\
&= \frac{1}{2}M\left[\frac{U^2+V^2+2UV}{4} + \frac{U^2+V^2-2UV}{12}\right] \\
&= \frac{M}{24}\left[3U^2+3V^2+6UV+U^2+V^2-2UV\right] \\
&= \frac{M}{24}\left[4U^2+4V^2+4UV\right] \\
&= \frac{M}{6}\left[U^2+V^2+UV\right].
\end{aligned}$$

► **Problem 4.11.5** *A heavy uniform rod, of length $2a$ is placed in a vertical plane with its ends in contact with a rough vertical wall and an equally rough horizontal plane, the coefficient of friction being $\tan\lambda$. Show that it will begin to slip down if its initial inclination to the vertical is greater than 2λ. Prove also that the inclination θ of the rod to the vertical at any time is given by*

$$\ddot{\theta}\left(K^2 + a^2\cos 2\lambda\right) - a^2\dot{\theta}^2\sin 2\lambda = ag\sin(\theta - 2\lambda).$$

Solution: Let AB be the rod of mass M and length $2a$. Let $G(\bar{x}, \bar{y})$ be the mid-point of the rod with respect to the axes Oy and Oy. Let S and R are the resultant reaction as A and B.

$$\begin{aligned}
\therefore\; \bar{x} &= a\sin\theta, & \dot{\bar{x}} &= a\cos\theta\cdot\dot{\theta}, & \ddot{\bar{x}} &= -a\sin\theta\cdot\dot{\theta}^2 + a\cos\theta\cdot\ddot{\theta} \\
\bar{y} &= a\cos\theta, & \dot{\bar{y}} &= -a\sin\theta\cdot\dot{\theta}, & \ddot{\bar{y}} &= -a\cos\theta\cdot\dot{\theta}^2 - a\sin\theta\cdot\ddot{\theta}.
\end{aligned}$$

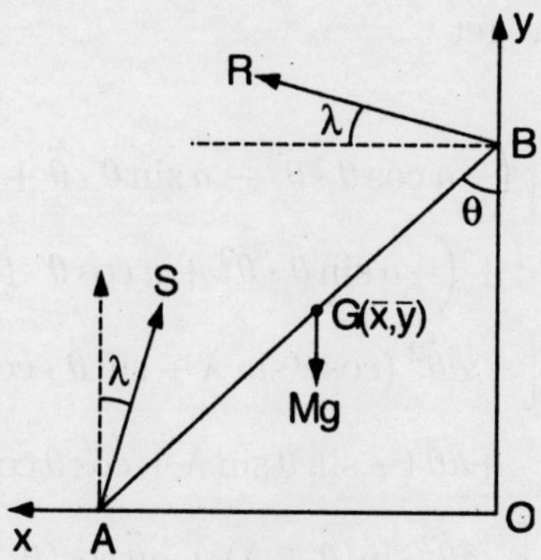

Fig 4.26

The equations of motion are

$$M\ddot{\bar{x}} = -S\cdot\sin\lambda + R\cos\lambda; \tag{4.11.9}$$

$$M\ddot{\bar{y}} = S\cos\lambda + R\sin\lambda - Mg \tag{4.11.10}$$

$$\text{and}\quad MK^2\ddot{\theta} = Sa\sin(\theta-\lambda) - Ra\cos(\theta-\lambda). \tag{4.11.11}$$

From (4.11.9); we get

$$-S\sin\lambda + R\cos\lambda = M\ddot{\bar{x}}. \tag{4.11.12}$$

From (4.11.10); we get

$$S\cos\lambda + R\sin\lambda = M\left(\ddot{\bar{y}} + g\right). \tag{4.11.13}$$

Multiplying (4.11.12) by $\sin\lambda$ and (4.11.13) by $\cos\lambda$ and subtracting, we get

$$S = M\left[-\ddot{\bar{x}}\sin\lambda + (\ddot{\bar{y}}+g)\cos\lambda\right]. \tag{4.11.14}$$

Multiplying (4.11.12) by $\cos\lambda$ and (4.11.13) by $\sin\lambda$ and adding, we get

$$R = M\left[(\ddot{\bar{y}}+g)\sin\lambda + \ddot{\bar{x}}\cos\lambda\right]. \tag{4.11.15}$$

From (4.11.14), we get

$$\begin{aligned} S &= M\Big[-\left(-a\sin\theta\cdot\dot{\theta}^2 + a\cos\theta\cdot\ddot{\theta}\right)\sin\lambda \\ &\qquad + \left(-a\cos\theta\cdot\dot{\theta}^2 - a\sin\theta\cdot\ddot{\theta}\right)\cos\lambda + g\cos\lambda\Big] \\ &= M\Big[-a\dot{\theta}^2\left(-\sin\theta\sin\lambda + \cos\theta\cos\lambda\right) \\ &\qquad -a\ddot{\theta}\left(\cos\theta\sin\lambda + \sin\theta\cos\lambda\right) + g\cos\lambda\Big] \\ &= M\left[-a\dot{\theta}^2\cos(\theta+\lambda) - a\ddot{\theta}\sin(\theta+\lambda) + g\cos\lambda\right]. \end{aligned}$$

From (4.11.15), we get

$$
\begin{aligned}
R &= M\Big[\Big(-a\cos\theta\cdot\dot{\theta}^2 - a\sin\theta\cdot\ddot{\theta} + g\Big)\sin\lambda \\
&\qquad + \Big(-a\sin\theta\cdot\dot{\theta}^2 + a\cos\theta\cdot\ddot{\theta}\Big)\cos\lambda\Big] \\
&= M\Big[-a\dot{\theta}^2(\cos\theta\sin\lambda + \sin\theta\cdot\cos\lambda) \\
&\qquad + a\ddot{\theta}(-\sin\theta\sin\lambda + \cos\theta\cos\lambda) + g\sin\lambda\Big] \\
&= M\Big[-a\dot{\theta}^2\sin(\theta+\lambda) + a\ddot{\theta}\cos(\theta+\lambda) + g\sin\lambda\Big].
\end{aligned}
$$

From (4.11.11), we get

$$
\begin{aligned}
MK^2\ddot{\theta} &= a\sin(\theta-\lambda)\cdot M\Big[-a\dot{\theta}^2\cos(\theta+\lambda) - a\ddot{\theta}\sin(\theta+\lambda) + g\cos\lambda\Big] \\
&\qquad - a\cos(\theta-\lambda)M\Big[-a\dot{\theta}^2\sin(\theta+\lambda) + a\ddot{\theta}\cos(\theta+\lambda) + g\sin\lambda\Big] \\
\text{or,}\quad K^2\ddot{\theta} &= -a^2\dot{\theta}^2\left[\sin(\theta-\lambda)\cos(\theta+\lambda) - \cos(\theta-\lambda)\sin(\theta+\lambda)\right] \\
&\qquad - a^2\ddot{\theta}\left[\sin(\theta-\lambda)\sin(\theta+\lambda) + \cos(\theta-\lambda)\cos(\theta+\lambda)\right] \\
&\qquad + aMg\left[\sin(\theta-\lambda)\cos\lambda - \cos(\theta-\lambda)\sin\lambda\right] \\
&= a^2\dot{\theta}^2\sin 2\lambda - a^2\ddot{\theta}\cos 2\lambda + ag\sin(\theta-2\lambda),
\end{aligned}
$$

$$
\text{or,}\quad (K^2 + a^2\cos 2\lambda)\,\ddot{\theta} - a^2\dot{\theta}^2\sin 2\lambda = ag\sin(\theta-2\lambda).
$$

▶ **Problem 4.11.6** *A uniform beam of mass M and length d stands upright on perfectly rough ground; on the top of it, which is flat, rests a weight of mass m, the coefficient of friction between the beam and the weight being μ. If the beam is allowed to fall to the ground, its inclination θ to the vertical when the weight slips given by*

$$
\left(\frac{4}{3}M + 3m\right)\cos\theta - \left(\frac{M}{6\mu}\right)\sin\theta = M + 2m.
$$

Solution: Let AB be the rod of length d and of mass M. Let the weight of mass m be placed at B and R and F be the normal reaction on the mass m and the frictional forces at B. Let at any time the beam makes an angle θ with the vertical.

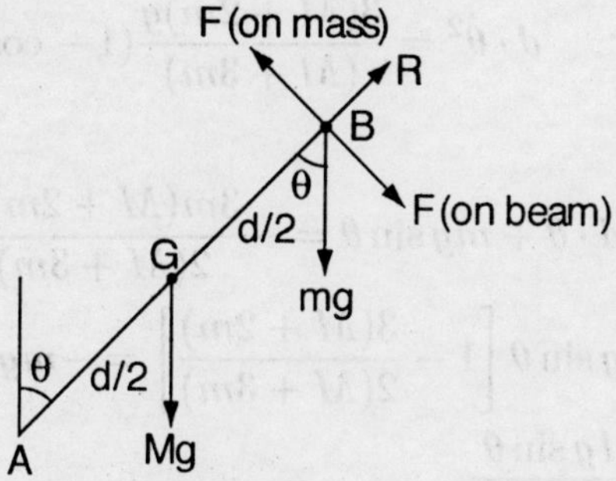

Fig 4.27

The equations of motion are for the beam

$$MK^2\ddot{\theta} = Mg \cdot \frac{d}{2} \sin\theta + F \cdot d$$

$$\text{or,}\quad M\left(\frac{d^2/4}{3} + \frac{d^2}{4}\right) \cdot \ddot{\theta} = Mg \cdot \frac{d}{2} \sin\theta + F \cdot d,$$

$$\text{or,}\quad M \cdot \frac{d^2}{3}\ddot{\theta} = Mg\frac{d}{2} \sin\theta + F \cdot d. \tag{4.11.16}$$

For the mass m

$$m\left(-d\dot{\theta}^2\right) = -mg\cos\theta + R,$$

$$\text{or,}\quad md \cdot \dot{\theta}^2 = mg\cos\theta - R \tag{4.11.17}$$

$$\text{and}\quad md \cdot \ddot{\theta} = mg\sin\theta - F. \tag{4.11.18}$$

Now eliminating F between (4.11.18) and (4.11.16), we get

$$M\frac{d}{3}\ddot{\theta} = \frac{M}{2}g\sin\theta - md\ddot{\theta} + mg\sin\theta,$$

$$\text{or,}\quad \frac{M+3m}{3}d \cdot \ddot{\theta} = \frac{(M+2m)}{2}g\sin\theta,$$

$$\text{or,}\quad d \cdot \ddot{\theta} = \frac{3(M+2m)g}{2(M+3m)} \sin\theta. \tag{4.11.19}$$

Integrating,

$$d \cdot \left(\dot{\theta}^2/2\right) = -\frac{3(M+2m)g}{2(M+3m)} \cos\theta + C.$$

Initially, $t = 0$, $\dot{\theta} = 0$, $\theta = 0$.

$$\therefore\quad C = \frac{3(M+2m)g}{2(M+3m)}$$

$$\therefore \quad d\cdot\dot{\theta}^2 = \frac{3(M+2m)g}{(M+3m)}(1-\cos\theta).$$

$$\begin{aligned}\therefore \quad F &= -md\cdot\ddot{\theta} + mg\sin\theta = -\frac{3m(M+2m)g}{2(M+3m)}\sin\theta + mg\sin\theta \\ &= -mg\sin\theta\left[1-\frac{3(M+2m)}{2(M+3m)}\right] = -mg\sin\theta\cdot\frac{-M}{2(M+3m)} \\ &= \frac{mMg\sin\theta}{2(M+3m)}. \qquad (4.11.20)\end{aligned}$$

$$\begin{aligned}\text{and} \quad R &= mg\cos\theta - md\cdot\dot{\theta}^2 = mg\cos\theta - \frac{3(M+2m)mg}{M+3m}(1-\cos\theta) \\ &= mg\cos\theta\left[1+\frac{3(M+2m)}{M+3m}\right] - \frac{3(M+2m)}{M+3m} \\ &= mg\cos\theta\frac{(4M+9m)}{M+3m} - \frac{3(M+2m)}{M+3m}.\end{aligned}$$

The mass m will slip if

$$\begin{aligned}\mu = \frac{F}{R} &= \frac{M\sin\theta}{2\left[(4M+9m)\cos\theta - 3(M+2m)\right]} \\ &= \frac{M\sin\theta}{6\left[\left(\frac{4}{3}M+3m\right)\cos\theta - (M+2m)\right]}\end{aligned}$$

$$6\mu\left[\left(\frac{4}{3}M+3m\right)\cos\theta - (M+2m)\right] = M\sin\theta,$$

$$\text{or,} \quad \left(\frac{4}{3}M+3m\right)\cos\theta - \frac{M}{6\mu}\sin\theta = (M+2m).$$

4.12 Motion of Uniform Solid Sphere, Uniform Spherical Shell, Uniform Solid Disc, Uniform Ring (Rolling Only)

A rough solid uniform sphere rolls down an inclined plane, rough enough to prevent any slipping. Investigate the motion. [CUH 1997,'01; KUH 2008]

Let O be initial point of contact of the sphere of mass M and radius a with the plane inclined at an angle α with the horizon. We take Ox and Oy as axes at O along and perpendicular to the plane.

Let $C(\bar{x}, a)$ be the centre, i.e., c.g. of the sphere. Let R and F be the normal reaction and frictional force at the point of contact B at any time t when the sphere rolls down a distance $\bar{x}$ along the plane. Let AC be a fixed straight line in the sphere makes an angle θ with a fixed line in space perpendicular to the plane at that time t.

$$\therefore \quad \bar{x} = BO = \text{Arc } AB = a\theta.$$

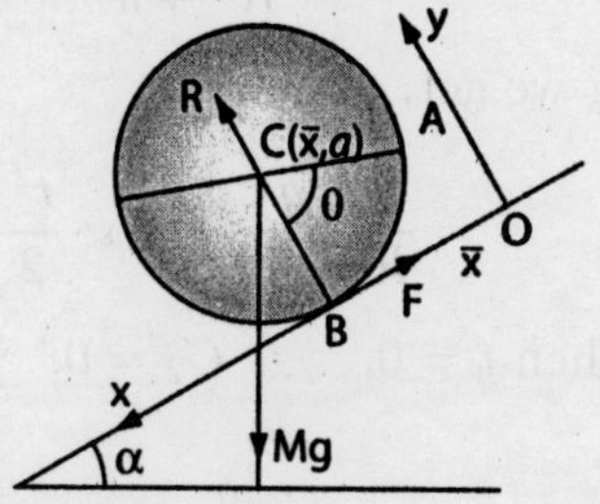

Fig 4.28

$$\left.\begin{aligned} \dot{\bar{x}} &= a\dot{\theta} \\ \text{or,} \quad \ddot{\bar{x}} &= a\ddot{\theta} \end{aligned}\right\}. \tag{4.12.1}$$

The equations of motion are

$$M\ddot{\bar{x}} = Mg\sin\alpha - F, \tag{4.12.2}$$

$$0 = M\ddot{\bar{y}} = R - Mg\cos\alpha \tag{4.12.3}$$

$$\text{and} \quad MK^2\ddot{\theta} = F \cdot a. \tag{4.12.4}$$

Using (4.12.2) in (4.12.4), we get

$$MK^2\ddot{\theta} = -(M\ddot{\bar{x}} - Mg\sin\alpha)\,a,$$

$$\text{or,} \quad K^2\ddot{\theta} + \ddot{\bar{x}}a = ga\sin\alpha,$$

$$\text{or,} \quad K^2\frac{\ddot{\bar{x}}}{a} + a\ddot{\bar{x}} = ga\sin\alpha \quad \text{[by (4.12.1)]}$$

$$\text{or,} \quad \left(\frac{K^2}{a^2} + 1\right)\ddot{\bar{x}} = g\sin\alpha$$

$$\text{or,} \quad \ddot{\bar{x}} = \frac{a^2 g}{K^2 + a^2}\sin\alpha. \tag{4.12.5}$$

Thus the sphere rolls down the plane with constant acceleration

$$\ddot{\bar{x}} = \frac{a^2 g}{K^2 + a^2}\sin\alpha.$$

Integrating, we get

$$\dot{\bar{x}} = \frac{a^2 g}{K^2 + a^2}\sin\alpha \cdot t + C_1.$$

Initially, $\dot{\bar{x}} = 0$ when $t = 0$. $\therefore\ C_1 = 0$.

$$\therefore \quad \dot{\bar{x}} = \frac{a^2 g}{K^2 + a^2} \sin\alpha \cdot t. \tag{4.12.6}$$

Again integrating we get,

$$\bar{x} = \frac{a^2 g}{K^2 + a^2} \sin\alpha \cdot \frac{t^2}{2} + C_2.$$

Initially, $\bar{x} = 0$ when $t = 0$. $\therefore\ C_2 = 0$.
Thus,

$$\bar{x} = \frac{a^2 g}{a^2 + K^2} \sin\alpha \cdot \frac{t^2}{2}. \tag{4.12.7}$$

Thus from (4.12.2),

$$\begin{aligned} F &= Mg\sin\alpha - M\ddot{\bar{x}} = Mg\sin\alpha - M \cdot \frac{a^2 g}{K^2 + a^2} \sin\alpha \\ &= Mg\left(1 - \frac{a^2}{K^2 + a^2}\right)\sin\alpha = Mg\frac{K^2}{K^2 + a^2}\sin\alpha. \end{aligned} \tag{4.12.8}$$

From (4.12.3);

$$R = Mg\cos\alpha. \tag{4.12.9}$$

For solid sphere we have

$$K^2 = \frac{2a^2}{5}.$$

$\therefore$ From (4.12.5), we get

$$\ddot{\bar{x}} = \frac{a^2 g \sin\alpha}{a^2 + \frac{2a^2}{5}} = \frac{5}{7} g \sin\alpha. \tag{4.12.10}$$

Thus the acceleration of the centre of the sphere is $\frac{5}{7} g \sin\alpha$ with which the sphere rolls down along the plane.

From (4.12.8),

$$F = Mg \cdot \frac{\frac{2a^2}{5}}{\frac{2a^2}{5} + a^2} \sin\alpha = \frac{2Mg}{7} \sin\alpha$$

and from (4.12.9) $R = Mg\cos\alpha$.

If the motion be purely rolling one, i.e., there is no sliding, then the coefficient of friction

$$\mu > \frac{F}{R} = \frac{2}{7}\tan\alpha. \tag{4.12.11}$$

In other words, the motion will be purely a rolling one if the coefficient of friction $\mu > \frac{2}{7}\tan\alpha$.

Remark 4.12.1 *If the body be a thin spherical shell, then we have* $K^2 = \frac{2a^2}{3}$.

Then, from (4.12.5) *we get the acceleration of the centre of the shell*

$$\ddot{x} = \frac{a^2}{\frac{2a^2}{3} + a^2} g \sin\alpha = \frac{3}{5} g \sin\alpha. \tag{4.12.12}$$

[BUH 2001]

Thus the acceleration of the centre of the Spherical shell is $\frac{3}{5} g \sin\alpha$ *with which the spherical shell rolls down along the plane.*

From (4.12.8), *we get*

$$F = Mg \cdot \frac{\frac{2a^2}{3}}{a^2 + \frac{2a^2}{3}} \cdot \sin\alpha = \frac{2Mg}{5} \sin\alpha \tag{4.12.13}$$

and from (4.12.9),

$$R = Mg\cos\alpha. \tag{4.12.14}$$

If the motion spherical shell be purely rolling, i. e., there is no sliding, then the coefficient of friction

$$\mu > \frac{F}{R} = \frac{2}{5} \sin\alpha.$$

Thus the motion of the spherical shell be purely rolling, then

$$\mu > \frac{2}{5} \sin\alpha. \tag{4.12.15}$$

[BUH 1997, 2002, 2004]

Remark 4.12.2 *If the body be a circular disc, then we have* $K^2 = a^2/2$.

In this case, from (4.12.5), *we get the acceleration of the centre of circular disc,*

$$\ddot{x} = \frac{a^2}{\frac{a^2}{2} + a^2} \cdot g \sin\alpha = \frac{2}{3} g \sin\alpha. \tag{4.12.16}$$

[BUH 2002]

Thus the circular disc rolls down the plane with constant acceleration of its centre $\frac{2}{3} g \sin\alpha$.

Again, from (4.12.8), *we get*

$$F = Mg \frac{\frac{a^2}{2}}{a^2 + \frac{a^2}{2}} \sin\alpha = \frac{Mg}{3} \sin\alpha.$$

and from (4.12.9),

$$R = Mg\cos\alpha.$$

Thus the motion of the circular disc be purely rolling, then coefficient of friction

$$\mu > \frac{F}{R}, \quad \text{i.e.,} \quad \mu > \frac{1}{3}\sin\alpha. \tag{4.12.17}$$

Remark 4.12.3 *If the body be a circular ring, then we have* $K^2 = a^2$.

In this case, from (4.12.5), *we get the acceleration of the centre of the ring is*

$$\ddot{\bar{x}} = \tfrac{1}{2}g\sin\alpha. \tag{4.12.18}$$

Thus the circular ring rolls down the plane with constant acceleration of its centre $\frac{1}{2}g\sin\alpha$.

Again from (4.12.8), *we get*

$$F = \frac{Mga^2}{a^2 + a^2}\sin\alpha = \frac{Mg}{2}\sin\alpha$$

$$\text{and} \quad R = Mg\cos\alpha.$$

Thus the motion of the circular ring be purely rolling, then the coefficient of friction

$$\mu > \frac{F}{R} = \frac{1}{2}\sin\alpha. \tag{4.12.19}$$

4.13 Solved Problems on Rolling Only

► **Problem 4.13.1** *A uniform solid cylinder is placed with its axis horizontal on a plane, whose inclination to the horizon is* α. *Show that the least coefficient of friction between it and the plane, so that it may roll and not slide, is* $\frac{1}{3}\tan\alpha$.

If the cylinder be hollow and of small thickness, the least value of coefficient of friction λ *is* $\frac{1}{2}\tan\alpha$. **[BUH 1997]**

Solution: Let initially the point A was at O. We take Ox and Oy as axes at O along and perpendicular to the plane, making an angle α with the horizon. Let AC be a fixed line, in a section of the cylinder of centre C, of mass M and radius a, makes an angle θ with a fixed line in space and perpendicular to the plane at any time t. Let in this time t, the cylinder rolls down a distance $\bar{x}$ along the plane. Thus the coordinates of C is $(\bar{x}, a)$.

Let R be the normal reaction and F be the frictional force at the point of contact B at time t.

$$\therefore \quad \bar{x} = OB = \text{arc}\, AB = a\theta. \quad \therefore\ \ddot{\bar{x}} = a\ddot{\theta}. \tag{4.13.1}$$

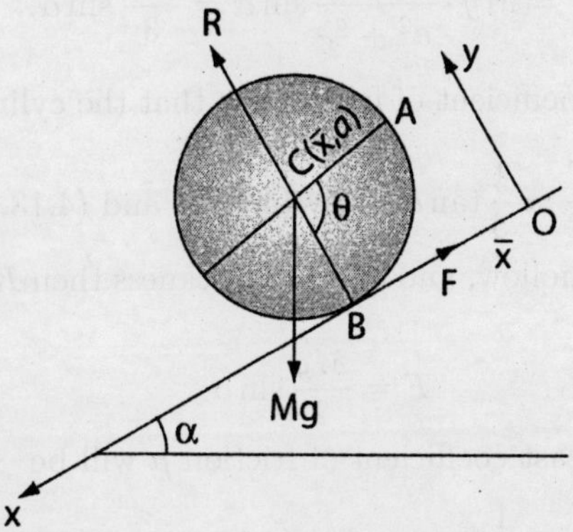

Fig 4.29

The equations of motion are

$$M\ddot{\bar{x}} = -F + Mg\sin\alpha \tag{4.13.2}$$

$$0 = M\ddot{\bar{y}} = R - Mg\cos\alpha \tag{4.13.3}$$

$$\text{and} \quad MK^2\ddot{\theta} = F \cdot a. \tag{4.13.4}$$

Using (4.13.2) in (4.13.4), we get

$$MK^2\ddot{\theta} = Mga\sin\alpha - M\ddot{\bar{x}}\cdot a, \quad \text{or,} \quad K^2\frac{\ddot{\bar{x}}}{a} + \ddot{x}a = ga\sin\alpha,$$

$$\text{or,} \quad \left(\frac{K^2}{a^2}+1\right)\ddot{\bar{x}} = g\sin\alpha.$$

$$\therefore \quad \ddot{\bar{x}} = \frac{a^2 g}{K^2+a^2}\sin\alpha. \tag{4.13.5}$$

$\therefore$ From (4.13.2), we get

$$\begin{aligned} F &= M(g\sin\alpha - \ddot{x}) \\ &= M\left(g\sin\alpha - \frac{a^2 g}{K^2+a^2}\sin\alpha\right) \\ &= Mg\cdot\frac{K^2}{K^2+a^2}\sin\alpha \end{aligned} \tag{4.13.6}$$

and from (4.13.3),

$$R = Mg. \tag{4.13.7}$$

If the body be solid cylinder, the section of the cylinder is a solid disc, then

$$K^2 = a^2/2.$$

$\therefore$ From (4.13.6),

$$F = Mg \cdot \frac{\frac{a^2}{2}}{a^2 + \frac{a^2}{2}} \sin\alpha = \frac{Mg}{3} \sin\alpha. \tag{4.13.8}$$

If μ be the least coefficient of friction, so that the cylinder will roll down then

$$\mu = \frac{F}{R} = \frac{1}{3}\tan\alpha. \quad \text{[by (4.13.8) and (4.13.7)]}$$

If the cylinder be hollow, and of small thickness then $K^2 = a^2$, then from (4.13.6),

$$F = \frac{Mg}{2} \sin\alpha. \tag{4.13.9}$$

In this case, the least coefficient of friction μ will be

$$\mu = \frac{F}{R} = \frac{1}{2}\tan\alpha. \quad \text{[by (4.13.9) and (4.13.7)]}$$

► **Problem 4.13.2** *A cylinder of mass M and radius R is held at rest in a plane inclined at an angle θ to the horizon. When the cylinder is released to moves down the slope. Discuss the case of pure rolling motion of the cylinder. Show that when the cylinder rolls down a length L of the slope, the velocity of its centre of mass is $\frac{4}{3}gL\sin\theta$.* **[CUH 2005]**

Solution: Let initially the point A was at O. We take Ox and Oy axes at O along and perpendicular to the plane. Let AC be the fixed line in a central section of the cylinder and makes an angle ϕ with a fixed line in space and perpendicular to the plane at any time t. Let also that the cylinder rolls down along the plane a distance $\bar{x}$ in this time t. Thus, $(\bar{x}, R)$ are the coordinates of the centre of the section C.

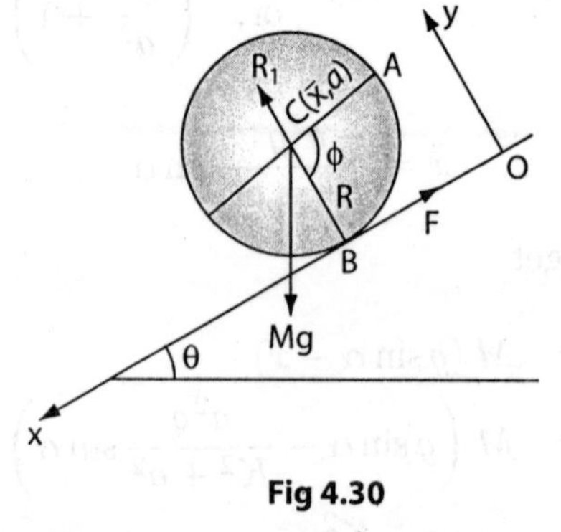

Fig 4.30

Let R_1 and F be the normal reaction and frictional force at the point of contact B in time t. As there is no sliding, where

$$\bar{x} = OB = \text{arc}\, AB = R \cdot \phi. \quad \therefore \quad \ddot{\bar{x}} = R \cdot \ddot{\phi}. \tag{4.13.10}$$

The equations of motion

$$M\ddot{\bar{x}} = -F + Mg\sin\theta \tag{4.13.11}$$

$$0 = M\ddot{\bar{y}} = R_1 - Mg\cos\theta \tag{4.13.12}$$

$$\text{and} \quad MK^2\ddot{\phi} = F \cdot R. \tag{4.13.13}$$

From (4.13.11) and (4.13.13) we get

$$MK^2\ddot{\phi} = R\left[Mg\sin\theta - M\ddot{\bar{x}}\right],$$

$$\text{or,} \quad \frac{R^2}{2} \cdot \frac{\ddot{\bar{x}}}{R} = Rg\sin\theta - R\ddot{\bar{x}} \quad [\text{by } 4.13.10]$$

$$\text{or,} \quad \left(\frac{1}{2} + 1\right)\ddot{\bar{x}} = g\sin\theta,$$

$$\text{or,} \quad \ddot{\bar{x}} = \frac{2g\sin\theta}{3}. \tag{4.13.14}$$

Thus the centre of mass of the cylinder rolls down the plane with constant acceleration $(2g/3)\sin\theta$.

Again from (4.13.11) we get

$$F = Mg\sin\theta - M\ddot{\bar{x}} = Mg\sin\theta - M \cdot \frac{2g}{3}\sin\theta$$

$$= \frac{Mg}{3}\sin\theta.$$

For rolling motion of the cylinder, the coefficient of friction μ,

$$\mu > \frac{F}{R} = \frac{1}{3}\sin\theta.$$

Now from (4.13.14), we get

$$\dot{\bar{x}}\frac{d}{d\bar{x}}(\dot{\bar{x}}) = \frac{2g}{3}\sin\theta, \quad \text{or,} \quad \dot{\bar{x}} \cdot d\dot{x} = \frac{2g}{3}\sin\theta \cdot d\bar{x}.$$

Let $\dot{\bar{x}} = v$ when $\bar{x} = L$

$$\therefore \int_0^v \dot{\bar{x}}\, d\dot{\bar{x}} = \frac{2g}{3}\sin\theta \cdot \int_0^L d\bar{x}, \quad \text{or,} \quad \frac{v^2}{2} = \frac{2g}{3}\sin\theta \cdot L$$

$$\text{or,} \quad v^2 = \frac{4g}{3}\sin\theta \cdot L.$$

$$\therefore \quad v = \left(\frac{4gL}{3} \cdot \sin\theta\right)^{1/2}.$$

► **Problem 4.13.3** *One end of a thread, which is wound on to a reel, is fixed, and the reel falls in a vertical line its axis being horizontal and the unwounded part of the thread being vertical. If the reel be a solid cylinder of radius a and weight W, show that the acceleration of the centre of the reel is* $\frac{2}{3}g$ *and the tension of the thread W/3.*

Solution: We take the axes as shown in the figure. $C(\bar{x}, \bar{y})$ be the centre of the reel. Let AC be a line fixed in the reel which makes an angle θ with a fixed line in space and perpendicular to x-axes at any time t. Let T be the tension in the thread. Let M be the mass of the reel.

$$\therefore \quad \bar{x} = a\theta, \quad \text{or,} \quad \ddot{\bar{x}} = a\ddot{\theta} \tag{4.13.15}$$

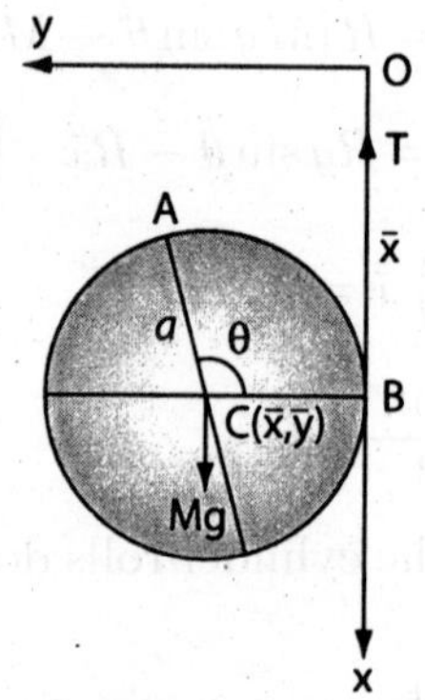

Fig 4.31

The equations of motion are

$$M\ddot{\bar{x}} = -T + Mg \tag{4.13.16}$$

$$0 = M\ddot{\bar{y}} = 0 \tag{4.13.17}$$

$$\text{and} \quad MK^2\ddot{\theta} = T \cdot a. \tag{4.13.18}$$

From (4.13.16) and (4.13.18), we get

$$MK^2\ddot{\theta} = a\left(Mg - M\ddot{\bar{x}}\right),$$

$$\text{or,} \quad K^2\frac{\ddot{\bar{x}}}{a} = ag - a\ddot{\bar{x}},$$

$$\text{or,} \quad \left[\frac{a^2}{2} \cdot \frac{1}{a} + a\right]\ddot{\bar{x}} = ag \quad \left[\because \quad K^2 = \frac{a^2}{2}\right],$$

$$\text{or,} \quad \frac{3a}{2}\ddot{\bar{x}} = ag, \quad \text{or,} \quad \ddot{\bar{x}} = \frac{2g}{3}.$$

Thus the acceleration of the centre of the reel $C = \frac{2g}{3}$.

Again from (4.13.16),

$$T = Mg - M\ddot{x} = Mg - M \cdot \frac{2g}{3} = \frac{Mg}{3} = \frac{W}{3}.$$

▶ **Problem 4.13.4** *A cylinder rolls down a smooth plane whose inclination to the horizon is α, unwrapping, as it goes, a fine string fixed to the highest point of the plane; find its acceleration and the tension of the string.*

Solution: In remark 4.12.2, replace the friction force F by T the tension in the string to get the corresponding results.

▶ **Problem 4.13.5** *A solid circular disc of radius a is rolling up a rough plane (along the line of greatest slope) inclined at an angle α to the horizontal. If V be the velocity of the centre of the disc at an instant, show that the disc ascends a further distance $\frac{3V^2}{4g\sin\alpha}$ along the plane, before coming to rest.*

[CUH 1998, 2004]

Solution: We take Ox and Oy as axes at O. Let $C(\bar{x}, a)$ be the centre of the circular disc of mass M and of radius a at any time t. Let initially the point A was at O. Let the line AC, fixed in the circular disc makes an angle θ with a fixed line in space and perpendicular to the plane making an angle α with the horizontal. Let R and F be the normal reaction and frictional force at the point of contact B at any time t.

$$\therefore \quad \bar{x} = OB = \text{arc } AB = a\theta. \quad \therefore \quad \ddot{\bar{x}} = a\ddot{\theta}. \tag{4.13.19}$$

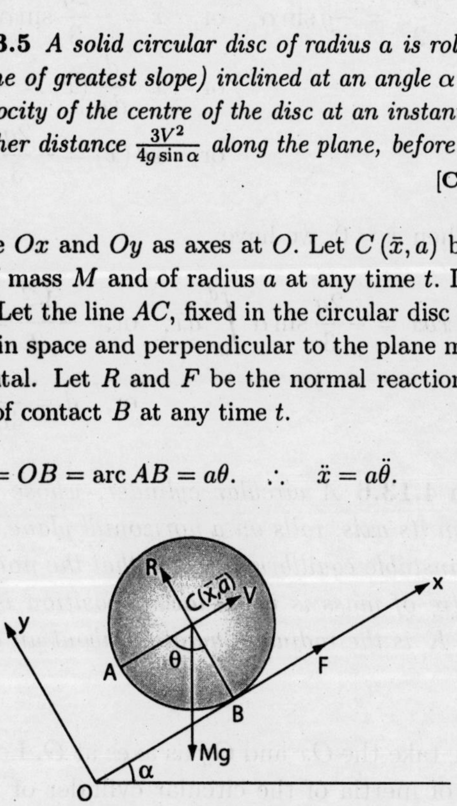

Fig 4.32

The equations of motion are

$$M\ddot{\bar{x}} = F - Mg\sin\alpha \tag{4.13.20}$$

$$0 = M\ddot{\bar{y}} = R - Mg\cos\alpha \tag{4.13.21}$$

$$\text{and} \quad MK^2\ddot{\theta} = -F \cdot a. \tag{4.13.22}$$

From (4.13.20) and (4.13.22), we get

$$MK^2\ddot{\theta} = -(M\ddot{\bar{x}} + Mg\sin\alpha)\cdot a$$

$$\text{or,}\quad K^2\cdot\frac{\ddot{\bar{x}}}{a} + \ddot{\bar{x}}a = -ga\sin\alpha \quad \text{[by (4.13.19)]}$$

$$\text{or,}\quad \left(\frac{a^2}{2}\cdot\frac{1}{a} + a\right)\ddot{\bar{x}} = -ga\sin\alpha. \quad \left[\text{as } K^2 = \frac{a^2}{2}\right]$$

$$\frac{3}{2}\ddot{\bar{x}} = -g\sin\alpha, \quad \text{or,} \quad \ddot{\bar{x}} = -\frac{2g}{3}\sin\alpha$$

$$\text{or,}\quad \dot{\bar{x}}\cdot\frac{d}{d\bar{x}}(\dot{\bar{x}}) = -\frac{2g}{3}\sin\alpha$$

$$\text{or,}\quad \dot{\bar{x}}d(\dot{\bar{x}}) = -\frac{2g}{3}\sin\alpha d\bar{x}.$$

If $\bar{x} = d$ when $\dot{\bar{x}} = 0$, we have

$$\int_V^0 \dot{\bar{x}}d\dot{\bar{x}} = -\frac{2g}{3}\sin\alpha\int_0^d d\bar{x}, \quad \text{or,} \quad -\frac{V^2}{2} = -\frac{2g}{3}\sin\alpha\cdot d$$

$$\text{or,}\quad d = \frac{3V^2}{4g\sin\alpha}.$$

▶ **Problem 4.13.6** *A circular cylinder, whose centre of inertia is at a distance c from its axis, rolls on a horizontal plane. If it be just started from a position of unstable equilibrium, show that the normal reaction of the plane when the centre of mass is in its lowest position is* $1 + \frac{4c^2}{(a-c)^2+K^2}$ *times its weight, where K is the radius of gyration about an axis through the centre of mass.*

Solution: We take the Ox and Oy as axes at O. Let $G(\bar{x}, \bar{y})$ be the position of the centre of inertia of the circular cylinder of mass M and radius a at any time t. Let initially the point A was at O. Let AO' be a fixed line in the central section of the cylinder which makes an angle θ with a fixed line in space perpendicular to Ox. Let R and F are the normal reaction and frictional force at the point of contact B.

$$\therefore \quad \bar{x} = OB + c\sin\theta = a\theta + c\sin\theta$$

$$\bar{y} = a + c\cos\theta.$$

$$\left.\begin{aligned} \therefore\quad &\dot{\bar{x}} = a\dot{\theta} + c\cos\theta\cdot\dot{\theta}; \quad \ddot{\bar{x}} = a\ddot{\theta} - c\sin\theta\cdot\dot{\theta}^2 + c\cos\theta\cdot\ddot{\theta} \\ &\dot{\bar{y}} = -c\sin\theta\cdot\dot{\theta}; \quad \ddot{\bar{y}} = -c\cos\theta\cdot\dot{\theta}^2 - c\sin\theta\cdot\ddot{\theta} \end{aligned}\right\}. \quad (4.13.23)$$

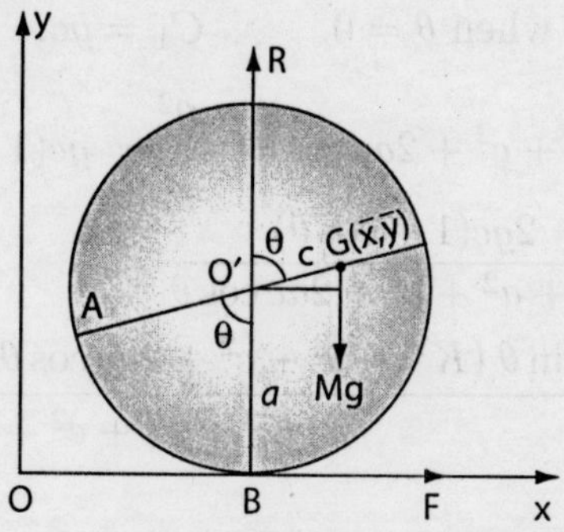

Fig 4.33

The equations of motion are

$$M\ddot{\bar{x}} = F \tag{4.13.24}$$

$$M\ddot{\bar{y}} = R - Mg \tag{4.13.25}$$

$$\text{and} \quad MK^2\ddot{\theta} = R \cdot c\sin\theta - F(a + c\cos\theta). \tag{4.13.26}$$

Eliminating R and F, we get from (4.13.26)

$$\begin{aligned}
MK^2\ddot{\theta} &= c\sin\theta\,(M\ddot{\bar{y}} + Mg) - (a + c\cos\theta)\cdot M\ddot{\bar{x}} \\
&= M\Big[c\sin\theta\left(-c\cos\theta\cdot\dot{\theta}^2 - c\sin\theta\cdot\ddot{\theta} + g\right) \\
&\qquad -(a + c\cos\theta)\left(a\ddot{\theta} - c\sin\theta\cdot\dot{\theta}^2 + c\cos\theta\cdot\ddot{\theta}\right)\Big]
\end{aligned}$$

$$\begin{aligned}
\text{or,} \quad K^2\ddot{\theta} &= -c^2\sin\theta\cos\theta\cdot\dot{\theta}^2 - c^2\sin^2\theta\cdot\ddot{\theta} + gc\sin\theta \\
&\qquad -(a + c\cos\theta)\left[(a + c\cos\theta)\ddot{\theta} - c\sin\theta\cdot\dot{\theta}^2\right] \\
&= -c^2\sin\theta\cos\theta\cdot\dot{\theta}^2 - c^2\sin^2\theta\cdot\ddot{\theta} + gc\sin\theta \\
&\qquad -(a + c\cos\theta)^2\ddot{\theta} + c\sin\theta(a + c\cos\theta)\dot{\theta}^2 \\
&= -\left[c^2\sin^2\theta + a^2 + c^2\cos^2\theta + 2ac\cos\theta\right]\ddot{\theta} \\
&\qquad + \left(ac\sin\theta + c^2\sin\theta\cos\theta - c^2\sin\theta\cos\theta\right)\dot{\theta}^2 + gc\sin\theta \\
&= -\left[a^2 + c^2 + 2ac\cos\theta\right]\ddot{\theta} + ac\sin\theta\cdot\dot{\theta}^2 + gc\sin\theta
\end{aligned}$$

$$\text{or,} \quad \left[K^2 + a^2 + c^2 + 2ac\cos\theta\right]\ddot{\theta} - ac\sin\theta\cdot\dot{\theta}^2 = gc\sin\theta$$

$$\text{or,} \quad \frac{d}{d\theta}\left[\left(K^2 + a^2 + c^2 + 2ac\cos\theta\right)\frac{\dot{\theta}^2}{2}\right] = gc\sin\theta.$$

Integrating, we get,

$$\left(K^2 + a^2 + c^2 + 2ac\cos\theta\right)\cdot\frac{\dot{\theta}^2}{2} = -gc\cos\theta + C_1.$$

Initially, $\dot{\theta} = 0$ when $\theta = 0$. $\quad \therefore \; C_1 = gc$.

$$\therefore \quad \left(K^2 + a^2 + c^2 + 2ac\cos\theta\right) \cdot \frac{\dot{\theta}^2}{2} = gc(1 - \cos\theta)$$

$$\text{or,} \quad \dot{\theta}^2 = \frac{2gc(1-\cos\theta)}{K^2 + a^2 + c^2 + 2ac\cos\theta} \tag{4.13.27}$$

$$\text{or,} \quad \ddot{\theta} = gc\left[\frac{\sin\theta\left(K^2 + a^2 + c^2 + 2ac\cos\theta\right) + (1-\cos\theta)\cdot 2ac\sin\theta}{\left(K^2 + a^2 + c^2 + 2ac\cos\theta\right)^2}\right]. \tag{4.13.28}$$

Now from (4.13.25), we get

$$\begin{aligned} R &= M\ddot{y} + Mg = M\left[g - c\cos\theta\cdot\dot{\theta}^2 - c\sin\theta\cdot\ddot{\theta}\right] \\ \text{or,} \quad R\big|_{\theta=\pi} &= M\left[g - c\cos\theta\cdot\dot{\theta}^2 - c\sin\theta\cdot\ddot{\theta}\right]_{\theta=\pi} \\ &= M\left[g + c\cdot\dot{\theta}^2\big|_{\theta=\pi}\right] \\ &= M\left[g + c\cdot\frac{4gc}{K^2 + a^2 + c^2 - 2ac}\right] \quad \text{[by (4.13.27)]} \\ &= Mg\left[1 + \frac{4c^2}{K^2 + (a-c)^2}\right]. \end{aligned}$$

▶ **Problem 4.13.7** *Two equal cylinders, each of mass m are bound together by an elastic string, whose tension T and roll with their axis horizontal down a rough plane of inclination α. Show that their acceleration is*

$$\frac{2}{3}g\sin\alpha\left[1 - \frac{2\mu T}{mg\sin\alpha}\right]$$

where μ is the coefficient of friction between the cylinders. **[VUH 2003]**

Solution: Let C_1 and C_2 be the centre of central section of the cylinders each of mass m and radius a. Let initially the point A_1 was at O. We take Ox and Oy as the axes at O, along and perpendicular to the plane of inclination α with the horizon. Let at any time t, the two fixed straight lines in both the cylinders make the same angle θ with fixed line in space and perpendicular to the plane. Let R_1 and F_1 are the normal reaction and frictional force at B_1 on the cylinder of centre C_1 and R_2 and F_2 are the normal reaction and frictional force at B_2 on the cylinder of centre C_2 and T be the tension in the string and R and μR are the reaction and the frictional force between the cylinder. Let in this time t, the cylinder of centre C_1 rolls down a distance $\bar{x}$ along the plane. Thus the coordinates of C_1 is $(\bar{x}, a)$ and that of C_2 is $(\bar{x} + 2a, a)$.

$$\therefore \quad \bar{x} = a\theta, \quad \dot{\bar{x}} = a\cdot\dot{\theta} \quad \text{and} \quad \ddot{\bar{x}} = a\ddot{\theta}. \tag{4.13.29}$$

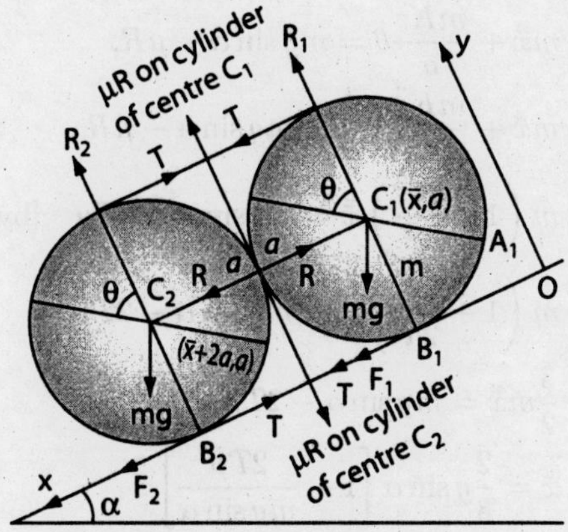

Fig 4.34

Now the equations of motion for the cylinder of centre C_1 is

$$m\ddot{\bar{x}} = mg\sin\alpha + 2T - R + F_1 \tag{4.13.30}$$

and for the cylinder of centre C_2 is

$$m\frac{d^2}{dt^2}(\bar{x}+2a) = m\ddot{\bar{x}} = mg\sin\alpha - 2T + R + F_2 \tag{4.13.31}$$

and

$$mK^2\ddot{\theta} = -F_1 a - \mu R\cdot a = -a(F_1 + \mu R) \tag{4.13.32}$$

for the cylinder of centre C_1 and

$$mK^2\ddot{\theta} = -F_2 a - \mu R\cdot a = -a(F_2 + \mu R) \tag{4.13.33}$$

for the cylinder of centre C_2.

From (4.13.32) and (4.13.33), we get

$$F_1 + \mu R = \mu_2 + \mu R \Rightarrow F_1 = F_2. \tag{4.13.34}$$

Again from (4.13.30) and (4.13.31) we get

$$mg\sin\alpha + 2T - R + F_1 = mg\sin\alpha - 2T + R + F_2$$

$$\text{or,}\quad 4T = 2R,\quad \text{or,}\quad R = 2T\quad [\text{As } F_1 = F_2]. \tag{4.13.35}$$

Thus from (4.13.30), we get

$$\begin{aligned} m\ddot{\bar{x}} &= mg\sin\alpha + 2T - 2T + F_1 \quad [\text{by (4.13.35)}] \\ &= mg\sin\alpha + F_1 = mg\sin\alpha - \frac{mK^2\ddot{\theta}}{a} - \mu R \quad [\text{by (4.13.32)}] \end{aligned}$$

$$\text{or,}\quad m\ddot{\bar{x}} + \frac{mK^2}{a}\ddot{\theta} = mg\sin\alpha - \mu R,$$

$$\text{or,}\quad m\ddot{\bar{x}} + \frac{mK^2}{a}\cdot\frac{\ddot{\bar{x}}}{a} = mg\sin\alpha - \mu R,$$

$$\text{or,}\quad m\left(1+\frac{K^2}{a^2}\right)\ddot{\bar{x}} = mg\sin\alpha - 2T\mu \quad [\text{by } (4.13.35)],$$

$$\text{or,}\quad m\left(1+\frac{a^2}{2a^2}\right)\ddot{\bar{x}} = mg\sin\alpha - 2T\mu,$$

$$\text{or,}\quad \frac{3}{2}m\ddot{\bar{x}} = mg\sin\alpha - 2T\mu,$$

$$\text{or,}\quad \ddot{\bar{x}} = \frac{2}{3}g\sin\alpha\left[1-\frac{2T\mu}{mg\sin\alpha}\right].$$

4.14 Two Dimensional Motion of a Body (Both Rolling and Sliding)

We have already consider the motion (Rolling only) of perfectly rough body having central circular section in the previous article. Our present work is to consider the motion of the imperfectly rough body having central circular section taking into account the sliding motion also.

To consider such type of motion we must have to reach to a definite idea that when a body rolls or slides. For this we know that if the velocity of the point of contact is not equal to zero, the body slides and when the velocity of the point of contact is zero, then the body may rolls.

Velocity of the point of contact for a body of central circular Section

Let a rough sphere moves on a plane. Let at any time t, the centre of mass G has the coordinates $(\bar{x}, a)$ referred to the Ox and Oy as axes and the sphere rotates an angle θ.

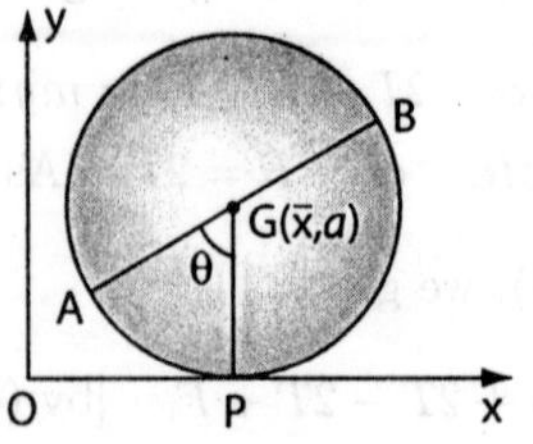

Fig 4.35

Thus the velocity of the point of contact of P at time t

$= $ velocity of centre of mass $G +$ velocity of P relative to G

$= \dot{\bar{x}} - a\dot{\theta}$. **[BUH 1999]**

4.15 Motion of an Imperfectly Rough Solid Sphere Down an Inclined Plane (Both Rolling and Sliding)

An imperfectly rough solid sphere move from rest down a plane making an angle α with horizon. Investigate the motion. [KUH '01, '03, '05, '06]

Initially the point A was at the point O. We take the Ox and Oy as axes at O along and perpendicular to the plane which makes an angle α with the horizon. Let at any time t, the centre of mass C describes a distance $\bar{x}$ along the plane. Let CA be a fixed line in the sphere which makes an θ with a fixed line in space and perpendicular to the plane. Let R and μR are the normal reaction and frictional force at the point of contact B at that time, where μ is the coefficient of friction.

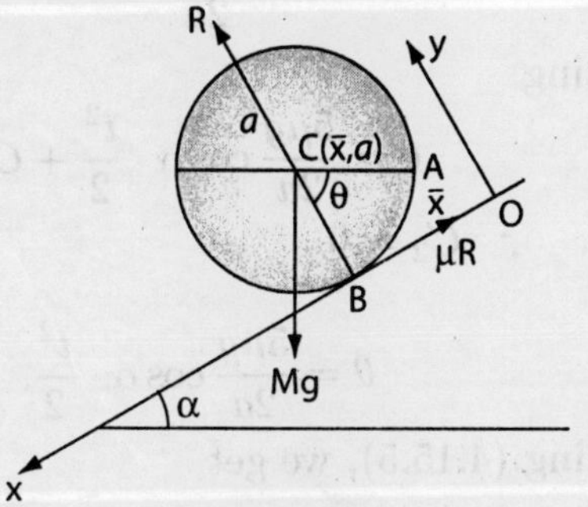

Fig 4.36

If the frictional force is not sufficient for pure rolling, then the sphere will slide as well as roll. Thus the maximum frictional force μR must act up the plane. Thus

$$\bar{x} = OB = \text{arc}\, AB = a\theta,$$

$$\text{or,} \quad \dot{\bar{x}} = a\dot{\theta} \quad \text{and} \quad \ddot{\bar{x}} = a \cdot \ddot{\theta}. \tag{4.15.1}$$

The equations of motion are

$$M\ddot{\bar{x}} = Mg\sin\alpha - \mu R \tag{4.15.2}$$

$$0 = M\ddot{\bar{y}} = R - Mg\cos\alpha \tag{4.15.3}$$

$$\text{and} \quad MK^2\ddot{\theta} = a \cdot \mu R. \tag{4.15.4}$$

From (4.15.2) and (4.15.3) we get

$$M\ddot{\bar{x}} = Mg\sin\alpha - Mg\mu\cos\alpha$$

$$\text{or,}\quad \ddot{\bar{x}} = g\sin\alpha - \mu g\cos\alpha = g(\sin\alpha - \mu\cos\alpha). \tag{4.15.5}$$

Thus the centre of mass moves along the plane with constant acceleration. Again using (4.15.3) in (4.15.4) we get

$$MK^2\ddot{\theta} = a\mu\cdot Mg\cos\alpha,\quad \text{or,}\quad \frac{2a^2}{5}\ddot{\theta} = a\mu g\cos\alpha,$$

$$\text{or,}\quad \ddot{\theta} = \frac{5\mu g}{2a}\cos\alpha. \tag{4.15.6}$$

Integrating, we get

$$\dot{\theta} = \frac{5\mu g\cos\alpha}{2a}\cdot t + C_1.$$

Initially, when $t = 0$, $\dot{\theta} = 0$. $\therefore\ C_1 = 0$.

$$\therefore\quad \dot{\theta} = \frac{5\mu g}{2a}\cdot\cos\alpha\cdot t. \tag{4.15.7}$$

Again integrating

$$\theta = \frac{5\mu g}{2a}\cos\alpha\cdot\frac{t^2}{2} + C_2$$

when $t = 0$, $\theta = 0$. $\therefore\ C_2 = 0$.

Thus

$$\theta = \frac{5\mu g}{2a}\cos\alpha\cdot\frac{t^2}{2}. \tag{4.15.8}$$

Again integrating (4.15.5), we get

$$\dot{\bar{x}} = g(\sin\alpha - \mu\cos\alpha)t + C_3$$

$$\text{and}\quad \bar{x} = g(\sin\alpha - \mu\cos\alpha)\frac{t^2}{2} + C_3t + C_4$$

when $t = 0$, $\dot{\bar{x}} = 0$, $\bar{x} = 0$. $\therefore\ C_3 = 0$ and $C_4 = 0$. Thus

$$\dot{\bar{x}} = g(\sin\alpha - \mu\cos\alpha)t \tag{4.15.9}$$

$$x = g(\sin\alpha - \mu\cos\alpha)\frac{t^2}{2}. \tag{4.15.10}$$

Now the velocity of the point of contact B, down the plane at the time t

$$\begin{aligned} = \dot{\bar{x}} - a\dot{\theta} &= g(\sin\alpha - \mu\cos\alpha)t - \frac{5}{2}\mu g\cos\alpha\cdot t \\ &= \frac{gt}{2}\left[2\sin\alpha - 2\mu\cos\alpha - 5\mu\cos\alpha\right] \\ &= \frac{gt}{2}\left[2\sin\alpha - 7\mu\cos\alpha\right]. \end{aligned} \tag{4.15.11}$$

From equation (4.15.11), it is clear that three important cases arises.

Case I. *If* $2\sin\alpha > 7\mu\cos\alpha$ *or* $\mu < \frac{2}{7}\tan\alpha$. **[VUH 2004]**

The sphere never rolls as the velocity of the point of contact is always positive for all values of t, hence the point of contact always slides down the plane and the maximum frictional force μR acts. All the equations of motion established above holds good throughout the entire motion.

Case II. *If* $2\sin\alpha = 7\mu\cos\alpha$ *or* $\mu = \frac{2}{7}\tan\alpha$.

In this case, the velocity of the point of contact is zero for all values of t and hence the sphere always rolls throughout the motion and the maximum frictional force μR acts. And all the equations of motion established above hold good.

Case III. *If* $2\sin\alpha < 7\mu\cos\alpha$ *or* $\mu > \frac{2}{7}\tan\alpha$. **[VUH 2006; KUH 2003]**

In this case, the velocity of the point of contact will be negative, i.e., the point of contact will slide up the plane which is impossible, if the maximum frictional force μR be allowed to act, since friction acts as a force in its maximum value to keep the point at rest. Thus in this case, the equations of motion established above do not hold good.

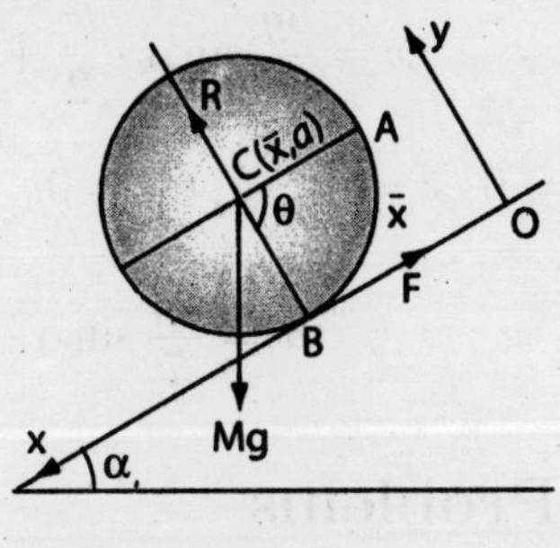

Fig 4.37

In this case, the equations of motion are

$$M\ddot{\bar{x}} = Mg\sin\alpha - F \tag{4.15.12}$$

$$0 = M\ddot{\bar{y}} = R - Mg\cos\alpha \tag{4.15.13}$$

$$MK^2\ddot{\theta} = F\cdot a. \tag{4.15.14}$$

Again the velocity of point of contact,

$$\dot{\bar{x}} - a\dot{\theta} = 0, \tag{4.15.15}$$

$$\text{or,}\quad \ddot{\bar{x}} = a\ddot{\theta}. \tag{4.15.16}$$

From (4.15.14) and (4.15.12), we get

$$MK^2\ddot{\theta} = a(Mg\sin\alpha - M\ddot{x}),$$

$$\text{or,}\quad \frac{2a^2}{5}\ddot{\theta} = ag\sin\alpha - a^2\ddot{\theta} \quad [\text{by } (4.15.16)],$$

$$\text{or,}\quad \frac{7a}{5}\cdot\ddot{\theta} = g\sin\alpha,$$

$$\text{or,}\quad \ddot{\bar{x}} = \frac{5g}{7}\sin\alpha \quad [\text{by } (4.15.16)]. \tag{4.15.17}$$

Integrating,

$$\dot{\bar{x}} = \frac{5g}{7}\sin\alpha\cdot t + C_4.$$

When $t = 0$, $\dot{\bar{x}} = 0$. $\therefore\ C_4 = 0$

$$\therefore\quad \dot{\bar{x}} = \frac{5g}{7}\sin\alpha\cdot t. \tag{4.15.18}$$

Again integrating, we get

$$\bar{x} = \frac{5g}{7}\sin\alpha\cdot\frac{t^2}{2} + C_5.$$

Initially, when $t = 0$, $\bar{x} = 0$. $\therefore\ C_5 = 0$.

$$\therefore\quad \bar{x} = \frac{5g}{7}\sin\alpha\cdot\frac{t^2}{2}. \tag{4.15.19}$$

4.16 Solved Problems

► **Problem 4.16.1** *A homogeneous sphere, of radius a, rotating with angular velocity ω about a horizontal diameter, is gently placed on a table whose coefficient of friction is μ. Show that there will be slipping at the point of contact for a time $\frac{2\omega a}{7\mu g}$, and that then the sphere will roll with constant angular velocity $\frac{2\omega}{7}$.* **[CUH 2002; BUH 2002, 2005; VUH 2001]**

Solution: Let $(\bar{x}, a)$ be the coordinates of the centre of gravity G of the sphere of mass M and radius a, referred to the axes Ox and Oy. Let AC be straight line fixed in the sphere makes an angle θ with a fixed line in space and perpendicular to Ox. Let R be the normal reaction at the point of contact at any time t and μR be the frictional force.

Initially $\dot{\bar{x}} = 0$.

Thus the initial velocity of point of contact $B = \dot{x} - a\dot{\theta} = -a\omega \neq 0$.

Hence initially the body slides.

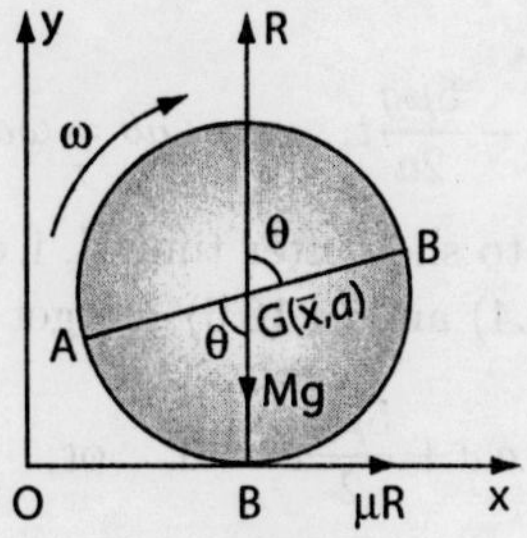

Fig 4.38

Now the equations of motion are

$$M\ddot{\bar{x}} = \mu R \tag{4.16.1}$$

$$0 = M\ddot{\bar{y}} = R - Mg \tag{4.16.2}$$

$$\text{and} \quad MK^2\ddot{\theta} = -\mu R \cdot a \tag{4.16.3}$$

From (4.16.1) and (4.16.2), we get

$$M\ddot{\bar{x}} = \mu \cdot Mg, \quad \text{or,} \quad \ddot{\bar{x}} = \mu g.$$

Integrating we get

$$\dot{\bar{x}} = \mu g t + C_1$$

when $t = 0$, $\dot{\bar{x}} = 0$, so $C_1 = 0$.

Thus,

$$\dot{\bar{x}} = \mu g t. \tag{4.16.4}$$

Again integrating

$$\bar{x} = \mu g \cdot \frac{t^2}{2} + C_2.$$

Initially $t = 0$, $\bar{x} = 0$. $\therefore$ $C_2 = 0$.

Thus,

$$\bar{x} = \mu g \cdot \frac{t^2}{2}. \tag{4.16.5}$$

Again from (4.16.3) and (4.16.2), we get

$$MK^2\ddot{\theta} = -\mu a \cdot Mg, \quad \text{or,} \quad K^2\ddot{\theta} = -\mu a g,$$

$$\text{or,} \quad \frac{2a^2}{5}\ddot{\theta} = -\mu a g \quad \left[\text{as for solid sphere } K^2 = \frac{2a^2}{5}\right]$$

$$\text{or,} \quad \ddot{\theta} = -\frac{5\mu g}{2a}.$$

Integrating

$$\dot{\theta} = -\frac{5\mu g}{2a} \cdot t + C_3$$

when $t = 0$, $\dot{\theta} = \omega$. $\therefore$ $C_3 = \omega$.

Thus

$$\dot{\theta} = \omega - \frac{5\mu g}{2a}t, \quad \text{or,} \quad a\dot{\theta} = \omega a - \frac{5\mu g}{2}t. \tag{4.16.6}$$

Let the sphere cases to slide after time t', i. e., at $t = t'$, $\dot{\bar{x}} - a\dot{\theta} = 0$. Therefore from (4.16.4) and (4.16.6), we get

$$\mu g t' - a\omega + \frac{5\mu g}{2}t' = 0, \quad \text{or,} \quad t' = \frac{2a\omega}{7\mu g}.$$

And the angular velocity of the sphere at $t = t'$, we get from (4.16.6)

$$\dot{\theta} = \omega - \frac{5\mu g}{2a}t' = \omega - \frac{5\mu g}{2a} \cdot \frac{2a\omega}{7\mu g} = \frac{2\omega}{7}.$$

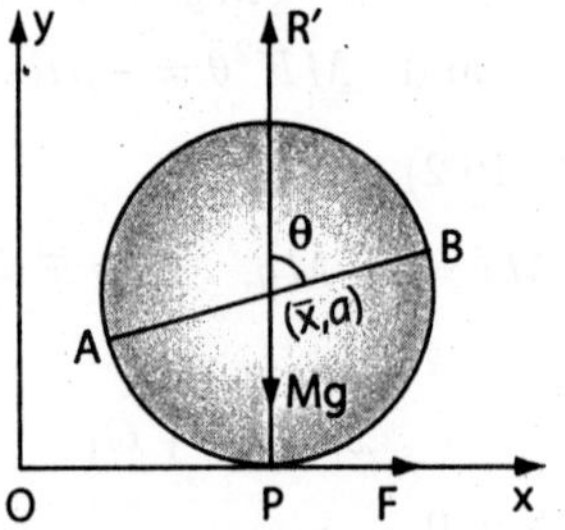

Fig 4.39

Thus after this time t', the sphere may roll. In this case the equations of motion are

$$M\ddot{\bar{x}} = F \tag{4.16.7}$$
$$0 = M\ddot{\bar{y}} = R' - Mg \tag{4.16.8}$$
$$MK^2\ddot{\phi} = -Fa. \tag{4.16.9}$$

Again,

$$\bar{x} = a\phi; \quad \dot{\bar{x}} = a\dot{\phi} \quad \text{or,} \quad \ddot{\bar{x}} = a\ddot{\phi}, \tag{4.16.10}$$

as there is no sliding.

From (4.16.7) and (4.16.9), we get

$$MK^2\ddot{\phi} = -M\ddot{\bar{x}}a, \quad \text{or,} \quad K^2\ddot{\phi} = -a\ddot{\bar{x}}, \quad \text{or,} \quad K^2\frac{\ddot{\bar{x}}}{a} = -\ddot{\bar{x}}a,$$

$$\text{or,} \quad \left(\frac{K^2}{a^2} + 1\right)\ddot{\bar{x}} = 0. \quad \therefore\ \ddot{\bar{x}} = 0. \quad \therefore\ F = 0 \quad [\text{by } (4.16.7)].$$

Thus

$$\ddot{\phi} = 0 \quad [\text{by } (4.16.9)]. \tag{4.16.11}$$

Thus for rolling, no frictional force is required throughout the motion. Integrating we get from (4.16.11)

$$\dot{\phi} = \text{constant} = \text{initial value} = \frac{2\omega}{7}.$$

Thus after time t', the sphere moves continuously with constant angular velocity $\frac{2\omega}{7}$.

► **Problem 4.16.2** *A uniform circular disc of radius a and mass M is projected along a rough horizontal plane, with velocity V and an initial angular velocity Ω about its centre. Investigate the motion.*

[BUH 2000; VUH 2003]

Solution:

Case I. *If* $V > a\Omega$.

In this case, the initial velocity of the point of contact $P = V - a\Omega > 0$. So the frictional force μR acts towards origin O. Let at any time t, the disc rotates an angle θ and c.g. G of the disc describe a distance $\bar{x}$.

$$\therefore \quad \bar{x} = a\theta, \quad \dot{\bar{x}} = a\dot{\theta}; \quad \ddot{\bar{x}} = a\ddot{\theta}.$$

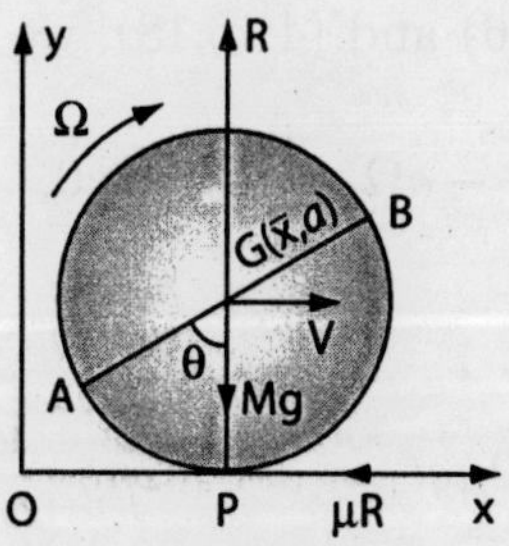

Fig 4.40

Now the equations of motion are

$$M\ddot{\bar{x}} = -\mu R \tag{4.16.12}$$
$$0 = M\ddot{\bar{y}} = R - Mg \tag{4.16.13}$$
$$MK^2\ddot{\theta} = \mu R \cdot a. \tag{4.16.14}$$

From (4.16.12) and (4.16.13), we get

$$M\ddot{\bar{x}} = -\mu Mg, \quad \text{or,} \quad \ddot{\bar{x}} = -\mu g. \tag{4.16.15}$$

Integrating, we get

$$\dot{x} = -\mu g t + C_1,$$

when $\dot{x} = V$, $t = 0$. $\therefore\ C_1 = V$.

Thus

$$\dot{x} = V - \mu g t. \tag{4.16.16}$$

From (4.16.14) and (4.16.13), we get

$$MK^2\ddot{\theta} = \mu a \cdot Mg, \quad \text{or,} \quad K^2\ddot{\theta} = \mu a g,$$

$$\text{or,} \quad \frac{a^2}{2}\ddot{\theta} = \mu a g,$$

$$\text{or,} \quad a\ddot{\theta} = 2\mu g.$$

Integrating, we get

$$a\dot{\theta} = 2\mu g t + C_2. \tag{4.16.17}$$

Initially $\dot{\theta} = \Omega$, $t = 0$. $\therefore\ C_2 = a\Omega$.

Thus

$$a\dot{\theta} = a\Omega + 2\mu g t. \tag{4.16.18}$$

If after time t_1, the disc begins rolling the velocity of the point of contact P, i.e., $\left(\dot{x} - a\dot{\theta}\right)\big|_{t=t_1} = 0$.

Thus using (4.16.16) and (4.16.18), we get

$$V - \mu g t_1 - a\Omega - 2\mu g t_1 = 0, \quad \text{or,} \quad t_1 = \frac{V - a\Omega}{3\mu g}.$$

From (4.16.18),

$$a\dot{\theta}\big|_{t=t_1} = a \cdot \Omega + 2\mu g t_1 = a\Omega + 2\mu g \cdot \frac{V - a\Omega}{3\mu g} = \frac{a\Omega + 2V}{3}. \tag{4.16.19}$$

Thus after time $t_1 = \frac{V - a\Omega}{3\mu g}$, the disc begins to rolls and at this time the velocity of the centre of disc is

$$\begin{aligned} \dot{x}\big|_{t=t_1} &= V - \mu g \cdot t_1 = V - \mu g \cdot \frac{V - a\Omega}{3\mu g} \\ &= \frac{3V - V + a\Omega}{3} = \frac{2V + a\Omega}{3}. \end{aligned} \tag{4.16.20}$$

When the disc rolls, the equations of motion are

$$M\ddot{x} = -F \tag{4.16.21}$$

$$0 = M\ddot{y} = R' - Mg \tag{4.16.22}$$

$$MK^2\ddot{\phi} = F \cdot a \tag{4.16.23}$$

$$\text{and} \quad \dot{x} - a\dot{\phi} = 0 \quad \text{or,} \quad \ddot{x} = a\ddot{\phi} \quad \text{[for rolling]} \tag{4.16.24}$$

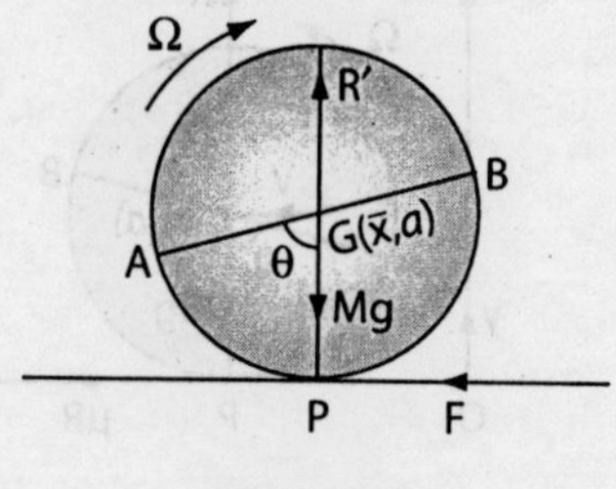

Fig 4.41

From (4.16.21) and (4.16.19), we get

$$MK^2\ddot{\phi} = -aM\ddot{\bar{x}}, \quad \text{or,} \quad K^2\ddot{\phi} = -a\ddot{\bar{x}},$$

$$\text{or,} \quad K^2 \cdot \frac{\ddot{\bar{x}}}{a} + a\ddot{\bar{x}} = 0 \quad [\text{by (4.16.24)}]$$

$$\text{or,} \quad \left(\frac{a^2}{2a} + a\right)\ddot{\bar{x}} = 0 \quad [\text{as for disc } K^2 = \tfrac{a^2}{2}]$$

$$\text{or,} \quad \frac{3a}{2}\ddot{\bar{x}} = 0$$

$\therefore\ \ddot{\bar{x}} = 0$. Thus from (4.16.21); $F = 0$.
Again from (4.16.23), $\ddot{\phi} = 0$.
Thus for rolling the equations (4.16.21) and (4.16.23) become as

$$\left.\begin{aligned} M\ddot{\bar{x}} &= 0 \\ MK^2\ddot{\phi} &= 0 \end{aligned}\right\}. \tag{4.16.25}$$

Thus $\ddot{\bar{x}} = 0$ or $\dot{\bar{x}}$ = constant = initial value = $\frac{2V+a\Omega}{3}$ and $\dot{\phi}$ = constant = initial value = $\frac{a\Omega+2V}{3a}$, [by (4.16.20)].
Thus the disc is continuously rolling with angular velocity $\frac{a\Omega+2V}{3a}$.

Case II. *If* $V < a\Omega$.

In this case, the initial velocity of the point of contact $P = V - a\Omega < 0$, thus the frictional force μR acts in the direction of x-axes. Let at any time t, the disc rotates an angle θ and the centre of gravity describe a distance $\bar{x}$

$$\therefore \quad \bar{x} = a\theta \quad \text{or,} \quad \dot{\bar{x}} = a\dot{\theta} \quad \text{and} \quad \ddot{\bar{x}} = a\ddot{\theta}. \tag{4.16.26}$$

The equations of motion are

$$M\ddot{\bar{x}} = \mu R \tag{4.16.27}$$

$$0 = M\ddot{\bar{y}} = R - Mg \tag{4.16.28}$$

$$\text{and} \quad MK^2\ddot{\theta} = -\mu R \cdot a. \tag{4.16.29}$$

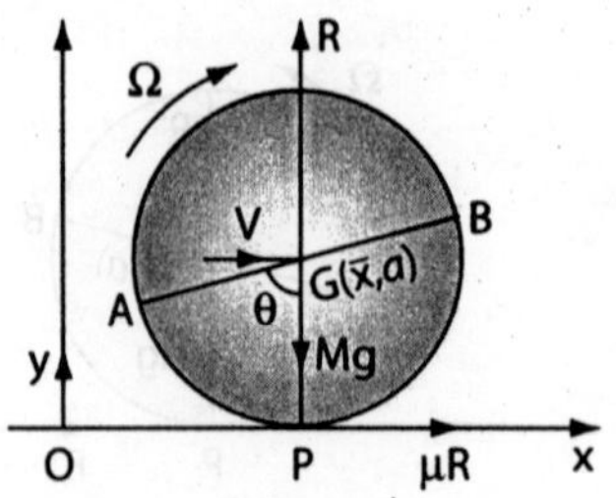

Fig 4.42

From (4.16.27) and (4.16.28), we get

$$M\ddot{\bar{x}} = \mu Mg \quad \text{or,} \quad \ddot{\bar{x}} = \mu g. \tag{4.16.30}$$

Integrating

$$\dot{\bar{x}} = \mu gt + C_3.$$

Initially, $\dot{\bar{x}} = V$, $t = 0$, then $C_3 = V$.

$$\dot{\bar{x}} = \mu gt + V. \tag{4.16.31}$$

Again from (4.16.29) and (4.16.28), we get

$$MK^2\ddot{\theta} = -\mu a \cdot Mg \quad \text{or,} \quad K^2\ddot{\theta} = -\mu ag$$

$$\text{or,} \quad \frac{a^2}{2}\ddot{\theta} = -\mu ag \quad \left[\text{as for disc } K^2 = \tfrac{a^2}{2}\right]$$

$$\text{or,} \quad a\ddot{\theta} = -2\mu g.$$

Integrating

$$a\dot{\theta} = -2\mu gt + C_4.$$

Initially $\dot{\theta} = \Omega$ when $t = 0$. $\therefore$ $C_4 = a\Omega$.

$$\therefore \quad a\dot{\theta} = a\Omega - 2\mu gt. \tag{4.16.32}$$

Let the disc begins to roll after time t_2, then the velocity of the point of contact at $t = t_2$, will be zero, i. e.,

$$\dot{\bar{x}} - a\dot{\theta}\big|_{t=t_2} = \mu gt_2 + V - a\Omega + 2\mu gt_2 = 0 \quad [\text{by (4.16.31) and (4.16.32)}]$$

$$\therefore \quad t_2 = \frac{a\Omega - V}{3\mu g}.$$

So, in this case, the disc begins to roll after time $t_2 = \frac{a\Omega - V}{3\mu\mu}$ and at this time, the velocity of the centre of the disc is

$$\begin{aligned} \dot{\bar{x}}\big|_{t=t_2} &= V + \mu g \cdot t_2 = V + \mu g \cdot \frac{a\Omega - V}{3\mu g} \quad [\text{by (4.16.31)}] \\ &= \frac{2V + a\Omega}{3} \end{aligned} \tag{4.16.33}$$

and the angular velocity about the centre is

$$\begin{aligned}\dot{\theta}\big|_{t=t_2} &= \Omega - \frac{2\mu g}{a}\cdot t_2 \quad \text{[by (4.16.32)]}\\ &= \Omega - \frac{2\mu g}{a}\cdot\frac{a\Omega - V}{3\mu g} = \frac{a\Omega + 2V}{3a}. \qquad (4.16.34)\end{aligned}$$

For rolling, the equations of motion are

$$M\ddot{x} = F \qquad (4.16.35)$$

$$0 = M\ddot{y} = R' - Mg \qquad (4.16.36)$$

$$MK^2\ddot{\phi} = -Fa \qquad (4.16.37)$$

$$\text{and} \quad \dot{x} - a\dot{\phi} = 0 \qquad (4.16.38)$$

$$\text{or,} \quad \ddot{x} = a\ddot{\phi}. \qquad (4.16.39)$$

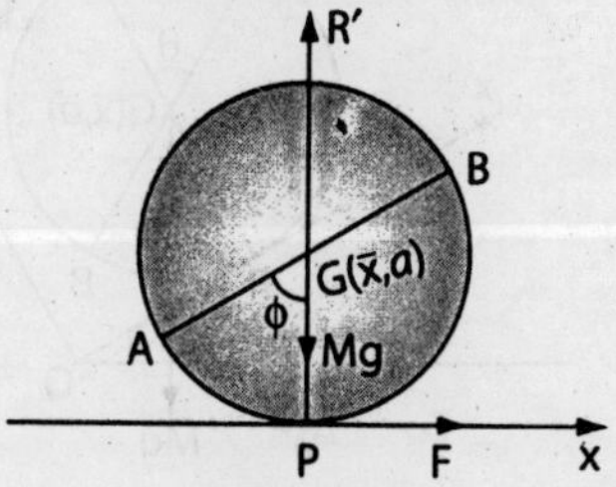

Fig 4.43

From (4.16.35) and (4.16.37) we have

$$\begin{aligned} MK^2\ddot{\phi} &= -aM\ddot{x}\\ \text{or,}\quad K^2\ddot{\phi} &= -a\ddot{x} = -a^2\ddot{\phi} \quad \text{[by (4.16.38)]}\\ \text{or,}\quad (K^2+a^2)\,\ddot{\phi} &= 0.\end{aligned}$$

Thus

$$\ddot{\phi} = 0, \qquad (4.16.40)$$

then by (4.16.37), $F = 0$.
Thus by (4.16.35), we get

$$\ddot{x} = 0. \qquad (4.16.41)$$

Integrating (4.16.40), we get

$$\dot{\phi} = \text{constant} = \text{initial value} = \frac{a\Omega + 2V}{3a}.$$

Thus, the disc rolls continuously with constant angular velocity $\frac{a\Omega+2V}{3a}$.

► **Problem 4.16.3** *A solid circular cylinder rotating about its axis is placed gently with its axis horizontal on a rough plane, whose inclination to the horizon is α. Initially the friction acts up the plane and the coefficient of friction is μ. Show that the cylinder will move upwards if $\mu > \tan\alpha$, and find the time that elapses before rolling takes place.*

Solution: We consider the Ox and Oy as axes along and perpendicular to the plane, making an angle α with the horizon. Let the cylinder is rotating with angular velocity ω and R be the normal reaction at the point of contact P and μR be the frictional force up the plane. Let at any time t, the centre of the central section of the cylinder of radius a describes a distance $\bar{x}$ along the plane and the cylinder rotates an angle θ in this time. Thus $(\bar{x}, a)$ be the coordinate of the centre of the central section G at any time.

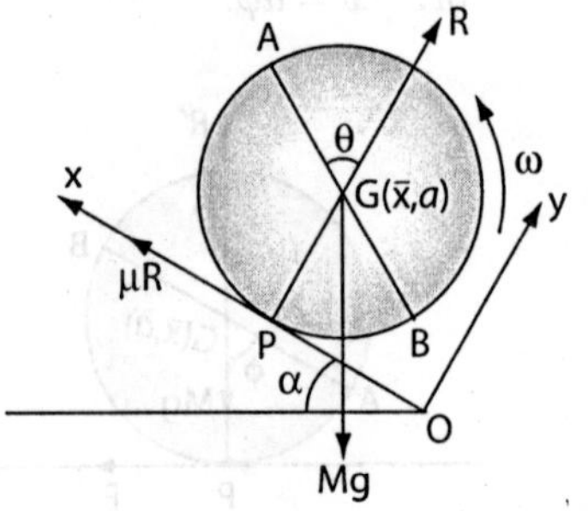

Fig 4.44

Initially, $\bar{x} = 0$, thus the initial velocity of the point of contact

$$\dot{\bar{x}} - a\dot{\theta} = -a\omega \neq 0.$$

Hence, initially the cylinder slides.
The equations of motion are

$$M\ddot{\bar{x}} = \mu R - Mg\sin\alpha \tag{4.16.42}$$

$$0 = M\ddot{\bar{y}} = R - Mg\cos\alpha \tag{4.16.43}$$

$$MK^2\ddot{\theta} = -\mu R \cdot a. \tag{4.16.44}$$

From (4.16.43) and (4.16.42), we get

$$M\ddot{\bar{x}} = \mu \cdot Mg\cos\alpha - Mg\sin\alpha$$

$$\text{or,}\quad \ddot{\bar{x}} = g(\mu\cos\alpha - \sin\alpha).$$

Integrating we get,

$$\dot{\bar{x}} = g(\mu\cos\alpha - \sin\alpha)t + C_1.$$

Initially, $t = 0$, $\dot{\bar{x}} = 0$. $\therefore\ C_1 = 0$
Thus,

$$\dot{\bar{x}} = g(\mu \cos\alpha - \sin\alpha) \cdot t. \tag{4.16.45}$$

Again from (4.16.44) and (4.16.43), we get

$$MK^2\ddot{\theta} = -\mu a \cdot Mg\cos\alpha, \quad \text{or,} \quad K^2\ddot{\theta} = -\mu g a \cos\alpha$$

$$\text{or,} \quad \frac{a^2}{2}\ddot{\theta} = -\mu g a\cos\alpha \quad [\because \text{ for disc } K^2 = a^2/2]$$

$$\text{or,} \quad \ddot{\theta} = -\frac{2\mu g}{a}\cos\alpha.$$

Integrating

$$a\dot{\theta} = -2\mu g\cos\alpha \cdot t + C_2.$$

Initially, $t = 0$, $\dot{\theta} = \omega$. $\therefore\ C_2 = a\omega$.
Thus,

$$a\dot{\theta} = a\omega - 2\mu g\cos\alpha \cdot t. \tag{4.16.46}$$

If the cylinder moves up the plane, then $\dot{\bar{x}} > 0$.
Thus from (4.16.45), we get

$$g(\mu\cos\alpha - \sin\alpha)t > 0, \quad \text{or,} \quad \mu > \tan\alpha.$$

If after time t', the cylinder begins to roll, then at this time t' the velocity of the point of contact is zero, i.e.,

$$\dot{\bar{x}} - a\dot{\theta}\big|_{t=t'} = 0.$$

Thus using (4.16.45) and (4.16.46), we get

$$g(\mu\cos\alpha - \sin\alpha)t' - a\omega + 2\mu g\cos\alpha \cdot t' = 0,$$

$$\text{or,} \quad t' = \frac{a\omega}{g(3\mu\cos\alpha - \sin\alpha)}.$$

► **Problem 4.16.4** *A circular homogeneous plate is projected up a rough inclined plane with velocity V with no rotation, the plane of the plate being in the plane of greatest slope. Show that the plate stops sliding after a time*

$$\frac{V}{g\{(3\mu\cos\alpha + \sin\alpha)\}}$$

where μ is the coefficient of friction and α is the inclination of the plane with the horizon. **[CUH 2003, 2007(O)]**

Solution: We take the Ox and Oy axes along and perpendicular to the plane, making an angle α with the horizon. Let at any time t, the circular plate of radius a and mass M, describes a distance $\bar{x}$ along the plane and the plate rotates an angle θ in this time. Thus, $(\bar{x}, a)$ be the coordinates of the c.g. G at that time. The initial velocity of the point of contact $= \dot{\bar{x}} - a\dot{\theta} = V > 0$. Thus the plate slides up

$$\therefore \quad \bar{x} = a\theta \quad \text{and} \quad \ddot{\bar{x}} = a\ddot{\theta}. \tag{4.16.47}$$

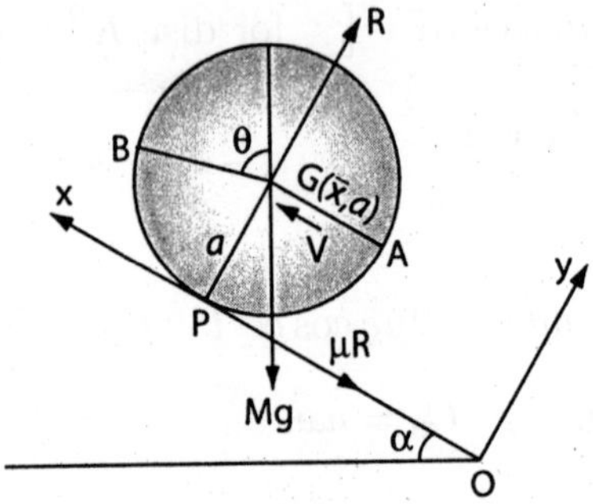

Fig 4.45

The equations of motion are

$$M\ddot{\bar{x}} = -\mu R - Mg\sin\alpha \tag{4.16.48}$$
$$0 = M\ddot{\bar{y}} = R - Mg\cos\alpha \tag{4.16.49}$$
$$MK^2\ddot{\theta} = \mu R \cdot a \tag{4.16.50}$$

From (4.16.48) and (4.16.49), we get

$$M\ddot{\bar{x}} = -\mu \cdot Mg\cos\alpha - Mg\sin\alpha$$
$$\text{or,} \quad \ddot{\bar{x}} = -g(\mu\cos\alpha + \sin\alpha).$$

Integrating,

$$\dot{\bar{x}} = -g(\mu\cos\alpha + \sin\alpha) \cdot t + C_1.$$

Initially, $t = 0$, $\dot{\bar{x}} = V$. $\therefore$ $C_1 = V$.
Thus,

$$\dot{\bar{x}} = V - g(\mu\cos\alpha + \sin\alpha) \cdot t. \tag{4.16.51}$$

Again from (4.16.50) and (4.16.49), we get

$$MK^2\ddot{\theta} = \mu a \cdot Mg\cos\alpha, \quad \text{or,} \quad K^2\ddot{\theta} = \mu ag\cos\alpha,$$
$$\text{or,} \quad \frac{a^2}{2}\ddot{\theta} = \mu ag\cos\alpha,$$
$$\text{or,} \quad a\ddot{\theta} = 2\mu g\cos\alpha.$$

Integrating,

$$a\dot{\theta} = 2\mu g \cos\alpha \cdot t + C_2.$$

Initially, $t = 0$, $\dot{\theta} = 0$. $\therefore\ C_2 = 0$.
Thus,

$$a\dot{\theta} = 2\mu a \cos\alpha \cdot t. \tag{4.16.52}$$

If at $t = t_1$, the circular plate stops sliding, then the velocity of the point of contact at $t = t_1$ is zero, i. e.,

$$\left(\dot{\bar{x}} - a\dot{\theta}\right)\Big|_{t=t_1} = 0.$$

Thus from (4.16.51) and (4.16.52) we get

$$V - g(\mu\cos\alpha + \sin\alpha)t_1 - 2\mu g\cos\alpha\, t_1 = 0,$$

$$\text{or,}\quad t_1 = \frac{V}{g[3\mu\cos\alpha + \sin\alpha]}.$$

► **Problem 4.16.5** *A uniform sphere rolls in a straight line on a rough horizontal plane and is acted upon by a force X at its centre in a direction opposite to the motion of the centre. Show that the centre of the sphere moves as it would if the mass were collected there and the force reduced to $\frac{5X}{7}$ and the friction is equal to $\frac{2X}{7}$ and is in a direction opposite to that of X.*

[CUH 1988, 1996]

Solution: Let Ox and Oy be taken as x and y axes. Let at any time t, the sphere rolls an angle θ and describes a distance $\bar{x}$ along Ox. Let R and F are the normal reaction and frictional force at the point of contact P. Thus, the coordinate of the centre of the sphere of mass M and radius a at that time t is $(\bar{x}, a)$

$$\therefore\quad \bar{x} = a\theta \quad \text{or,}\quad \ddot{\bar{x}} = a\ddot{\theta}. \tag{4.16.53}$$

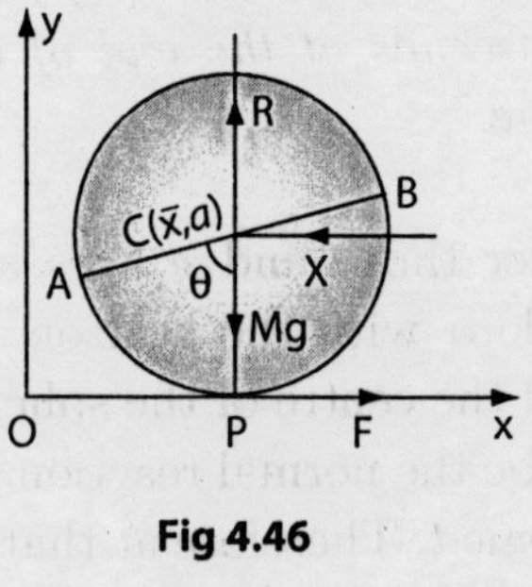

Fig 4.46

The equations of motion are

$$M\ddot{\bar{x}} = F - X \tag{4.16.54}$$

$$0 = M\ddot{\bar{y}} = R - Mg \tag{4.16.55}$$

$$\text{and } MK^2\ddot{\theta} = -Fa. \tag{4.16.56}$$

From (4.16.54) and (4.16.56), we get

$$M\ddot{\bar{x}} = -\frac{MK^2\ddot{\theta}}{a} - X$$

$$\text{or, } M\left(1 + \frac{K^2}{a^2}\right)\ddot{\bar{x}} = -X \quad [\text{by } (4.16.53)]$$

$$\text{or, } M\left(1 + \frac{2}{5}\frac{a^2}{a^2}\right)\ddot{\bar{x}} = -X \quad \left[\text{as } K^2 = \frac{2a^2}{5} \text{ for sphere}\right]$$

$$\text{or, } M\ddot{\bar{x}} = -\frac{5}{7}X \tag{4.16.57}$$

which is equivalent to the motion of the mass M collected at the centre under the action of the force $\frac{5X}{7}$ which acts in the opposite direction to that of the motion of the sphere, i.e., X reduced to $\frac{5X}{7}$.

From (4.16.54), we get

$$F = M\ddot{\bar{x}} + X = -\frac{5}{7}X + X \quad [\text{by } (4.16.57)]$$

$$= \frac{2X}{7}.$$

▶ **Problem 4.16.6** *A sphere, of radius a is projected up an inclined plane with velocity V and angular velocity Ω in the sense which would cause it to roll up; it $V > a\Omega$ and the coefficient of friction $> \frac{2}{7}\tan\alpha$, show that the sphere will cease to ascends at the end of a time $\frac{5V+2a\Omega}{5g\sin\alpha}$; where α is the inclination of the plane.* **[CUH 1984, 1989; BUH 1999]**

Solution: We consider the x and y axes along and perpendicular to the plane making an angle α with the horizon. Let at any time t, the sphere rotates an angle θ and the centre of the sphere moves a distance $\bar{x}$ along the plane. Let R and μR be the normal reaction and frictional force at the point of contact P at that time t. Therefore at that time t, $(\bar{x}, a)$ be the coordinate of the centre of the sphere of radius a and mass M. Initially the velocity of the point of contact $\dot{\bar{x}} - a\dot{\theta} = V - a\Omega > 0$. So initially the sphere slides up the plane. Thus

$$\bar{x} = a\theta \quad \text{or,} \quad \ddot{\bar{x}} = a\ddot{\theta} \tag{4.16.58}$$

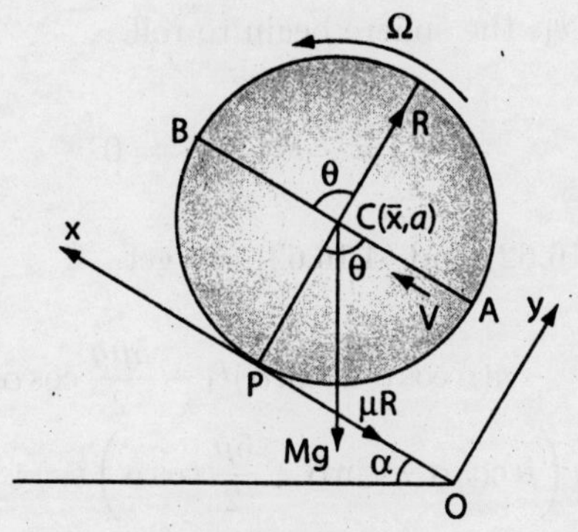

Fig 4.47

The equations of motion are

$$M\ddot{\bar{x}} = -\mu R - Mg\sin\alpha \tag{4.16.59}$$

$$0 = M\ddot{\bar{y}} = R - Mg\cos\alpha \tag{4.16.60}$$

$$\text{and} \quad MK^2\ddot{\theta} = \mu R\cdot a. \tag{4.16.61}$$

From (5.16.59) and (5.16.60), we get

$$M\ddot{\bar{x}} = -\mu Mg\cos\alpha - Mg\sin\alpha = -Mg[\mu\cos\alpha + \sin\alpha]$$

$$\text{or,} \quad \ddot{\bar{x}} = -g[\mu\cos\alpha + \sin\alpha]$$

Integrating,

$$\dot{\bar{x}} = -g[\mu\cos\alpha + \sin\alpha]\cdot t + C_1.$$

Initially $t = 0$, $\dot{\bar{x}} = V$. $\therefore\ C_1 = V$.
Thus,

$$\dot{\bar{x}} = V - g[\mu\cos\alpha + \sin\alpha]t. \tag{4.16.62}$$

From (4.16.61) and (4.16.60), we get

$$MK^2\ddot{\theta} = \mu a\cdot Mg\cos\alpha, \quad \text{or,} \quad K^2\ddot{\theta} = \mu ag\cos\alpha,$$

$$\text{or,} \quad \frac{2a^2}{5}\ddot{\theta} = \mu ag\cos\alpha \quad \left(\text{as } K^2 = \tfrac{2a^2}{5} \text{ for sphere}\right),$$

$$\text{or,} \quad a\ddot{\theta} = \frac{5\mu g}{2}\cos\alpha.$$

Integrating,

$$a\dot{\theta} = \frac{5\mu g}{2}\cos\alpha\cdot t + C_2.$$

Initially, $t = 0$, $\dot{\theta} = \Omega$. $\therefore\ C_2 = a\Omega$.
Thus

$$a\dot{\theta} = a\Omega + \frac{5\mu g}{2}\cos\alpha\cdot t. \tag{4.16.63}$$

Let after time t_1, the sphere begin to roll.
Thus

$$\left(\dot{\bar{x}} - a\dot{\theta}\right)\Big|_{t=t_1} = 0.$$

Thus from (4.16.62) and (4.16.63), we get

$$V - g(\mu\cos\alpha + \sin\alpha)t_1 - \frac{5\mu g}{2}\cos\alpha\, t_1 - a\Omega = 0$$

$$g\left(\mu\cos\alpha + \sin\alpha + \frac{5\mu}{2}\cos\alpha\right)t_1 = V - a\Omega$$

$$\text{or,}\quad t_1 = \frac{2(V - a\Omega)}{(7\mu\cos\alpha + 2\sin\alpha)g} \tag{4.16.64}$$

and from (4.16.62),

$$\begin{aligned}\dot{\bar{x}}\Big|_{t=t_1} &= V - g(\mu\cos\alpha + \sin\alpha)\cdot\frac{2(V - a\Omega)}{(7\mu\cos\alpha + 2\sin\alpha)g} \\ &= V - \frac{2(\mu\cos\alpha + \sin\alpha)(V - a\Omega)}{(7\mu\cos\alpha + 2\sin\alpha)}.\end{aligned} \tag{4.16.65}$$

Thus after time t_1, the motion of the sphere be rolling one.
The equations of motion, in this case are

$$M\ddot{\bar{x}} = -F - Mg\sin\alpha \tag{4.16.66}$$

$$0 = M\ddot{\bar{y}} = R' - Mg\cos\alpha \tag{4.16.67}$$

$$MK^2\ddot{\phi} = F\cdot a \tag{4.16.68}$$

$$\text{and}\quad \dot{\bar{x}} - a\dot{\phi} = 0 \quad\text{and}\quad \ddot{\bar{x}} = a\ddot{\phi}. \tag{4.16.69}$$

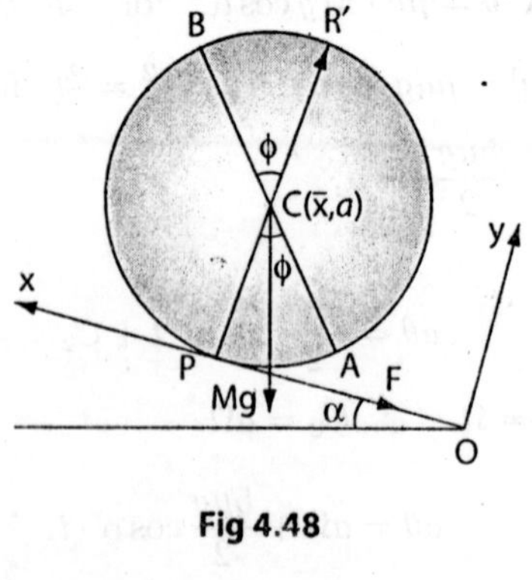

Fig 4.48

From (5.16.66) and (5.16.68), we get

$$M\ddot{x} = -\frac{MK^2\ddot{\phi}}{a} - Mg\sin\alpha,$$

$$\text{or,}\quad \ddot{x} = -\frac{K^2\ddot{x}}{a^2} - g\sin\alpha \quad \text{[by (4.16.69)]},$$

$$\text{or,}\quad \left(1 + \frac{K^2}{a^2}\right)\ddot{x} = -g\sin\alpha,$$

$$\text{or,}\quad \left(1 + \frac{2a^2}{5a^2}\right)\ddot{x} = -g\sin\alpha,$$

$$\text{or,}\quad \ddot{x} = -\frac{5g}{2}\sin\alpha. \tag{4.16.70}$$

$\therefore$ From (4.16.66),

$$F = -M\ddot{x} - Mg\sin\alpha = M\left(\frac{5g}{7}\sin\alpha - g\sin\alpha\right)$$

$$= -\frac{2Mg}{7}\sin\alpha. \tag{4.16.71}$$

Now

$$|\mu| = \left|\frac{F}{R'}\right| = \left|\frac{\frac{2Mg}{7}\sin\alpha}{Mg\cos\alpha}\right| = \frac{2}{7}\tan\alpha \quad \text{[by (4.16.67)]}.$$

Thus as $\mu > \frac{2}{7}\tan\alpha$ (given) then the rolling motion is possible.
Again integrating (4.16.70), we get

$$\dot{x} = -\frac{5g}{7}\sin\alpha \cdot t + C_3.$$

The initial values are, $t = t_1$, $\dot{x} = \dot{x}\big|_{t=t_1}$

$$\therefore \quad C_3 = \dot{x}\big|_{t=t_1} + \frac{5g}{7}\sin\alpha \cdot t_1.$$

$$\begin{aligned}\therefore\ C_3 &= V - \frac{2(\mu\cos\alpha + \sin\alpha)(V - a\Omega)}{(7\mu\cos\alpha + 2\sin\alpha)} \\ &\quad + \frac{5g}{7}\sin\alpha \cdot \frac{2(V - a\Omega)}{(7\mu\cos\alpha + 2\sin\alpha)g} \\ &\qquad\qquad \text{[by (4.16.64) and (4.16.65)]} \\ &= V + \frac{2(V - a\Omega)}{(7\mu\cos\alpha + 2\sin\alpha)}\left[-\mu\cos\alpha - \sin\alpha + \frac{5\sin\alpha}{7}\right] \\ &= V - \frac{2(V - a\Omega)}{(7\mu\cos\alpha + 2\sin\alpha)}\left[\frac{7\mu\cos\alpha + 2\sin\alpha}{7}\right] \\ &= \frac{7V - 2(V - a\Omega)}{7} = \frac{5V + a\Omega}{7}.\end{aligned}$$

Thus

$$\dot{\bar{x}} = -\frac{5g}{7}\sin\alpha \cdot t + \frac{5V + 2a\Omega}{7}.$$

Let at time $t = t_2$, the sphere ceases to ascends and at this time $\dot{\bar{x}} = 0$.

$$\therefore \quad -\frac{5g}{7}\sin\alpha \cdot t_2 + \frac{5V + 2a\Omega}{7} = 0, \quad \text{or,} \quad t_2 = \frac{5V + 2a\Omega}{5g\sin\alpha}.$$

▶ **Problem 4.16.7** *A sphere is projected with an underhand twist down a rough inclined plane; show that it will turn back in its course of its motion if* $2a\omega(\mu - \tan\alpha) > 5\mu u$, *where* u, ω *are the initial linear and angular velocities of the sphere,* μ *is the coefficient of friction and* α *is the inclination of the plane to the horizon.* **[BUH 2003]**

Solution: We take the x and y axes along and perpendicular to the plane, making an angle α with the horizon. Let R and μR be the normal reaction and frictional force at the point of contact P. Let at any time t, the sphere rotates an angle θ and centre of the sphere describes a distance $\bar{x}$ along the plane. Initial velocity of the point of contact $= \dot{\bar{x}} - a\dot{\theta} = u + a\omega > 0$. So the sphere slides down the plane and hence friction acts along up the plane.

$$\therefore \quad \bar{x} = a\theta \quad \text{or,} \quad \ddot{\bar{x}} = a\ddot{\theta}. \tag{4.16.72}$$

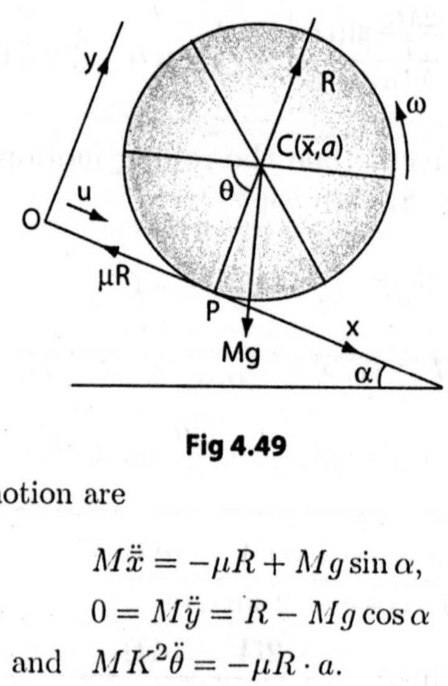

Fig 4.49

The equations of motion are

$$M\ddot{\bar{x}} = -\mu R + Mg\sin\alpha, \tag{4.16.73}$$

$$0 = M\ddot{\bar{y}} = R - Mg\cos\alpha \tag{4.16.74}$$

$$\text{and} \quad MK^2\ddot{\theta} = -\mu R \cdot a. \tag{4.16.75}$$

From (4.16.73) and (4.16.74), we get

$$M\ddot{\bar{x}} = -\mu Mg\cos\alpha + Mg\sin\alpha$$

$$\text{or,} \quad \ddot{\bar{x}} = g(\sin\alpha - \mu\cos\alpha).$$

Integrating

$$\dot{\bar{x}} = g(\sin\alpha - \mu\cos\alpha)\cdot t + C_1.$$

Initially, $t = 0$, $\dot{\bar{x}} = u$. $\therefore\ C_1 = u$.

$$\therefore\quad \dot{\bar{x}} = u + g(\sin\alpha - \mu\cos\alpha)t. \tag{4.16.76}$$

From (4.16.74) and (4.16.75), we get

$$M\cdot\frac{2a^2}{5}\ddot{\theta} = -\mu Mga\cos\alpha \quad \left[\text{as } K^2 = \frac{2a^2}{5}\right]$$

$$\text{or,}\quad a\ddot{\theta} = -\frac{5\mu g}{2}\cos\alpha.$$

Integrating,

$$a\dot{\theta} = -\frac{5\mu g}{2}\cos\alpha\cdot t + C_2.$$

Initially, $t = 0$, $\dot{\theta} = \omega$. $\therefore\ C_2 = a\omega$.

$$\text{or,}\quad a\dot{\theta} = a\omega - \frac{5\mu g}{2}\cos\alpha\cdot t. \tag{4.16.77}$$

The sphere will turn back when $\dot{\bar{x}} = 0$ and the velocity of the point of contact must be negative.

Let at $t = t_1$, $\dot{\bar{x}} = 0$.

$\therefore$ From (4.16.76), we get

$$u + g(\sin\alpha - \mu\cos\alpha)\cdot t_1 = 0$$

$$\text{or,}\quad t_1 = -\frac{u}{g(\sin\alpha - \mu\cos\alpha)} \tag{4.16.78}$$

and the velocity of point of contact at $t = t_1$,

$$\left(\dot{\bar{x}} - a\dot{\theta}\right)_{t=t_1} < 0, \quad \text{or,}\quad 0 - a\dot{\theta}\big|_{t=t_1} < 0,$$

$$\text{or,}\quad -a\omega + \frac{5\mu g}{2}\cos\alpha\cdot t_1 < 0,$$

$$\text{or,}\quad -a\omega - \frac{5\mu g}{2}\cos\alpha\cdot\frac{u}{g(\sin\alpha - \mu\cos\alpha} < 0 \quad \text{[by (4.16.78)]}$$

$$\text{or,}\quad \frac{5\mu u\cos\alpha}{2(\sin\alpha - \mu\cos\alpha)} < -a\omega, \quad \text{or,}\quad \frac{5\mu u}{2(\tan\alpha - \mu)} < -a\omega,$$

$$\text{or,}\quad 5\mu u < -2a\omega(\tan\alpha - \mu),$$

$$\text{or,}\quad 5\mu u < 2a\omega(\mu - \tan\alpha).$$

► **Problem 4.16.8** *If a sphere be project up an inclined plane, for which $\mu = \frac{1}{7}\tan\alpha$, with velocity V and an initial angular velocity Ω (in the direction in which it would roll up) and if $V > a\Omega$ show that friction acts downwards at first and upwards afterwards, and prove that the whole time during which the sphere rises is $\frac{17V+4a\Omega}{18g\sin\alpha}$.* **[CUH 1988]**

Solution: We take x and y axes along and perpendicular to the plane, making an angle α with the horizon. Let C be the centre of the sphere of mass M and radius a at any time. Let the sphere rotates an angle θ and the centre of the sphere describe a distance $\bar{x}$ in that time t.

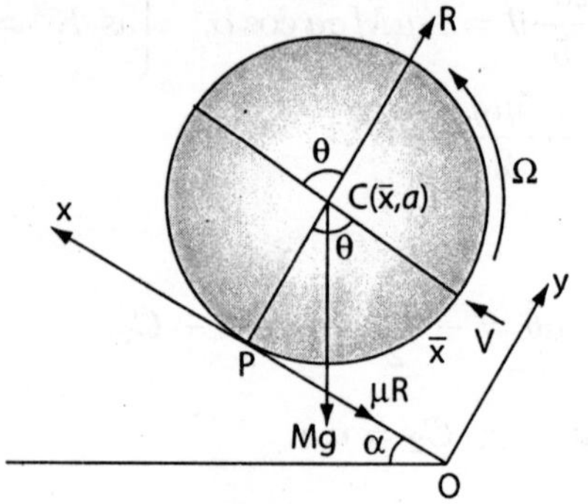

Fig 4.50

The initial velocity of the point of contact $= \dot{\bar{x}} - a\dot{\theta} = V - a\Omega > 0$ as $V > a\Omega$. Hence the sphere initially slides up. So friction μR acts downwards, where R is the normal reaction. Therefore

$$\bar{x} = a\theta \quad \text{or,} \quad \ddot{\bar{x}} = a\ddot{\theta}. \tag{4.16.79}$$

The equations of motion are

$$M\ddot{\bar{x}} = -\mu R - Mg\sin\alpha \tag{4.16.80}$$

$$0 = M\ddot{\bar{y}} = R - Mg\cos\alpha \tag{4.16.81}$$

$$\text{and} \quad MK^2\ddot{\theta} = \mu R \cdot a. \tag{4.16.82}$$

From (4.16.80) and (4.16.81), we have

$$M\ddot{\bar{x}} = -\mu Mg\cos\alpha - Mg\sin\alpha$$

$$\text{or,} \quad \ddot{\bar{x}} = -g(\mu\cos\alpha + \sin\alpha).$$

Integrating

$$\dot{\bar{x}} = -g(\mu\cos\alpha + \sin\alpha)\cdot t + C_1.$$

Initially $t = 0$, $\dot{\bar{x}} = V$. $\therefore\ C_1 = V$.

Thus

$$\dot{\bar{x}} = V - g(\mu \cos\alpha + \sin\alpha) \cdot t. \tag{4.16.83}$$

From (4.16.81) and (4.16.82), we get

$$MK^2\ddot{\theta} = \mu a \cdot Mg\cos\alpha, \quad \text{or,} \quad \frac{2a^2}{5}\ddot{\theta} = \mu ag\cos\alpha \qquad \left[\text{as } K^2 = \frac{2a^2}{5}\right]$$

$$\text{or,} \quad a\ddot{\theta} = \frac{5\mu g}{2}\cos\alpha.$$

Integrating,

$$a\dot{\theta} = \frac{5\mu g}{2}\cos\alpha \cdot t + C_2.$$

Initially, $t = 0$, $\dot{\theta} = \Omega$. $\therefore\ C_2 = a\Omega$.
Thus,

$$a\dot{\theta} = a\Omega + \frac{5\mu g}{2}\cos\alpha \cdot t. \tag{4.16.84}$$

Let after time t', the velocity of the point of contact is zero, i. e.,

$$\left(\dot{\bar{x}} - a\dot{\theta}\right)\Big|_{t=t'} = 0$$

then by (4.16.83) and (4.16.84), we get

$$V - g(\mu\cos\alpha + \sin\alpha)\cdot t' - a\Omega - \frac{5\mu g}{2}\cos\alpha \cdot t' = 0$$

$$\text{or,} \quad (V - a\Omega) - g\left(\mu\cos\alpha + \sin\alpha + \frac{5\mu\cos\alpha}{2}\right)t' = 0$$

$$\text{or,} \quad t' = \frac{2(V - a\Omega)}{(7\mu\cos\alpha + 2\sin\alpha)g}$$

$$= \frac{2(V - a\Omega)}{(\tan\alpha\cos\alpha + 2\sin\alpha)g} \qquad \left[\because \quad \mu = \tfrac{1}{7}\tan\alpha\right]$$

$$= \frac{2}{3}\frac{V - a\Omega}{g\sin\alpha}. \tag{4.16.85}$$

Thus after time t', the motion of the sphere may be rolling one.

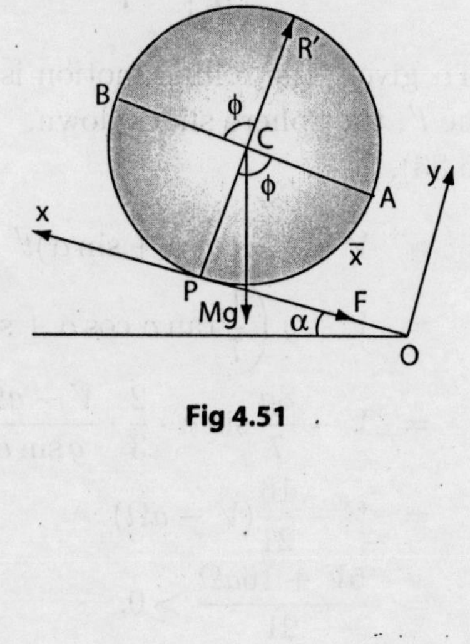

Fig 4.51

The equations of motion after t' are

$$M\ddot{\bar{x}} = -F - Mg\sin\alpha \tag{4.16.86}$$

$$0 = M\ddot{\bar{y}} = R' - Mg\cos\alpha \tag{4.16.87}$$

$$MK^2\ddot{\phi} = F\cdot a \tag{4.16.88}$$

$$\text{and}\quad \bar{x} = a\phi;\quad \dot{\bar{x}} = a\dot{\phi}\quad \text{or,}\quad \ddot{\bar{x}} = a\ddot{\phi}. \tag{4.16.89}$$

From (4.16.86) and (4.16.88), we have

$$MK^2\ddot{\phi} = -a\left(M\ddot{\bar{x}} + Mg\sin\alpha\right),$$

$$\text{or,}\quad K^2\ddot{\phi} = -a\left(\ddot{\bar{x}} + g\sin\alpha\right),$$

$$\text{or,}\quad K^2\frac{\ddot{\bar{x}}}{a^2} = -\ddot{\bar{x}} - g\sin\alpha,$$

$$\text{or,}\quad \left(\frac{2}{5}+1\right)\ddot{\bar{x}} = -g\sin\alpha,$$

$$\text{or,}\quad \ddot{\bar{x}} = -\frac{5g}{7}\sin\alpha. \tag{4.16.90}$$

From (4.16.86)

$$\begin{aligned} F &= -M\left(\ddot{\bar{x}} + g\sin\alpha\right) = -M\left(-\frac{5g}{7}\sin\alpha + g\sin\alpha\right) \\ &= -\frac{2Mg}{7}\sin\alpha \end{aligned}$$

and $R' = Mg\cos\alpha$.

The rolling motion is only possible if

$$\mu > \left|\frac{F}{R'}\right| = \frac{2}{7}\tan\alpha.$$

Since $\mu = \frac{1}{7}\tan\alpha$ given, the rolling motion is not possible after time t'. Hence after time t', the sphere slides down.
Now from (4.16.83),

$$\begin{aligned} \text{or,}\quad \dot{\bar{x}}\Big|_{t=t'} &= V - g(\mu\cos\alpha + \sin\alpha)t' \\ &= V - g\left(\frac{1}{7}\tan\alpha\cos\alpha + \sin\alpha\right)\cdot\frac{2}{3}\cdot\frac{V - a\Omega}{g\sin\alpha} \\ &= V - \frac{8g}{7}\sin\alpha\cdot\frac{2}{3}\cdot\frac{V - a\Omega}{g\sin\alpha} \\ &= V - \frac{16}{21}(V - a\Omega) \\ &= \frac{5V + 16a\Omega}{21} > 0. \end{aligned} \tag{4 16.91}$$

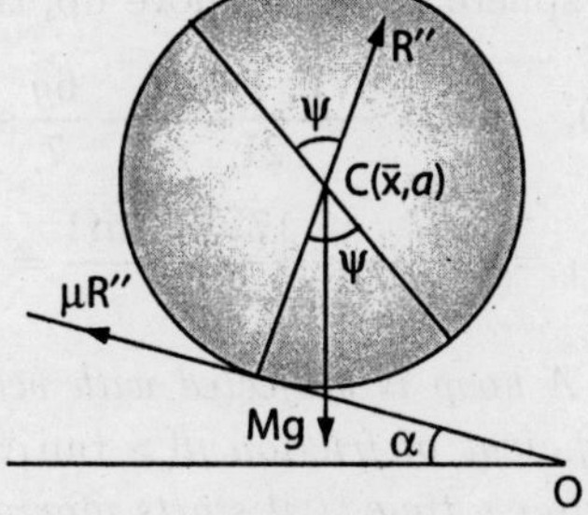

Fig 4.52

Hence after time t', the centre of the sphere moves up the plane, and the point of contact slides down. Hence the friction acts upwards.

Thus the equations of motion after time t' are

$$M\ddot{x} = \mu R'' - Mg\sin\alpha \tag{4.16.92}$$

$$0 = M\ddot{y} = R'' - Mg\cos\alpha \tag{4.16.93}$$

$$\text{and}\quad MK^2\ddot{\psi} = -\mu R''\cdot a. \tag{4.16.94}$$

From (4.16.92) and (4.16.93) we get

$$\begin{aligned} M\ddot{x} &= \mu\cdot Mg\cos\alpha - Mg\sin\alpha \\ \text{or,}\quad \ddot{x} &= g(\mu\cos\alpha - \sin\alpha) \\ &= g\left(\frac{1}{7}\tan\alpha\cdot\cos\alpha - \sin\alpha\right) \quad [\text{as } \mu = \tfrac{1}{7}\tan\alpha] \\ &= -\frac{6g}{7}\sin\alpha. \end{aligned}$$

Integrating,

$$\dot{x} = -\frac{6g}{7}\sin\alpha\cdot t + C_3.$$

$$\begin{aligned} \therefore\quad C_3 &= \dot{x}\Big|_{t=t'} + \frac{6g}{7}\sin\alpha\cdot t' \\ &= \frac{5V+16a\Omega}{21} + \frac{6g}{7}\sin\alpha\cdot\frac{2}{3}\frac{V-a\Omega}{g\sin\alpha} \quad [\text{by (4.16.91) and (4.16.85)}] \\ &= \frac{5V+16a\Omega}{21} + \frac{12}{21}(V-a\Omega) \\ &= \frac{17V+4a\Omega}{21} \\ \therefore\quad \dot{x} &= \frac{17V+4a\Omega}{21} - \frac{6g}{7}\sin\alpha\cdot t. \end{aligned}$$

If after time t_1 the sphere ceases to move up, then

$$\dot{\bar{x}}\big|_{t=t_1} = 0, \quad \text{i.e.,} \quad \frac{17V + 4a\Omega}{21} - \frac{6g}{7}\sin\alpha \cdot t_1 = 0$$

$$\text{or,} \quad t_1 = \frac{7}{6g\sin\alpha} \cdot \frac{17V + 4a\Omega}{21} = \frac{17V + 4a\Omega}{18g\sin\alpha}.$$

▶ **Problem 4.16.9** *A hoop is projected with velocity V down a plane of inclination α, the coefficient of friction $\mu(> \tan\alpha)$. If has initially such a backward spin Ω that after a time t_1 it starts moving up hill and continue to do so for a time t_2 after which it once move descends. The motion being in a vertical plane at right angles to the given inclined plane, show that*

$$(t_1 + t_2)g\sin\alpha = a\Omega - V.$$

Solution: We take the x and y-axes along and perpendicular to the plane, making an angle α to the horizon. Let at any time the hoop rotates an angle θ and the centre of the hoop moves up a distance $\bar{x}$ along the plane. Let R and μR are the normal reaction and friction act at P, the point of contact, at time t.

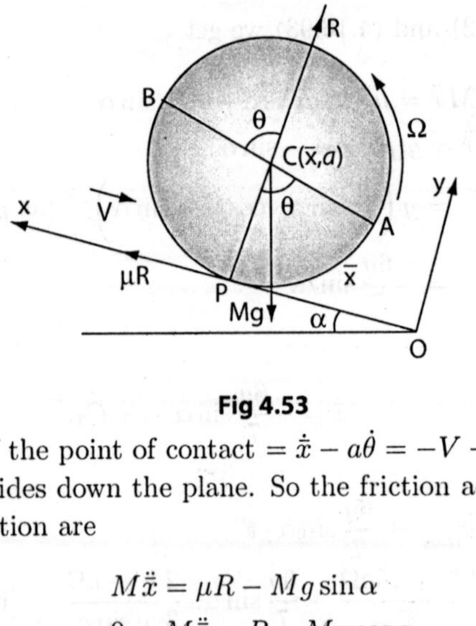

Fig 4.53

Initial velocity of the point of contact $= \dot{\bar{x}} - a\dot{\theta} = -V - a\Omega < 0$.

Thus the hoop slides down the plane. So the friction acts up the plane. The equations of motion are

$$M\ddot{\bar{x}} = \mu R - Mg\sin\alpha \tag{4.16.95}$$

$$0 = M\ddot{\bar{y}} = R - Mg\cos\alpha \tag{4.16.96}$$

$$\text{and} \quad MK^2\ddot{\theta} = -\mu R \cdot a. \tag{4.16.97}$$

From (4.16.95) and (4.16.96) we get

$$M\ddot{\bar{x}} = \mu \cdot Mg\cos\alpha - Mg\sin\alpha$$

$$\text{or,} \quad \ddot{\bar{x}} = g(\mu\cos\alpha - \sin\alpha).$$

Integrating,

$$\dot{x} = g(\mu\cos\alpha - \sin\alpha)t + C_1.$$

Initially, $t = 0$, $\dot{x} = -V$. $C_1 = -V$.
Thus,

$$\dot{x} = g(\mu\cos\alpha - \sin\alpha)t - V. \tag{4.16.98}$$

If the hoop ceases to move down at $t = t_1$ then

$$\dot{x}\big|_{t=t_1} = 0, \quad \text{or,} \quad g(\mu\cos\alpha - \sin\alpha)t_1 - V = 0$$

$$\text{or,} \quad t_1 = \frac{V}{g(\mu\cos\alpha - \sin\alpha)}. \tag{4.16.99}$$

Again from (4.16.96) and (4.16.97) we get,

$$\begin{aligned} MK^2\ddot{\theta} &= -\mu a \cdot Mg\cos\alpha, \\ \text{or,} \quad K^2\ddot{\theta} &= -\mu ag\cos\alpha, \\ \text{or,} \quad a^2\ddot{\theta} &= -\mu ag\cos\alpha, \\ \text{or,} \quad a^2\ddot{\theta} &= -\mu g\cos\alpha. \end{aligned}$$

Integrating

$$a\dot{\theta} = -\mu g\cos\alpha \cdot t + C_2.$$

Initially, $t = 0$, $\dot{\theta} = \Omega$. $\therefore\ C_2 = a\Omega$.
Thus

$$a\dot{\theta} = a\Omega - \mu g\cos\alpha \cdot t. \tag{4.16.100}$$

If the hoop moves up the plane after time t_1, then the velocity of the point of contact will be greater than zero at $t = t_1$, i. e.,

$$\dot{x}\big|_{t=t_1} - a\dot{\theta}\big|_{t=t_1} > 0$$

$$0 - a\Omega + \mu g\cos\alpha \cdot t_1 > 0 \quad \left[\text{as } \dot{x}\big|_{t=t_1} = 0\right]$$

$$\text{or,} \quad \mu g\cos\alpha \cdot \frac{V}{g(\mu\cos\alpha - \sin\alpha)} > a\Omega$$

$$\text{or,} \quad \frac{V\mu\cos\alpha}{\mu\cos\alpha - \sin\alpha} > a\Omega.$$

Now the velocity of the point of contact at $t = t_1$

$$= \dot{x}\big|_{t=t_1} - a\dot{\theta}\big|_{t=t_1} = -a\dot{\theta} \quad \text{at } t = t_1$$

acts downwards the plane. Therefore the friction acts upwards and the hoop moves up the plane.

At time t_1, $\dot{\bar{x}}\Big|_{t=t_1} = 0$

$$\left.\begin{aligned} \text{and} \quad a\dot{\theta}\Big|_{t=t_1} &= -\mu g\cos\alpha\cdot t_1 + a\Omega \\ &= a\Omega - \mu g\cos\alpha\cdot\frac{V}{g(\mu\cos\alpha - \sin\alpha)} \\ &= a\Omega - \frac{\mu V\cos\alpha}{\mu\cos\alpha - \sin\alpha} \end{aligned}\right\} \tag{4.16.101}$$

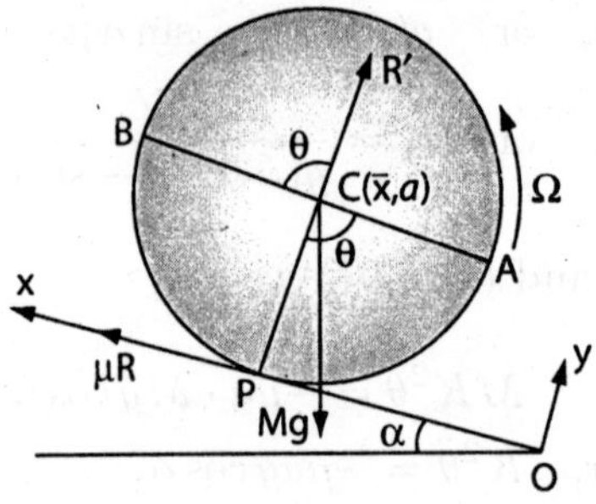

Fig 4.54

After time t_1, the equations of motion are

$$M\ddot{\bar{x}} = \mu R' - Mg\sin\alpha \tag{4.16.102}$$

$$0 = M\ddot{\bar{y}} = R' - Mg\cos\alpha \tag{4.16.103}$$

$$MK^2\ddot{\theta} = -\mu R'\cdot a \tag{4.16.104}$$

and we consider the time is measured from the instant when the body begins to move up.

From (4.16.102) and (4.16.103) we get

$$\begin{aligned} M\ddot{\bar{x}} &= \mu Mg\cos\alpha - Mg\sin\alpha, \\ \text{or,} \quad \ddot{\bar{x}} &= g(\mu\cos\alpha - \sin\alpha). \end{aligned}$$

Integrating

$$\dot{\bar{x}} = g(\mu\cos\alpha - \sin\alpha)t + C_3,$$

Initially $\dot{\bar{x}} = 0$, when $t = 0$ we get $C_3 = 0$,

$$\therefore \quad \dot{\bar{x}} = g(\mu\cos\alpha - \sin\alpha)t. \tag{4.16.105}$$

From (4.16.104) and (4.16.103), we get

$$\begin{aligned} MK^2\ddot{\theta} &= -\mu a\cdot Mg\cos\alpha \\ a\ddot{\theta} &= -\mu g\cos\alpha \quad [\text{As } K^2 = a^2 \text{ for the hoop}] \\ a\dot{\theta} &= -\mu g\cos\alpha\cdot t + C_4. \end{aligned}$$

Initially, using (4.16.101) we get,

$$C_4 = a\Omega - \frac{\mu V \cos\alpha}{\mu\cos\alpha - \sin\alpha}.$$

Thus,

$$a\dot{\theta} = a\Omega - \frac{\mu V \cos\alpha}{\mu\cos\alpha - \sin\alpha} - \mu g\cos\alpha \cdot t. \tag{4.16.106}$$

Now let the hoop begins to roll after time t', then

$$\dot{\bar{x}} - a\dot{\theta}\big|_{t=t'} = 0,$$

i. e.,

$$g(\mu\cos\alpha - \sin\alpha)t' - a\Omega + \mu g\cos\alpha \cdot t' + \frac{\mu V\cos\alpha}{\mu\cos\alpha - \sin\alpha} = 0,$$

$$\text{or,}\quad t' = \frac{a\Omega - \frac{\mu V\cos\alpha}{\mu\cos\alpha - \sin\alpha}}{g(2\mu\cos\alpha - \sin\alpha)}. \tag{4.16.107}$$

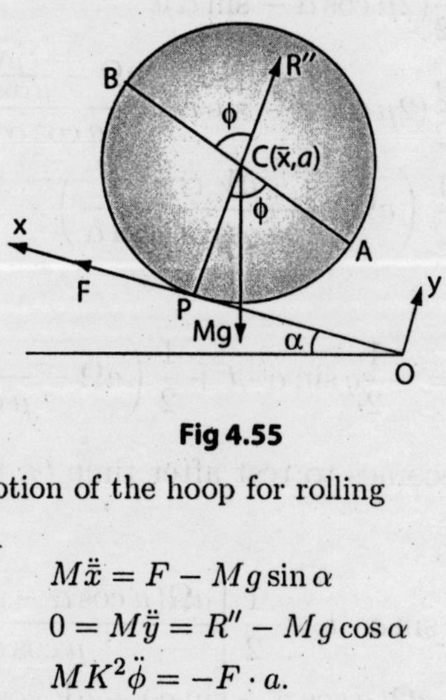

Fig 4.55

The equations of motion of the hoop for rolling

$$M\ddot{\bar{x}} = F - Mg\sin\alpha \tag{4.16.108}$$
$$0 = M\ddot{\bar{y}} = R'' - Mg\cos\alpha \tag{4.16.109}$$
$$MK^2\ddot{\phi} = -F\cdot a. \tag{4.16.110}$$

and for rolling

$$\dot{\bar{x}} = a\dot{\phi}. \tag{4.16.111}$$

From (4.16.108) and (4.16.110) we get

$$\begin{aligned} M\ddot{x} &= -\frac{MK^2\ddot{\phi}}{a} - Mg\sin\alpha \\ &= -\frac{MK^2\ddot{x}}{a^2} - Mg\sin\alpha \quad \text{[by (4.16.111)]} \\ \text{or,}\quad \ddot{x}\left(1+\frac{K^2}{a^2}\right) &= -g\sin\alpha \\ \text{or,}\quad 2\ddot{x} &= -g\sin\alpha \quad [\text{as } K^2 = a^2 \text{ for hoop}] \\ \text{or,}\quad \ddot{x} &= -\frac{g}{2}\sin\alpha. \end{aligned}$$

Integrating

$$\dot{x} = -\frac{g}{2}\sin\alpha\cdot t + C_5.$$

$$\begin{aligned} \therefore\quad C_5 &= \dot{x}\big|_{t=t'} + \frac{g}{2}\sin\alpha\cdot t' \\ &= g(\mu\cos\alpha - \sin\alpha)t' + \frac{g}{2}\sin\alpha\cdot t' \quad \text{[by (4.16.105)]} \\ &= \frac{g}{2}(2\mu\cos\alpha - \sin\alpha)t' \\ &= \frac{g}{2}(2\mu\cos\alpha - \sin\alpha)\frac{a\Omega - \frac{\mu V\cos\alpha}{\mu\cos\alpha - \sin\alpha}}{g(2\mu\cos\alpha - \sin\alpha)} \quad \text{[by (4.16.107)]} \\ &= \frac{1}{2}\left(a\Omega - \frac{\mu V\cos\alpha}{\mu\cos\alpha - \sin\alpha}\right). \end{aligned}$$

Thus,

$$\dot{x} = -\frac{1}{2}g\sin\alpha\cdot t + \frac{1}{2}\left(a\Omega - \frac{\mu V\cos\alpha}{\mu\cos\alpha - \sin\alpha}\right)$$

If the body becomes to rest after time t_2, then $\dot{x}\big|_{t=t_2} = 0$.
Thus,

$$-\frac{1}{2}g\sin\alpha\cdot t_2 + \frac{1}{2}\left[\frac{a\Omega(\mu\cos\alpha - \sin\alpha) - \mu V\cos\alpha}{\mu\cos\alpha - \sin\alpha}\right] = 0,$$

$$\text{or,}\quad t_2 = \frac{a\Omega(\mu\cos\alpha - \sin\alpha) - \mu V\cos\alpha}{g\sin\alpha(\mu\cos\alpha - \sin\alpha)}.$$

$$\begin{aligned} \therefore\quad t_1 + t_2 &= \frac{V}{g(\mu\cos\alpha - \sin\alpha)} + \frac{a\Omega(\mu\cos\alpha - \sin\alpha) - \mu V\cos\alpha}{g\sin\alpha(\mu\cos\alpha - \sin\alpha)} \\ &= \frac{V\sin\alpha + a\Omega(\mu\cos\alpha - \sin\alpha) - \mu V\cos\alpha}{g\sin\alpha(\mu\cos\alpha - \sin\alpha)}. \end{aligned}$$

$$\text{or,}\quad (t_1+t_2)g\sin\alpha = \frac{a\Omega(\mu\cos\alpha-\sin\alpha)-V(\mu\cos\alpha-\sin\alpha)}{\mu\cos\alpha-\sin\alpha}$$
$$= a\Omega - V.$$

► **Problem 4.16.10** *A uniform sphere, of radius a, is rotating about a horizontal diameter with angular velocity Ω and is gently placed on a rough plane which is inclined at an angle α to the horizontal, the sense of rotation being such as to tend to cease the sphere to move up the plane along the line of greatest slope. Show that, if the coefficient of friction be* $\tan\alpha$, *the centre of sphere will remains at rest for a time* $\frac{2a\Omega}{5g\sin\alpha}$, *and will then move downwards with acceleration* $\frac{5}{7}g\sin\alpha$.

If the body be a thin hoop instead of a sphere,show that the time is $\frac{a\Omega}{g\sin\alpha}$ *and the acceleration* $\frac{1}{2}g\sin\alpha$. **[BUH 2005]**

Solution: We consider the x and y axes at O along and perpendicular to the plane, which makes an angle α with the horizon. Let at any time t, the sphere rotates an angle θ and the centre C of the sphere of mass M and radius a describes a distance $\bar{x}$ along the plane. Thus the coordinates of the centre C is $(\bar{x}, a)$.

The initial velocity of the point of contact $= \dot{\bar{x}} - a\dot{\theta} = -a\Omega < 0$. Thus the sphere moves down and the friction μR acts up the plane, where R is the normal reaction.

Now the equation of motion are

$$M\ddot{\bar{x}} = \mu R - Mg\sin\alpha \tag{4.16.112}$$
$$0 = M\ddot{\bar{y}} = R - Mg\cos\alpha \tag{4.16.113}$$
$$\text{and}\quad MK^2\ddot{\theta} = -\mu R\cdot a. \tag{4.16.114}$$

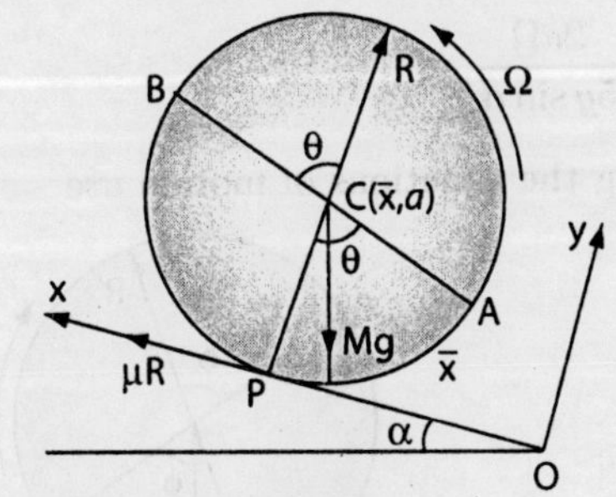

Fig 4.56

From (4.16.112) and (4.16.113), we have

$$M\ddot{\bar{x}} = \mu\cdot Mg\cos\alpha - Mg\sin\alpha$$
$$\text{or,}\quad \ddot{\bar{x}} = g(\mu\cos\alpha-\sin\alpha) = g(\tan\alpha\cos\alpha-\sin\alpha) = 0 \quad [\text{as } \mu=\tan\alpha]$$

Thus, integrating

$$\dot{x} = \text{constant} = \text{initial value} = 0. \tag{4.16.115}$$

Hence the motion of the sphere is such that the centre of the sphere remains fixed for certain time.

From (4.16.114) and (4.16.113) we get

$$MK^2\ddot{\theta} = -\mu a \cdot Mg\cos\alpha, \quad \text{or,} \quad \frac{2a^2}{5}\ddot{\theta} = -\mu a g\cos\alpha$$

$$\text{or,} \quad a\ddot{\theta} = -\frac{5\mu g}{2}\cos\alpha = -\frac{5}{2}g\cdot\tan\alpha\cdot\cos\alpha \quad [\text{as } \mu = \tan\alpha]$$

$$= -\frac{5}{2}g\sin\alpha.$$

Integrating,

$$a\dot{\theta} = -\frac{5}{2}g\sin\alpha\cdot t + C_1.$$

Initially $t = 0$, $\dot{\theta} = \Omega$. $\therefore\ C_1 = a\Omega$.

Thus,

$$a\dot{\theta} = a\Omega - \frac{5}{2}g\sin\alpha\cdot t. \tag{4.16.116}$$

It is clear that the centre of the sphere will remain fixed till the rolling starts. Let after time t_1, the rolling of the sphere starts. Thus the velocity of the point of contact must vanish at $t = t_1$, i.e.,

$$\left(\dot{x} - a\dot{\theta}\right)\Big|_{t=t_1} = 0$$

$$0 + \frac{5}{2}g\sin\alpha\cdot t_1 - a\Omega = 0 \quad [\text{by (4.16.115) and (4.16.116)}]$$

$$\text{or,} \quad t_1 = \frac{2a\Omega}{5g\sin\alpha}. \tag{4.16.117}$$

Now for rotting the equations of motion are

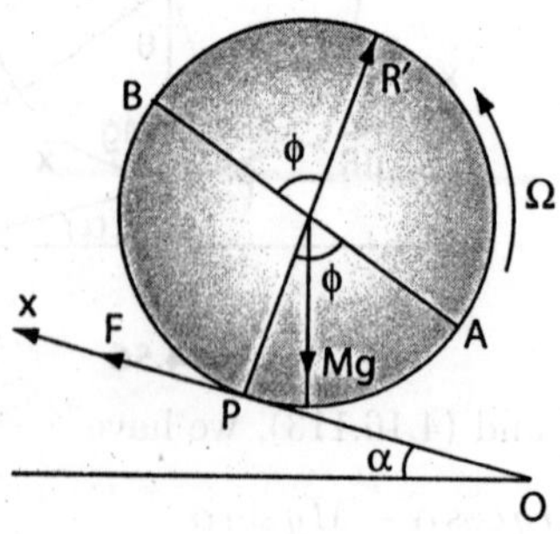

Fig 4.57

$$M\ddot{\bar{x}} = F - Mg\sin\alpha \quad (4.16.118)$$

$$0 = M\ddot{\bar{y}} = R' - Mg\cos\alpha \quad (4.16.119)$$

$$MK^2\ddot{\phi} = -F \cdot a \quad (4.16.120)$$

$$\text{and} \quad \dot{\bar{x}} = a\dot{\phi} \quad \text{and} \quad \ddot{\bar{x}} = a\ddot{\phi}. \quad (4.16.121)$$

From (4.16.118) and (4.16.120), we get

$$M\ddot{\bar{x}} = -\frac{MK^2\ddot{\phi}}{a} - Mg\sin\alpha,$$

$$\text{or,} \quad \ddot{\bar{x}} + \frac{K^2}{a^2}\ddot{\bar{x}} = -g\sin\alpha \quad \text{[by (4.16.121)]},$$

$$\text{or,} \quad \left(1 + \frac{K^2}{a^2}\right)\ddot{\bar{x}} = -g\sin\alpha,$$

$$\text{or,} \quad \left(1 + \frac{2}{5}\right)\ddot{\bar{x}} = -g\sin\alpha \left[\text{As } K^2 = \frac{2a^2}{5}\right],$$

$$\therefore \quad \ddot{\bar{x}} = -\frac{5g}{7}\sin\alpha.$$

Thus, the downwards acceleration of the centre $= \frac{5g}{7}\sin\alpha$.

When the body is circular hoop.

The equations of motion are

$$M\ddot{\bar{x}} = \mu R' - Mg\sin\alpha \quad (4.16.122)$$

$$0 = M\ddot{\bar{y}} = R' - Mg\cos\alpha \quad (4.16.123)$$

$$MK^2\ddot{\theta} = -\mu R' \cdot a \quad (4.16.124)$$

From (4.16.122) and (4.16.123), we get

$$M\ddot{\bar{x}} = \mu \cdot Mg\cos\alpha - Mg\sin\alpha$$

$$\text{or,} \quad \ddot{\bar{x}} = g(\mu\cos\alpha - \sin\alpha) = g(\tan\alpha\cos\alpha - \sin\alpha) = 0 \quad [\text{as } \mu = \tan\alpha]$$

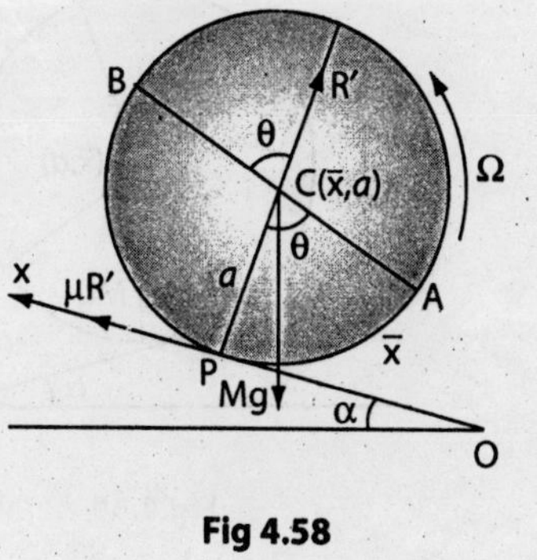

Fig 4.58

Integrating

$$\dot{\bar{x}} = \text{constant} = \text{initial value} = 0. \tag{4.16.125}$$

Thus, the centre of the hoop remains fixed for certain time, till the rolling starts.

From (4.16.124) and (4.16.123), we get

$$MK^2\ddot{\theta} = -\mu a \cdot Mg\cos\alpha$$

or, $K^2\ddot{\theta} = -ag\mu\cos\alpha = -ag\tan\alpha\cdot\cos\alpha = -ag\sin\alpha$ [as $\mu = \tan\alpha$]

or, $a^2\ddot{\theta} = -ag\sin\alpha$ [as $K^2 = a^2$ for hoop]

or, $a\ddot{\theta} = -g\sin\alpha$.

Integrating,

$$a\dot{\theta} = -g\sin\alpha \cdot t + C_2.$$

Initially, $t = 0$, $\dot{\theta} = \Omega$. $\therefore$ $C_2 = a\Omega$.

Thus,

$$a\dot{\theta} = a\Omega - g\sin\alpha \cdot t. \tag{4.16.126}$$

Let after time t_2, the hoop starts rolling

$$\text{i.e.,} \quad \left(\dot{\bar{x}} - a\dot{\theta}\right)\Big|_{t=t_2} = 0$$

$$\text{or,} \quad 0 - a\Omega + g\sin\alpha \cdot t_2 = 0 \quad \text{or,} \quad t_2 = \frac{a\Omega}{g\sin\alpha}.$$

The equations of motion for rolling are

$$M\ddot{\bar{x}} = F - Mg\sin\alpha \tag{4.16.127}$$

$$0 = M\ddot{y} = R'' - Mg\cos\alpha \tag{4.16.128}$$

$$MK^2\ddot{\phi} = -F\cdot a \tag{4.16.129}$$

$$\text{and} \quad \dot{\bar{x}} = a\dot{\phi} \quad \text{or,} \quad \ddot{\bar{x}} = a\ddot{\phi} \tag{4.16.130}$$

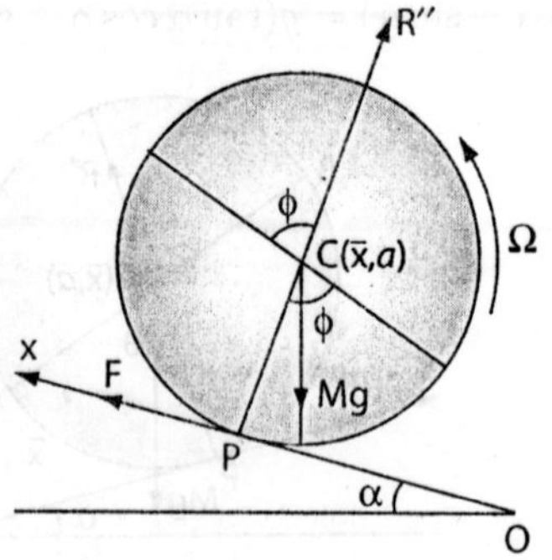

Fig 4.59

From (4.16.127) and (4.16.129) we have

$$M\ddot{\bar{x}} = -\frac{MK^2\ddot{\phi}}{a} - Mg\sin\alpha,$$

$$\text{or,}\quad \ddot{\bar{x}} = -\frac{K^2}{a^2}\ddot{\bar{x}} - g\sin\alpha \quad \text{[by (4.16.130)]}$$

$$\text{or,}\quad \left(1+\frac{K^2}{a^2}\right)\ddot{\bar{x}} = -g\sin\alpha,$$

$$\text{or,}\quad 2\ddot{\bar{x}} = -g\sin\alpha \quad [\text{As } K^2 = a^2 \text{ for hoop}],$$

$$\text{or,}\quad \ddot{\bar{x}} = -\frac{g\sin\alpha}{2}.$$

Thus the centre of the hoop moves downwards with acceleration $\frac{g\sin\alpha}{2}$.

► **Problem 4.16.11** *A uniform heavy circular cylinder rolls down certain distance along a perfectly rough inclined plane in t seconds find the time taken by a hollow circular cylinder of same radius to roll down the same distance along the plane.* **[BUH 1998, 2000]**

Solution: We consider Ox and Oy as axes along and perpendicular to inclined plane, making an angle α with horizontal. Let at any time t fixed line AC in the heavy cylinder of radius a and mass M, makes an angle θ with a fixed line in space (perpendicular to the plane). Let R and F are the normal reaction and friction at the point of contact P. Let at time t, the coordinate of the centre of central section of the cylinder are $(\bar{x}, a)$.

$$\therefore \quad \bar{x} = a\theta \quad \text{or,} \quad \ddot{\bar{x}} = a\ddot{\theta}. \tag{4.16.131}$$

The equations of motion are,

$$M\ddot{\bar{x}} = -F + Mg\sin\alpha \tag{4.16.132}$$

$$0 = M\ddot{y} = R - Mg\cos\alpha \tag{4.16.133}$$

$$\text{and}\quad MK^2\ddot{\theta} = F\cdot a \tag{4.16.134}$$

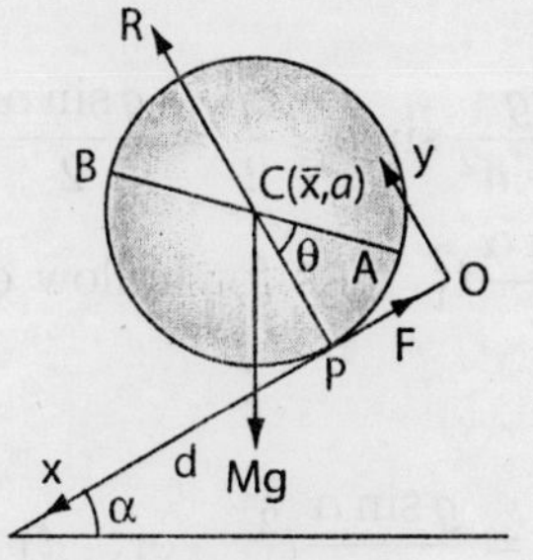

Fig 4.60

From (4.16.134) and (4.16.132), we get

$$M\ddot{\bar{x}} = -\frac{MK^2}{a}\ddot{\theta} + Mg\sin\alpha = -\frac{MK^2\ddot{\bar{x}}}{a^2} + Mg\sin\alpha \quad \text{[by (4.16.131)]}$$

$$\text{or,} \quad \left(1+\frac{K^2}{a^2}\right)\ddot{\bar{x}} = g\sin\alpha, \quad \text{or,} \quad \ddot{\bar{x}} = \frac{a^2 g}{K^2+a^2}\sin\alpha.$$

Integrating,

$$\dot{\bar{x}} = \frac{a^2 g}{K^2+a^2}\sin\alpha\cdot t + C_1.$$

Initially, $\dot{\bar{x}} = 0$, when $t = 0$. $\therefore\ C_1 = 0$.
Thus,

$$\dot{\bar{x}} = \frac{a^2 g}{K^2+a^2}\sin\alpha\cdot t.$$

Again integrating

$$\bar{x} = \frac{a^2 g}{K^2+a^2}\sin\alpha\cdot\frac{t^2}{2} + C_2, \quad \text{as } \bar{x} = 0, t = 0, \quad \therefore\ C_2 = 0$$

$$\text{or,} \quad \bar{x} = \frac{a^2 g}{K^2+a^2}\sin\alpha\cdot\frac{t^2}{2}.$$

For solid cylinder $K^2 = \frac{a^2}{2}$.
Let the solid cylinder takes t sec. to describe the distance d along the plane.

$$\therefore\quad d = \frac{a^2 g}{\frac{a^2}{2}+a^2}\sin\alpha\cdot\frac{t^2}{2} = \frac{g\sin\alpha}{3}\cdot t^2.$$

If t_1 be the time taken by hollow cylinder to move this distance d along the plane, we have

$$\begin{aligned} d &= \frac{a^2 g}{a^2+a^2}\sin\alpha\cdot\frac{t_1^2}{2} = \frac{g\sin\alpha}{2}\cdot\frac{t_1^2}{2} \\ &= \frac{g\sin\alpha}{4}t_1^2 \quad \text{[As for hollow cylinder } K^2 = a^2\text{]} \end{aligned}$$

Thus,

$$\frac{g\sin\alpha}{3}t^2 = \frac{g\sin\alpha}{4}t_1^2, \quad \text{or,} \quad t_1 = \sqrt{\frac{4}{3}}t = \frac{2t}{\sqrt{3}}.$$

▶ **Problem 4.16.12** *A uniform solid cylinder of mass M and radius a is rolled with its axis horizontal up a rough inclined plane by means of constant couple L. Show that for this to be possible, the coefficient of friction must exceed $\frac{1}{3}\tan\alpha + \frac{2}{3}\frac{L\sec\alpha}{Mga}$.* **[CUH 1964]**

Solution: We take the x and y axes along and perpendicular to the plane which makes an angle α with the horizon. Let at any time t the cylinder rotates an angle θ and the centre of central section of the cylinder C describes a distance $\bar{x}$ along the plane. R and F are the normal reaction and friction at the point of contact P. Thus the coordinates of C is $(\bar{x}, a)$.

For rolling, the velocity of the point of contact

$$= \dot{\bar{x}} - a\dot{\theta} = 0$$

$$\text{or,} \quad \dot{\bar{x}} = a\dot{\theta} \quad \text{and} \quad \ddot{\bar{x}} = a\ddot{\theta}. \tag{4.16.135}$$

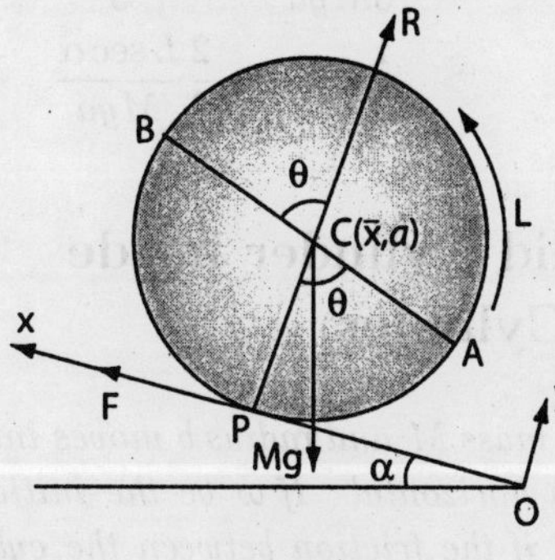

Fig 4.61

The equations of motion are

$$M\ddot{\bar{x}} = F - Mg\sin\alpha \tag{4.16.136}$$

$$0 = M\ddot{\bar{y}} = R - Mg\cos\alpha \tag{4.16.137}$$

$$\text{and} \quad MK^2\ddot{\theta} = L - F \cdot a \tag{4.16.138}$$

From (4.16.138), we get

$$M \cdot \frac{a^2}{2}\ddot{\theta} = L - Fa \quad [\text{For cylinder } K^2 = \tfrac{a^2}{2}]$$

$$\text{or,} \quad Ma\ddot{\theta} = \frac{2L}{a} - 2F. \tag{4.16.139}$$

From (4.16.135) and (4.16.136), we get

$$Ma\ddot{\theta} = F - Mg\sin\alpha. \tag{4.16.140}$$

Thus,

$$F - Mg\sin\alpha = \frac{2L}{a} - 2F,$$

$$\text{or,} \quad 3F = \frac{2L}{a} + Mg\sin\alpha$$

$$\text{or,} \quad F = \frac{2}{3} \cdot \frac{L}{a} + \frac{Mg}{3}\sin\alpha \tag{4.16.141}$$

and from (4.16.137),

$$R = Mg\cos\alpha. \tag{4.16.142}$$

Thus for rolling coefficient of friction $\mu > \frac{F}{R}$,

$$\text{or,}\quad \mu > \frac{\frac{2L}{3a} + \frac{Mg}{3}\sin\alpha}{Mg\sin\alpha}$$

$$> \frac{2L}{3Mga}\sec\alpha + \frac{1}{3}\tan\alpha$$

$$> \frac{1}{3}\tan\alpha + \frac{2}{3}\frac{L\sec\alpha}{Mga}.$$

4.17 Rough Solid Cylinder inside a Hollow Cylinder

A rough solid cylinder of mass M and radius b moves inside a hollow cylinder of radius a with its axis horizontal. If ω be the initial angular velocity of the moving cylinder and if the friction between the cylinder be sufficient to prevent any sliding, investigate the motion.

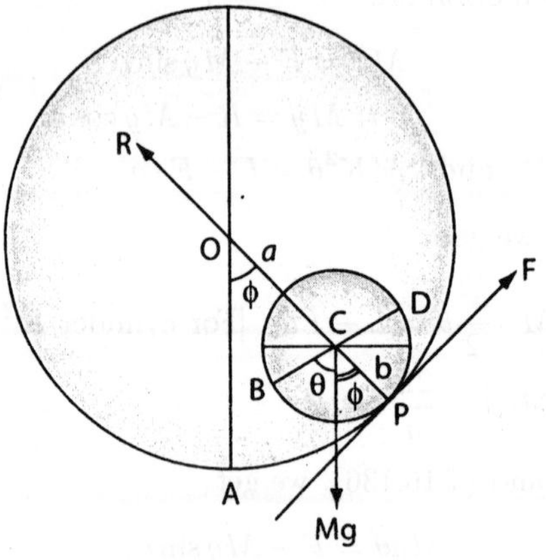

Fig 4.62

Let O be the centre of the fixed cylinder of radius a. Let C be the centre of the cylinder of mass M and of radius b. Initially the point B was in contact with the fixed cylinder at A, the lowest point. Let at any time, the diameter BD makes an angle θ with the vertical. Let the inside cylinder rolls to the P, so that $\angle AOP = \phi$. Let the initial angular velocity $\dot{\phi} = \omega$. Thus for pure rolling

$$\text{arc } AP = \text{arc } BP, \quad \text{or}, \quad a\phi = b(\theta + \phi),$$
$$\text{or}, \quad (a-b)\phi = b\theta, \quad \text{or}, \quad (a-b)\dot{\phi} = b\dot{\theta},$$
$$\text{or}, \quad (a-b)\ddot{\phi} = b\ddot{\theta}. \tag{4.17.1}$$

Let R and F be the normal reaction and friction at P, then the equations of motion are

$$M\left[-(a-b)\dot{\phi}^2\right] = -R + Mg\cos\phi,$$
$$\text{or}, \quad M(a-b)\dot{\phi}^2 = R - Mg\cos\phi, \tag{4.17.2}$$
$$M(a-b)\ddot{\phi} = F - Mg\sin\phi. \tag{4.17.3}$$

and

$$MK^2\ddot{\theta} = -F \cdot b \tag{4.17.4}$$
$$\text{or}, \quad M \cdot \frac{b^2}{2}\ddot{\theta} = -F \cdot b \quad \left[\text{as } K^2 = \frac{b^2}{2}\right],$$
$$\text{or}, \quad F = -\frac{Mb\ddot{\theta}}{2} = -\frac{Mb}{2} \cdot \frac{(a-b)\ddot{\phi}}{b} = -\frac{M(a-b)}{2}\ddot{\phi} \quad [\text{by } (4.17.1)]. \tag{4.17.5}$$

Using (4.17.5) in (4.17.3) we get

$$M(a-b)\ddot{\phi} = -\frac{M(a-b)}{2}\ddot{\phi} - Mg\sin\phi$$
$$\text{or}, \quad \frac{3(a-b)}{2}\ddot{\phi} = -g\sin\phi$$
$$\text{or}, \quad \frac{3(a-b)}{2}\frac{d}{d\phi}\left(\frac{\dot{\phi}^2}{2}\right) = -g\sin\phi.$$

Integrating

$$\frac{3}{4}(a-b)\dot{\phi}^2 = g\cos\phi + C_1.$$

Initially, $\dot{\phi} = \omega$ when $\phi = 0$. $\therefore\ C_1 = \frac{3}{4}(a-b)\omega^2 - g$.
Thus,

$$\begin{aligned}\frac{3}{4}(a-b)\dot{\phi}^2 &= g\cos\phi + \frac{3}{4}(a-b)\omega^2 - g\\ &= -g(1-\cos\phi) + \frac{3}{4}(a-b)\omega^2\\ \text{or}, \quad (a-b)\dot{\phi}^2 &= (a-b)\omega^2 - \frac{4}{3}g(1-\cos\phi).\end{aligned} \tag{4.17.6}$$

Differentially with respect to t, we get

$$(a-b)2\cdot\dot{\phi}\cdot\ddot{\phi}=-\frac{4g}{3}\sin\phi\cdot\dot{\phi}.$$

$$\text{or,}\quad (a-b)\ddot{\phi}=-\frac{2g}{3}\sin\phi. \tag{4.17.7}$$

Thus using (4.17.6) in (4.17.2) we get

$$\begin{aligned}
R &= M\left[(a-b)\omega^2-\frac{4}{3}g(1-\cos\phi)\right]+Mg\cos\phi\\
&= M(a-b)\omega^2-\frac{4Mg}{3}\left(1-\cos\phi-\frac{3\cos\phi}{4}\right)\\
&= M(a-b)\omega^2-\frac{Mg}{3}(4-7\cos\phi)\quad .\\
&= M(a-b)\omega^2+\frac{Mg}{3}(7\cos\phi-4). \qquad (4.17.8)
\end{aligned}$$

Using (4.17.7) in (4.17.3) we get,

$$\begin{aligned}
F &= M\left(-\frac{2g}{3}\sin\phi\right)+Mg\sin\phi\\
&= Mg\sin\phi\left(1-\frac{2}{3}\right)=\frac{Mg}{3}\sin\phi. \qquad (4.17.9)
\end{aligned}$$

Particular Remarks

Remark 4.17.1 *When the moving cylinder will leave the fixed cylinder. In this case $R=0$, thus from (4.17.8) we get*

$$M(a-b)\omega^2+\frac{Mg}{3}(7\cos\phi-4)=0,$$

$$\textit{or,}\quad 3(a-b)\omega^2+g(7\cos\phi-4)=0,$$

$$\textit{or,}\quad \omega^2=\frac{g(4-7\cos\phi)}{3(a-b)},$$

$$\textit{or,}\quad 7g\cos\phi=4g-3(a-b)\omega^2,$$

$$\textit{or,}\quad \phi=\cos^{-1}\left[\frac{4g-3(a-b)\omega^2}{7g}\right]. \tag{4.17.10}$$

Thus the moving cylinder may leave the fixed cylinder after rotating an angle ϕ to the vertical given by (4.17.10).

Remark 4.17.2 *When the moving cylinder may just complete the revolution.*

In this case $R = 0$ *when* $\phi = \pi$.
Thus from (4.17.8) *we get*

$$0 = M(a-b)\omega^2 - \frac{Mg}{3}(4+7) = M(a-b)\omega^2 - \frac{11}{3}Mg.$$

$$\text{or,}\quad \omega^2 = \frac{11g}{3(a-b)} \quad \text{or,}\quad \omega = \sqrt{\frac{11g}{3(a-b)}}.$$

Remark 4.17.3 *When the moving cylinder makes small oscillation.*

[**BUH 1996**]

In this case, ϕ *is very small, i. e.,* $\sin\phi = \phi$.
Thus from (4.17.7) *we get*

$$\ddot{\phi} = -\frac{2g}{3(a-b)}\phi.$$

Thus the time of small oscillation is

$$T' = \frac{2\pi}{\sqrt{\frac{2g}{3(a-b)}}} = 2\pi\sqrt{\frac{3(a-b)}{2g}} = \pi\sqrt{\frac{6(a-b)}{g}}.$$

4.18 Solved Problems

► **Problem 4.18.1** *A circular plate rolls down the inner circumference of a rough circle under the action of gravity, the planes of both plate and circle being vertical. When the line joining their centres is inclined at an angle* θ *to the vertical, show that the friction between the bodies is* $\frac{\sin\theta}{3}\times$ *the weight of the plate.*

Solution: Let O be the centre of the fixed circle of radius a. Let C be the centre of the moving circular plate of mass M and radius b. Let AB be a fixed diameter in the circular plate which makes an angle ϕ with the vertical at any time t and let the line joining the centres makes an angle θ with the vertical at that time. Let R and F are the normal reaction and friction at the point of contact P at that time.

For rolling the tangential velocity of the point of contact $P = 0$, i.e.,

$$-(a-b)\dot{\theta} - b\dot{\phi} = 0, \quad \text{or,}\quad (a-b)\dot{\theta} = -b\dot{\phi},$$
$$\text{or,}\quad (a-b)\ddot{\theta} = -b\ddot{\phi}. \tag{4.18.1}$$

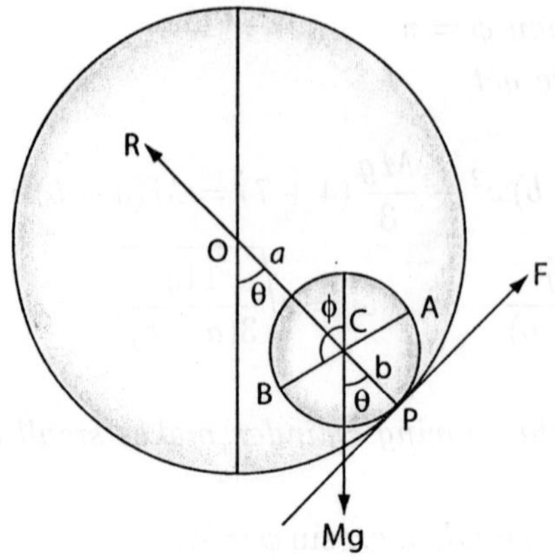

Fig 4.63

The equations of motion of the circular plate along and perpendicular to OC are

$$M\left\{-(a-b)\dot{\theta}^2\right\} = -R + Mg\cos\theta,$$

$$\text{or,}\quad M(a-b)\dot{\theta}^2 = R - Mg\cos\theta \qquad (4.18.2)$$

$$M(a-b)\ddot{\theta} = F - Mg\sin\theta \qquad (4.18.3)$$

$$\text{and}\quad MK^2\ddot{\phi} = F\cdot b. \qquad (4.18.4)$$

$$\therefore\quad F = \frac{MK^2\ddot{\phi}}{b} = \frac{M}{b}\cdot\frac{b^2}{2}\ddot{\phi} = -\frac{Mb}{2}\cdot\frac{(a-b)}{b}\ddot{\theta}$$

$$\left[\text{by (4.18.1) and as } K^2 = \tfrac{b^2}{2}\right]$$

$$\text{or,}\quad 2F = -M(a-b)\ddot{\theta} = -F + Mg\sin\theta,$$

$$\text{or,}\quad 3F = Mg\sin\theta,$$

$$\text{or,}\quad F = \frac{\sin\theta}{3}\times \text{ weight of the moving plate.}$$

► **Problem 4.18.2** *A solid homogeneous sphere is rolling on the inside of a fixed hollow sphere, the two centres being in the same vertical plane. Show that the smaller sphere will make complete revolutions if, when it is in its lowest position, the pressure on it is greater that $\frac{34}{7}$ times of its own weight.*

Solution: Let a and b be the radii of the large and small spheres respectively. Let at any time t, the line joining the centres of the spheres make an angle ϕ with the vertical and a fixed line in the small solid sphere makes an angle θ with a fixed line (vertical) in space in this time. Let the solid small sphere of mass moves up and the initial angular velocity of the moving sphere $\dot{\phi} = \Omega$ when $\phi = 0$ and $t = 0$.

Now for rolling, the tangential velocity of the point of contact P is zero, i. e.,

$$(a-b)\dot{\phi} - b\dot{\theta} = 0, \quad \text{or,} \quad (a-b)\ddot{\phi} = b\ddot{\theta}. \tag{4.18.5}$$

The equations of motion along and perpendicular to OC are

$$M\left[-(a-b)\dot{\phi}^2\right] = -R + Mg\cos\phi$$

$$\text{or,} \quad M(a-b)\dot{\phi}^2 = R - Mg\cos\phi \tag{4.18.6}$$

$$M(a-b)\ddot{\phi} = -F - Mg\sin\phi \tag{4.18.7}$$

and

$$MK^2\ddot{\theta} = F\cdot b. \tag{4.18.8}$$

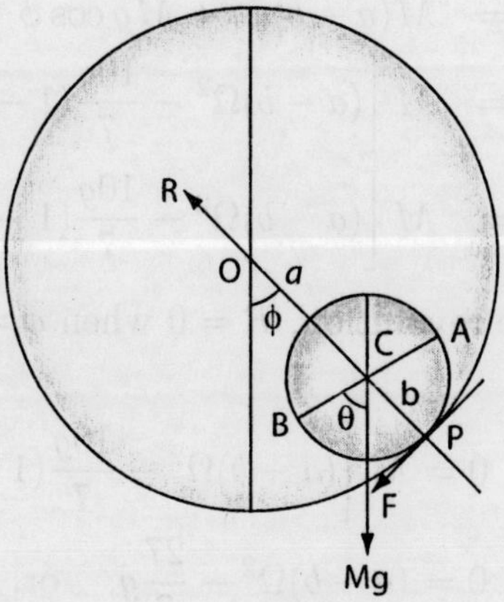

Fig 4.64

From (4.18.8),

$$F = \frac{MK^2\ddot{\theta}}{b} = \frac{M}{b}\cdot\frac{2b^2}{5}\ddot{\theta} \quad \left[\text{as } K^2 = \frac{2b^2}{5}\right]$$

$$= \frac{2}{5}Mb\ddot{\theta} = \frac{2}{5}M(a-b)\ddot{\phi} \quad [\text{by (4.18.5)}] \tag{4.18.9}$$

From (4.18.9) and (4.18.7), we get

$$M(a-b)\ddot{\phi} = -\frac{2}{5}M(a-b)\ddot{\phi} - Mg\sin\phi$$

$$\text{or,} \quad \frac{7}{5}(a-b)\ddot{\phi} = -g\sin\phi$$

$$\text{or,} \quad \frac{7}{5}(a-b)\frac{d}{d\phi}\left(\frac{\dot{\phi}^2}{2}\right) = -g\sin\phi.$$

Integrating,

$$\frac{7}{5}(a-b)\cdot\frac{\dot{\phi}^2}{2} = g\cos\phi + C_1, \quad \text{or,} \quad \dot{\phi}^2 = \frac{10g\cos\phi}{7(a-b)} + C_2.$$

Initially, $t = 0$, $\phi = 0$ and $\dot{\phi} = \Omega$. $\therefore\ C_2 = \Omega^2 - \frac{10g}{7(a-b)}$.
Thus

$$\dot{\phi}^2 = \frac{10g}{7(a-b)}\cos\phi + \Omega^2 - \frac{10g}{7(a-b)},$$

$$\text{or,} \quad (a-b)\dot{\phi}^2 = (a-b)\Omega^2 - \frac{10g}{7}(1-\cos\phi). \tag{4.18.10}$$

Using (4.18.10) in (4.18.6) we get

$$\begin{aligned} R &= M(a-b)\dot{\phi}^2 + Mg\cos\phi \\ &= M\left[(a-b)\Omega^2 - \frac{10g}{7}(1-\cos\phi)\right] + Mg\cos\phi \\ &= M\left[(a-b)\Omega^2 - \frac{10g}{7}(1-\cos\phi) + g\cos\phi\right]. \end{aligned} \tag{4.18.11}$$

For complete revolution, $R = 0$ when $\phi = \pi$.
Thus

$$0 = M\left[(a-b)\Omega^2 - \frac{10g}{7}(1+1) - g\right]$$

$$0 = (a-b)\Omega^2 - \frac{27}{7}g, \quad \text{or,} \quad (a-b)\Omega^2 = \frac{27}{7}g.$$

Thus from (4.18.11) we get

$$R = M\left[\frac{27}{7}g + g\cos\phi - \frac{10}{7}g(1-\cos\phi)\right]. \tag{4.18.12}$$

Therefore, the pressure at the lowest position, i. e., when $\phi = 0$ from (4.18.12), we get

$$R\big|_{\phi=0} = M\left[\frac{27}{7}g + g\right] = \frac{34}{7}Mg = \frac{34}{7} \times \text{ weight of the moving sphere.}$$

▶ **Problem 4.18.3** *A disc rolls on the inside of a fixed hollow circular cylinder whose axis horizontal, the plane of the disc being vertical and perpendicular to the axis of the cylinder; if, when in its lowest position, its centre is moving with a velocity* $\sqrt{\frac{8g(a-b)}{3}}$, *show that the centre of the disc will describe an angle* ϕ *about the centre of the cylinder in time*

$$\sqrt{\frac{3(a-b)}{2g}}\cdot\log\tan\left(\frac{\pi}{4}+\frac{\phi}{4}\right).$$

Solution: Let a and b be the radii of the hollow cylinder and the disc respectively. Let the line joining the centre of the disc and the centre of the central section of the hollow cylinder makes an angle ϕ with the vertical at any time t. Let the disc moves up and P be the point of contact at that time t and a fixed line in the disc makes angle θ with fixed line in space (vertical).

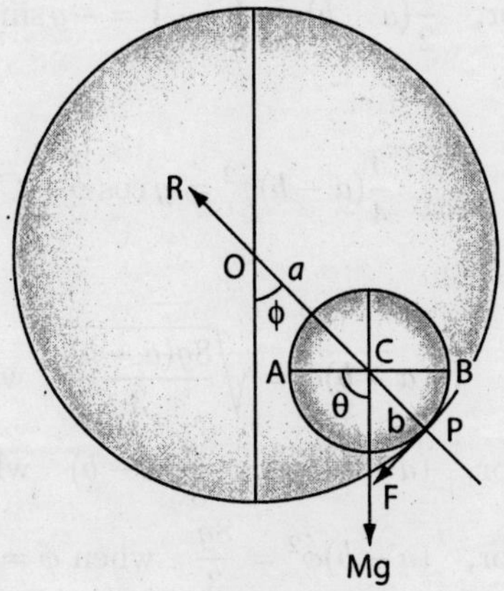

Fig 4.65

Let R and F be the normal reaction and friction at P in that time.

For rolling the tangential velocity of the point of contact P is zero, i.e.,

$$(a-b)\dot{\phi} - b\dot{\theta} = 0, \quad \text{or,} \quad (a-b)\ddot{\phi} = b\ddot{\theta}. \tag{4.18.13}$$

The equations of motion, along and perpendicular to OC are

$$M\left[-(a-b)\dot{\phi}^2\right] = -R + Mg\cos\phi,$$

$$\text{or,} \quad M(a-b)\dot{\phi}^2 = R - Mg\cos\phi \tag{4.18.14}$$

$$M(a-b)\ddot{\phi} = -F - Mg\sin\phi \tag{4.18.15}$$

$$\text{and} \quad MK^2\ddot{\theta} = F\cdot b. \tag{4.18.16}$$

From (4.18.16), we have

$$F = \frac{MK^2\ddot{\theta}}{b} = \frac{M}{b}\cdot\frac{b^2}{2}\ddot{\theta} = \frac{M}{2}b\ddot{\theta} \quad \left[\text{as } K^2 = \tfrac{b^2}{2}\right]$$

$$= \frac{M}{2}(a-b)\ddot{\phi} \quad [\text{by (4.18.13)}]. \tag{4.18.17}$$

From (4.18.17) and (4.18.15), we have

$$M(a-b)\ddot{\phi} = -\frac{M}{2}(a-b)\ddot{\phi} - Mg\sin\phi,$$

$$\text{or,}\quad \frac{3}{2}(a-b)\ddot{\phi} = -g\sin\phi,$$

$$\text{or,}\quad \frac{3}{2}(a-b)\frac{d}{d\phi}\left(\frac{\dot{\phi}^2}{2}\right) = -g\sin\phi.$$

Integrating,

$$\frac{3}{4}(a-b)\dot{\phi}^2 = g\cos\phi + C_1. \tag{4.18.18}$$

Initially, $t = 0$,

$$(a-b)\dot{\phi} = \sqrt{\frac{8g(a-b)}{3}} \quad \text{when } \phi = 0,$$

$$\text{or,}\quad (a-b)^2\dot{\phi}^2 = \frac{8g}{3}(a-b) \quad \text{when } \phi = 0,$$

$$\text{or,}\quad (a-b)\dot{\phi}^2 = \frac{8g}{3} \quad \text{when } \phi = 0.$$

Thus from (4.18.18) we have

$$C_1 = \frac{3}{4}\cdot\frac{8g}{3} - g = 2g - g = g.$$

Thus

$$\frac{3}{4}(a-b)\dot{\phi}^2 = g\cos\phi + g,$$

$$\text{or,}\quad (a-b)\dot{\phi}^2 = \frac{4g}{3}(1+\cos\phi) = \frac{8g}{3}\cos^2\frac{\phi}{2}.$$

$$\therefore\quad \dot{\phi} = \sqrt{\frac{8g}{3(a-b)}}\cos\frac{\phi}{2}, \quad \text{or,}\quad \sec\frac{\phi}{2}\,d\phi = \sqrt{\frac{8g}{3(a-b)}}\,dt.$$

Integrating, we have

$$\int dt = \sqrt{\frac{3(a-b)}{8g}}\int \sec\frac{\phi}{2}d\phi + C_2$$

$$\text{or,}\quad t = \frac{1}{2}\sqrt{\frac{3(a-b)}{2g}}\cdot\log\tan\left(\frac{\pi}{4}+\frac{\phi}{4}\right)\cdot 2 + C_2$$

$$= \sqrt{\frac{3(a-b)}{2g}}\log\tan\left(\frac{\pi}{4}+\frac{\phi}{4}\right) + C_2.$$

Initially, $t = 0$, $\phi = 0$. $\therefore$ $C_2 = 0$.
Thus,

$$t = \sqrt{\frac{3(a-b)}{2g}} \log \tan\left(\frac{\pi}{4} + \frac{\phi}{4}\right).$$

► **Problem 4.18.4** *A perfectly rough ball is a rest within a hollow cylindrical roller and the roller is then drawn along a level path with uniform velocity V. If $V^2 > \frac{27}{7}g(b-a)$, show that the ball will roll completely round the inside of the roller, a and b being the radii of the ball and the roller.*

Solution: Let a and b be radii of the rough ball and the roller respectively and C and O be the centres of ball and the roller. Let at any time t, CO makes an angle ϕ with the vertical. Let also a line fixed in the ball makes an angle θ with a line fixed in space (vertical) in that time and M be the mass of the ball. Let R and F be the normal reaction and friction at the point of contact P.

Thus initially,

$$t = 0, \phi = 0 \quad \text{and} \quad V = (b-a)\dot{\phi}. \tag{4.18.19}$$

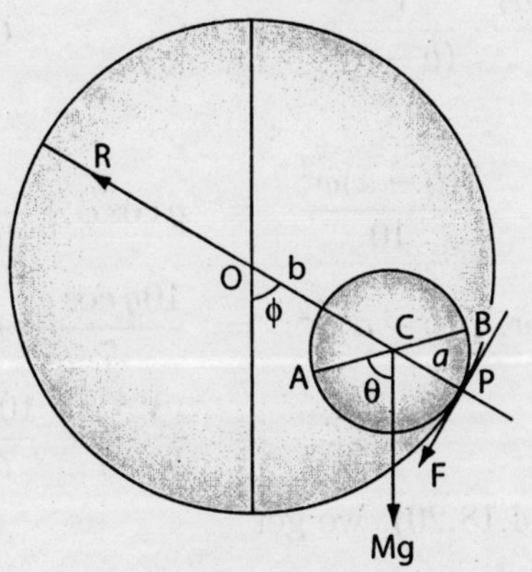

Fig 4.66

The equations of motion of the roller along and perpendicular to OC are

$$M\left[-(b-a)\dot{\phi}^2\right] = -R + Mg\cos\phi$$

$$\text{or,} \quad M(b-a)\dot{\phi}^2 = R - Mg\cos\phi \tag{4.18.20}$$

$$M(b-a)\ddot{\phi} = -F - Mg\sin\phi \tag{4.18.21}$$

$$\text{and} \quad MK^2\ddot{\theta} = F\cdot a. \tag{4.18.22}$$

Again for rolling, the tangential velocity of the point of contact P is zero, i.e.,

$$(b-a)\dot{\phi} - a\dot{\theta} = 0 \quad \text{or,} \quad (b-a)\ddot{\phi} = a\ddot{\theta}. \tag{4.18.23}$$

From (4.18.22), we have

$$F = \frac{MK^2\ddot{\theta}}{a} = \frac{M}{a} \cdot \frac{2}{5}a^2 \cdot \ddot{\theta} \quad \left[\text{as } K^2 = \tfrac{2a^2}{5}\right],$$

$$\text{or,} \quad F = \frac{2a}{5}M\ddot{\theta} = \frac{2M}{5} \cdot (b-a)\ddot{\phi} \quad [\text{by (4.18.23)}]$$

From (4.18.21), we get

$$M(b-a)\ddot{\phi} = -\frac{2M}{5}(b-a)\ddot{\phi} - Mg\sin\phi$$

$$\text{or,} \quad \frac{7}{5}(b-a)\ddot{\phi} = -g\sin\phi, \quad \text{or,} \quad \frac{7}{5}(a-b)\frac{d}{d\phi}\left(\frac{\dot{\phi}^2}{2}\right) = -g\sin\phi,$$

$$\text{or,} \quad \frac{7(b-a)}{10}\dot{\phi}^2 = g\cos\phi + C_1.$$

Initially, $t = 0$, $\phi = 0$ and $V = (b-a)\dot{\phi}$.

$$\frac{7(b-a)}{10} \cdot \frac{V^2}{(b-a)^2} = g + C_1 \quad \therefore \quad C_1 = \frac{7V^2}{10(b-a)} - g.$$

Thus

$$\begin{aligned} \frac{7(b-a)\dot{\phi}^2}{10} &= g\cos\phi + \frac{7V^2}{10(b-a)} - g, \\ \text{or,} \quad (b-a)\dot{\phi}^2 &= \frac{10g\cos\phi}{7} + \frac{V^2}{b-a} - \frac{10g}{7} \\ &= \frac{V^2}{b-a} - \frac{10g}{7}(1-\cos\phi). \end{aligned}$$

Again from (4.18.20), we get

$$\begin{aligned} R &= M\left[(b-a)\dot{\phi}^2 + g\cos\phi\right] \\ &= M\left[\frac{V^2}{b-a} - \frac{10g}{7}(1-\cos\phi) + g\cos\phi\right] \quad [\text{by (4.18.19)}]. \end{aligned}$$

If the ball rolls completely, then $R > 0$ when $\phi = \pi$, i.e.,

$$R\big|_{\theta=\pi} > 0,$$

$$\text{or,} \quad \frac{V^2}{b-a} - \frac{10g}{7}(1+1) - g > 0,$$

$$\text{or,} \quad \frac{V^2}{b-a} - \frac{10g}{7} > 0,$$

$$\text{or,} \quad V^2 > \frac{27g}{7}(b-a).$$

► **Problem 4.18.5** *A solid uniform sphere, resting on the top of another fixed sphere is slightly displaced and begins to roll down. Show that it will slip when the common normal makes with the vertical an angle given by*

$$2\sin\phi = \mu(17\cos\phi - 10),$$

where μ is the coefficient of friction.

Solution: Let O and C be the centries of the fixed sphere and moving sphere of mass M respectively. Let at any time t, OC makes an angle ϕ with the vertical, and in this time the moving sphere rolls angle θ. Let R and F are the normal reaction and friction at the point of contact P and a and b be the radii of the fixed and moving sphere respectively. For rolling the tangential velocity of P is zero, i. e.,

$$(a+b)\dot{\phi} - b\dot{\theta} = 0, \quad \text{or,} \quad (a+b)\ddot{\phi} = b\ddot{\theta}. \qquad (4.18.24)$$

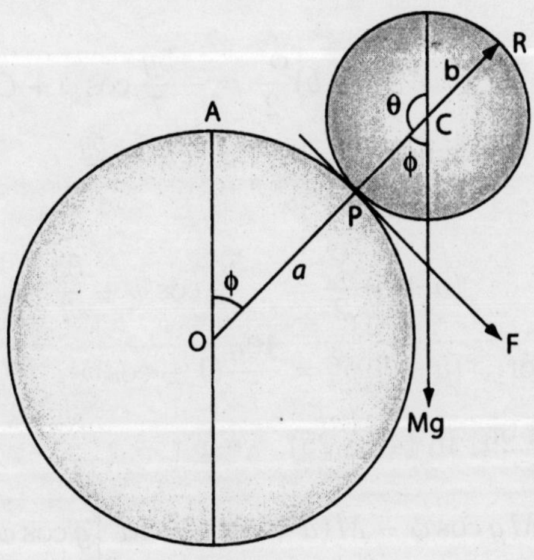

Fig 4.67

The equations of motion of the moving sphere are, along and perpendicular to OC,

$$M\left[-(a+b)\dot{\phi}^2\right] = R - Mg\cos\phi$$

$$\text{or,} \quad M(a+b)\dot{\phi}^2 = -R + Mg\cos\phi \qquad (4.18.25)$$

$$M(a+b)\ddot{\phi} = F + Mg\sin\phi \qquad (4.18.26)$$

$$\text{and} \quad MK^2\ddot{\theta} = -F\cdot b. \qquad (4.18.27)$$

Using (4.18.27) in (4.18.26), we get

$$M(a+b)\ddot{\phi} = -\frac{MK^2\ddot{\theta}}{b} + Mg\sin\phi$$

$$= -\frac{MK^2(a+b)\ddot{\phi}}{b^2} + Mg\sin\phi \quad \text{[by (4.18.24)]}$$

$$\text{or,} \quad \left(1+\frac{K^2}{b^2}\right)(a+b)\ddot{\phi} = g\sin\phi$$

$$\text{or,} \quad \left(1+\frac{2b^2}{5b^2}\right)(a+b)\ddot{\phi} = g\sin\phi \quad \left[\because\ K^2 = \tfrac{2b^2}{5}\right]$$

$$\text{or,} \quad (a+b)\ddot{\phi} = \frac{5g}{7}\sin\phi, \tag{4.18.28}$$

$$\text{or,} \quad (a+b)\frac{d}{d\phi}\left(\frac{\dot{\phi}^2}{2}\right) = \frac{5g}{7}\sin\phi.$$

Integrating,

$$(a+b)\frac{\dot{\phi}^2}{2} = -\frac{5g}{7}\cos\phi + C_1.$$

Initially, $\dot{\phi} = 0$ when $\phi = 0$. $\therefore\ C_1 = \frac{5g}{7}$.
Thus,

$$(a+b)\frac{\dot{\phi}^2}{2} = -\frac{5g}{7}\cos\phi + \frac{5g}{7} = \frac{5g}{7}(1-\cos\phi)$$

$$\text{or,} \quad (a+b)\dot{\phi}^2 = \frac{10g}{7}(1-\cos\phi). \tag{4.18.29}$$

Using (4.18.29) in (4.18.25), we get

$$R = Mg\cos\phi - M(a+b)\dot{\phi}^2 = M\left[g\cos\phi - \frac{10g}{7}(1-\cos\phi)\right]$$

$$= \frac{Mg}{7}(17\cos\phi - 10).$$

Again substituting (4.18.28) in (4.18.26), we get

$$F = M(a+b)\ddot{\phi} - Mg\sin\phi = M\left[\frac{5g}{7}\sin\phi - g\sin\phi\right]$$

$$= -\frac{2Mg}{7}\sin\phi.$$

Thus the sphere will slip if

$$\mu = \left|\frac{F}{R}\right| = \left|\frac{-\frac{2Mg}{7}\sin\phi}{\frac{Mg}{7}(17\cos\phi - 10)}\right| = \frac{2\sin\phi}{17\cos\phi - 10},$$

$$\text{or,} \quad 2\sin\phi = \mu(17\cos\phi - 10).$$

▶ **Problem 4.18.6** *A rough solid circular cylinder rolls down a second rough cylinder which is fixed with its axis horizontal. If the plane through their axis makes an angle α with the vertical when the first cylinder is at rest, show that the cylinders will separate when this angle of elevation is* $\cos^{-1}\left(\frac{4}{7}\cos\alpha\right)$.

Solution: Let O and C be the centres of the central section of the cylinders of radius a and b respectively. Let at any time t the line joining the centres makes an angle ϕ with the vertical OL and in that time a fixed diameter AB of the moving cylinder makes an angle θ with the vertical. Let M be the mass of the moving cylinder and R and F are the normal reaction and friction at the point of contact P.

Thus initially, $t = 0$, $\dot{\phi} = 0$, $\phi = \alpha$.

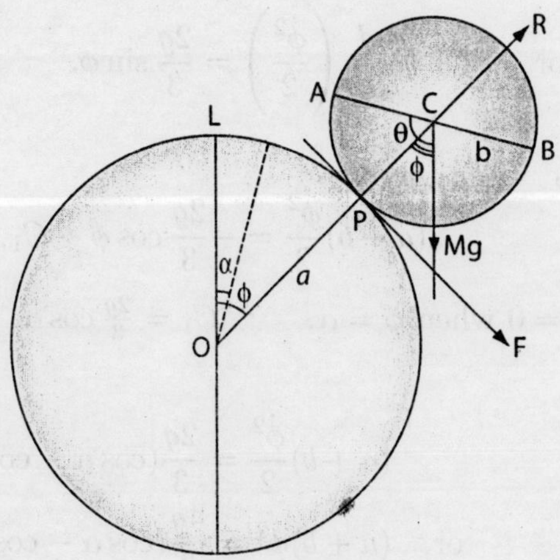

Fig 4.68

For rolling, the tangential velocity of the point of contact is zero, i. e.,

$$(a+b)\dot{\phi} - b\dot{\theta} = 0, \quad \text{or,} \quad (a+b)\ddot{\phi} = b\ddot{\theta}. \tag{4.18.30}$$

The equations of motion moving cylinder are

$$M\left[-(a+b)\dot{\phi}^2\right] = R - Mg\cos\phi,$$

$$\text{or,} \quad M(a+b)\dot{\phi}^2 = -R + Mg\cos\phi \tag{4.18.31}$$

$$M(a+b)\ddot{\phi} = F + Mg\sin\phi \tag{4.18.32}$$

$$\text{and} \quad MK^2\ddot{\theta} = -F\cdot b. \tag{4.18.33}$$

Using (4.18.33) in (4.18.32), we get

$$
\begin{aligned}
M(a+b)\ddot{\phi} &= -\frac{MK^2\ddot{\theta}}{b} + Mg\sin\phi \\
&= -\frac{MK^2(a+b)\ddot{\phi}}{b^2} + Mg\sin\phi \quad \text{[by (4.18.30)]},
\end{aligned}
$$

$$\text{or,}\quad \left(1+\frac{K^2}{b^2}\right)(a+b)\ddot{\phi} = Mg\sin\phi,$$

$$\text{or,}\quad \left(1+\frac{1}{2}\right)(a+b)\ddot{\phi} = g\sin\phi \quad \left[\text{as } K^2 = \tfrac{b^2}{2}\right],$$

$$\text{or,}\quad (a+b)\ddot{\phi} = \frac{2g}{3}\sin\phi, \tag{4.18.34}$$

$$\text{or,}\quad (a+b)\frac{d}{d\phi}\left(\frac{\dot{\phi}^2}{2}\right) = \frac{2g}{3}\sin\phi.$$

Integrating,

$$(a+b)\frac{\dot{\phi}^2}{2} = -\frac{2g}{3}\cos\phi + C_1.$$

Initially, $\dot{\phi} = 0$ when $\phi = \alpha$. $\therefore\ C_1 = \frac{2g}{3}\cos\alpha.$
Thus

$$(a+b)\frac{\dot{\phi}^2}{2} = \frac{2g}{3}(\cos\alpha - \cos\phi),$$

$$\text{or,}\quad (a+b)\dot{\phi}^2 = \frac{4g}{3}(\cos\alpha - \cos\phi). \tag{4.18.35}$$

Thus from (4.18.31),

$$
\begin{aligned}
R &= Mg\cos\phi - M(a+b)\dot{\phi}^2 \\
&= M\left[g\cos\phi - \frac{4g}{3}(\cos\alpha - \cos\phi)\right] \\
&= \frac{Mg}{3}[7\cos\phi - 4\cos\alpha].
\end{aligned}
$$

The cylinders will separate when $R = 0$.
Thus

$$7\cos\phi - 4\cos\alpha = 0, \quad \text{or,}\quad \cos\phi = \frac{4}{7}\cos\alpha.$$

$$\therefore\quad \phi = \cos^{-1}\left(\frac{4}{7}\cos\alpha\right).$$

► **Problem 4.18.7** *A solid uniform sphere, resting on the top of another fixed sphere is slightly displaced and begins to roll down. If the plane through* *their axis of rotation makes an angle α with the vertical when the first sphere is at rest, show that it will stop when the common normal makes with the vertical an angle ϕ, given by*

$$K^2 \sin\phi = \mu\left[(K^2 + 4b^2)\cos\phi - 2b^2\cos\alpha\right],$$

where b is the radius of the moving sphere and K is the radius of gyration. The upper sphere will leave the fixed sphere if

$$\phi = \cos^{-1}\left(\frac{2b^2\cos\alpha}{K^2 + 3b^2}\right).$$

Solution: Let O and C be the centres of the fixed and movable spheres having radii a and b respectively. Let M be the mass of the movable sphere. Let at any time t, the line OC makes an angle ϕ with the vertical and a fixed diameter AB of the movable sphere makes an angle θ with the vertical. Let R and F are the normal reaction and friction at the point of contact P.

For rolling, the tangential velocity of the point contact is zero, i. e.,

$$(a+b)\dot{\phi} - b\dot{\theta} = 0 \quad \text{or,} \quad (a+b)\ddot{\phi} = b\ddot{\theta}. \tag{4.18.36}$$

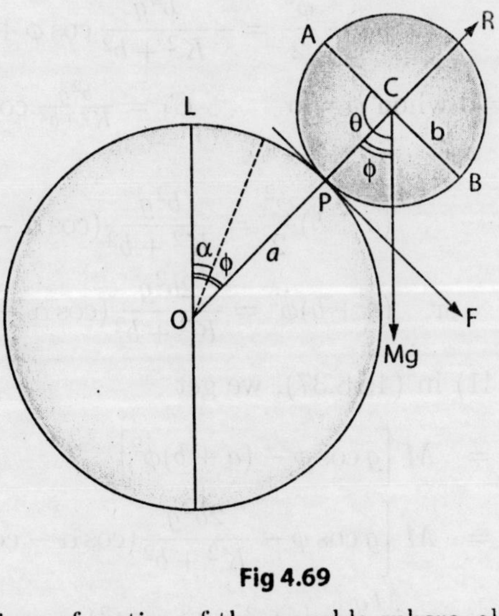

Fig 4.69

Thus the equations of motion of the movable sphere, along and perpendicular to OC, are

$$M\left[-(a+b)\dot{\phi}^2\right] = R - Mg\cos\phi,$$

$$\text{or,} \quad M(a+b)\dot{\phi}^2 = -R + Mg\cos\phi \tag{4.18.37}$$

$$M(a+b)\ddot{\phi} = F + Mg\sin\phi \tag{4.18.38}$$

$$\text{and} \quad MK^2\ddot{\theta} = -F\cdot b. \tag{4.18.39}$$

Using (4.18.39) in (4.18.38), we get

$$M(a+b)\ddot{\phi} = -\frac{MK^2\ddot{\theta}}{b} + Mg\sin\phi,$$

$$\text{or,} \quad (a+b)\ddot{\phi} = -\frac{K^2}{b^2}(a+b)\ddot{\phi} + g\sin\phi \quad \text{[by (4.18.36)]}$$

$$\text{or,} \quad \left(1+\frac{K^2}{b^2}\right)(a+b)\ddot{\phi} = g\sin\phi,$$

$$\text{or,} \quad (a+b)\ddot{\phi} = \frac{b^2 g}{K^2+b^2}\sin\phi, \tag{4.18.40}$$

$$\text{or,} \quad (a+b)\frac{d}{d\phi}\left(\frac{\dot{\phi}^2}{2}\right) = \frac{b^2 g}{K^2+b^2}\sin\phi.$$

Integrating,

$$(a+b)\cdot\frac{\dot{\phi}^2}{2} = -\frac{b^2 g}{K^2+b^2}\cos\phi + C_1.$$

Initially, $\dot{\phi}=0$ when $\phi=\alpha$. $\quad\therefore\ C_1 = \frac{b^2 g}{K^2+b^2}\cos\alpha$.
Thus

$$(a+b)\frac{\dot{\phi}^2}{2} = \frac{b^2 g}{K^2+b^2}(\cos\alpha - \cos\phi),$$

$$\text{or,} \quad (a+b)\dot{\phi}^2 = \frac{2b^2 g}{K^2+b^2}(\cos\alpha-\cos\phi). \tag{4.18.41}$$

Using (4.18.41) in (4.18.37), we get

$$\begin{aligned} R &= M\left[g\cos\phi - (a+b)\dot{\phi}^2\right] \\ &= M\left[g\cos\phi - \frac{2b^2 g}{K^2+b^2}(\cos\alpha-\cos\phi)\right] \\ &= \frac{Mg}{K^2+b^2}\left[\left(K^2+b^2+2b^2\right)\cos\phi - 2b^2\cos\alpha\right] \\ &= \frac{Mg}{K^2+b^2}\left[\left(K^2+3b^2\right)\cos\phi - 2b^2\cos\alpha\right]. \end{aligned} \tag{4.18.42}$$

Using (4.18.40) in (4.18.38), we get

$$\begin{aligned} F &= M(a+b)\ddot{\phi} - Mg\sin\phi = M\left[\frac{b^2 g}{K^2+b^2}\sin\phi - g\sin\phi\right] \\ &= \frac{Mg}{K^2+b^2}\left[-K^2\sin\phi\right] = -\frac{MgK^2}{K^2+b^2}\sin\phi \end{aligned}$$

when the rolling stops

$$\mu = \left|\frac{F}{R}\right| = \frac{\frac{MgK^2\sin\phi}{K^2+b^2}}{\frac{Mg}{K^2+b^2}\left[(K^2+3b^2)\cos\phi - 2b^2\cos\alpha\right]},$$

$$= \frac{K^2\sin\phi}{(K^2+3b^2)\cos\phi - 2b^2\cos\alpha}$$

$$\text{or,}\quad K^2\sin\phi = \mu\left[(K^2+3b^2)\cos\phi - 2b^2\cos\alpha\right].$$

The upper sphere leave the fixed sphere if $R = 0$.
Thus from (4.18.42) we get

$$(K^2+3b^2)\cos\phi = 2b^2\cos\alpha, \quad \text{or,} \quad \phi = \cos^{-1}\left[\frac{2b^2\cos\alpha}{(K^2+3b^2)}\right].$$

► **Problem 4.18.8** *A rough cylinder, of mass M, is capable of motion about its axis which is horizontal, a particle of mass m is placed it vertically above the axis and the system is slightly disturbed. Show that the particle will slip on the cylinder when it has moved through an angle θ given by $\mu(M+6m)\cos\theta - M\sin\theta = 4m\mu$, where μ is the coefficient of friction.*

Solution: Let O be the centre of the central section of the rough cylinder of mass M and radius a. Let at any time t, P be the position of the particle of mass m. R and F are the normal reaction and friction on the particle at that time t. Let in that time, the cylinder turns an angle θ, so that $\angle AOP = \theta$.

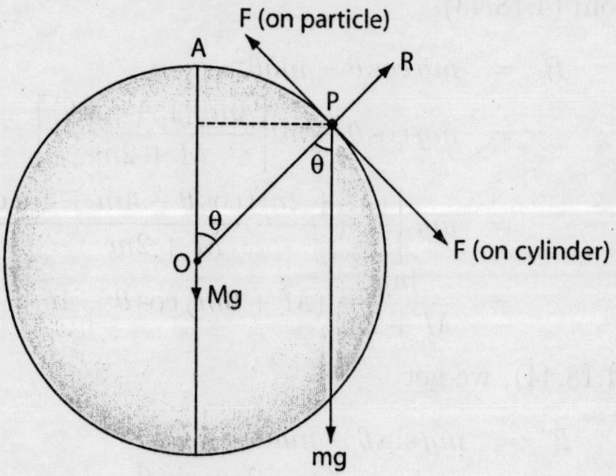

Fig 4.70

The equations of the motion of the particle along and perpendicular to OP are,

$$m\left(-a\dot{\theta}^2\right) = R - mg\cos\theta \qquad (4.18.43)$$

$$ma\ddot{\theta} = -F + mg\sin\theta. \qquad (4.18.44)$$

Considering the motion of the cylinder, we get

$$MK^2\ddot{\theta} = F \cdot a. \tag{4.18.45}$$

Using (4.18.44) in (4.18.45) we get

$$MK^2\ddot{\theta} = a\left[mg\sin\theta - ma\ddot{\theta}\right],$$

or, $(MK^2 + ma^2)\,\ddot{\theta} = agm\sin\theta,$

or, $\left(M \cdot \dfrac{a^2}{2} + ma^2\right)\ddot{\theta} = agm\sin\theta \qquad \left[\because\ K^2 = \dfrac{a^2}{2} \text{ for cylinder}\right]$

or, $(M + 2m)\,a\ddot{\theta} = 2gm\sin\theta, \qquad (4.18.46)$

or, $(M + 2m)\,a \cdot \dfrac{d}{d\theta}\left(\dfrac{\dot{\theta}^2}{2}\right) = 2gm\sin\theta.$

Integrating

$$(M + 2m)a\dot{\theta}^2 = -4gm\cos\theta + C_1.$$

Initially, $\dot{\theta} = 0$ when $\theta = 0$. $\therefore\ C_1 = 4mg.$
Thus

$$(M + 2m)a\dot{\theta}^2 = 4mg(1 - \cos\theta). \tag{4.18.47}$$

Thus from (4.18.43),

$$\begin{aligned} R &= mg\cos\theta - ma\dot{\theta}^2 \\ &= mg\cos\theta - m\left[\frac{4mg(1-\cos\theta)}{M+2m}\right] \\ &= mg\left[\frac{(M+2m)\cos\theta - 4m + 4m\cos\theta}{M+2m}\right] \\ &= \frac{mg}{M+2m}[(M+6m)\cos\theta - 4m]. \end{aligned}$$

From (4.18.44), we get

$$\begin{aligned} F &= mg\sin\theta - ma\ddot{\theta} \\ &= mg\sin\theta - m\cdot\frac{2gm\sin\theta}{M+2m} \quad \text{[by (4.18.46)]} \\ &= mg\sin\theta\left[\frac{M+2m-2m}{M+2m}\right] = \frac{Mmg\sin\theta}{M+2m}. \end{aligned}$$

The particle slips from cylinder, if

$$\mu = \left|\frac{F}{R}\right| = \frac{\frac{Mmg\sin\theta}{M+2m}}{\frac{mg}{M+2m}[(M+6m)\cos\theta - 4m]} = \frac{M\sin\theta}{(M+6m)\cos\theta - 4m},$$

or, $\mu\big[(M+6m)\cos\theta - 4m\big] = M\sin\theta,$

or, $\mu(M+6m)\cos\theta - M\sin\theta = 4\mu m.$

▶ **Problem 4.18.9** *The mass of a sphere $\frac{1}{5}$ that of another sphere of the same material which is free to move about its centre as a fixed point, the first sphere rolls down the second from rest at the highest point, the coefficient of friction being μ. Prove that sliding will begin when the angle ϕ which the line of centres makes with the vertical is given by*

$$\sin\phi = 2\mu(5\cos\phi - 3).$$

Solution: Let O and C be the centres of the lower and upper spheres of masses m and $5m$ respectively. Let OB and CD be two fixed line in the lower and upper spheres. Let at any time t, OC, OB and CD make angles ϕ, ψ and θ with the fixed line in space (vertical). Let

$$K_1 = \frac{2b^2}{5} \quad \text{and} \quad K_2^2 = \frac{2a^2}{5}$$

are the radius of gyration of the upper and lower spheres, where b and a are the radii of the upper and lower spheres. Let R and F are the normal reaction and friction at the point of contact P.

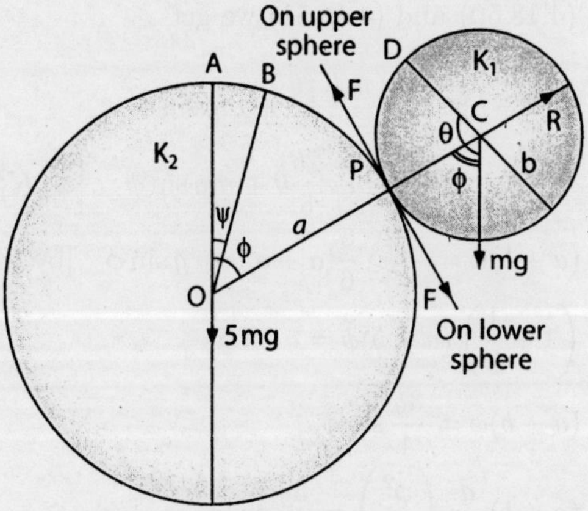

Fig 4.71

As there is no slipping, Arc PB = Arc PD

$$\therefore \quad a(\phi - \psi) = b(\theta - \phi),$$
$$\text{or,} \quad a(\ddot{\phi} - \ddot{\psi}) = b(\ddot{\theta} - \ddot{\phi}),$$
$$\text{or,} \quad (a+b)\ddot{\phi} = b\ddot{\theta} + a\ddot{\psi}. \qquad (4.18.48)$$

The equations of motion, along and perpendicular to OC, are for the upper sphere,

$$m\left[-(a+b)\dot{\phi}^2\right] = R - mg\cos\phi,$$

$$\text{or,}\quad m(a+b)\dot{\phi}^2 = -R + mg\cos\phi \tag{4.18.49}$$

$$m(a+b)\ddot{\phi} = -F + mg\sin\phi \tag{4.18.50}$$

and

$$mK_1^2\ddot{\theta} = F\cdot b. \tag{4.18.51}$$

Again, for the lower sphere

$$5mK_2^2\ddot{\psi} = F\cdot a. \tag{4.18.52}$$

From (4.18.52) and (4.18.51), we have

$$\frac{5K_2^2\ddot{\psi}}{K_1^2\ddot{\theta}} = \frac{a}{b},\quad \text{or,}\quad \frac{5\cdot\frac{2a^2}{5}\ddot{\psi}}{\frac{2b^2}{5}\ddot{\theta}} = \frac{a}{b},\quad \text{or,}\quad 5a\ddot{\psi} = b\ddot{\theta},$$

$$\text{or,}\quad \frac{a\ddot{\psi}}{1} = \frac{b\ddot{\theta}}{5} = \frac{a\ddot{\psi}+b\ddot{\theta}}{6} = \frac{(a+b)\ddot{\phi}}{6}\quad \text{[by (4.18.48)]}. \tag{4.18.53}$$

Again from (4.18.50) and (4.18.51) we get

$$m(a+b)\ddot{\phi} = -\frac{mK_1^2\ddot{\theta}}{b} + mg\sin\phi$$

$$= -m\cdot\frac{2b}{5}\ddot{\theta} + mg\sin\phi \quad \left[\text{as } K_1^2 = \tfrac{2b^2}{5}\right],$$

$$\text{or,}\quad (a+b)\ddot{\phi} = -\frac{2}{5}\cdot\frac{5}{6}(a+b)\ddot{\phi} + g\sin\phi \quad \text{[by (4.18.53)]},$$

$$\text{or,}\quad \left(1+\frac{1}{3}\right)(a+b)\ddot{\phi} = g\sin\phi,$$

$$\text{or,}\quad (a+b)\ddot{\phi} = \frac{3g}{4}\sin\phi, \tag{4.18.54}$$

$$\text{or,}\quad (a+b)\frac{d}{d\phi}\left(\frac{\dot{\phi}^2}{2}\right) = \frac{3g}{4}\sin\phi.$$

Integrating, we have

$$(a+b)\dot{\phi}^2 = -\frac{3g}{2}\cos\phi + C.$$

Initially, $\dot{\phi}=0$, when $\phi = 0$. $\therefore\ C = \frac{3g}{2}$.
Thus,

$$(a+b)\dot{\phi}^2 = \frac{3g}{2}(1-\cos\phi). \tag{4.18.55}$$

Thus from (4.18.49) and (4.18.55), we have

$$\begin{aligned} R &= mg\cos\phi - m\cdot\frac{3g}{2}(1-\cos\phi) = \frac{mg}{2}[2\cos\phi - 3 + 3\cos\phi] \\ &= \frac{mg}{2}[5\cos\phi - 3] \end{aligned}$$

and from (4.18.50) and (4.18.54) we get

$$F = mg\sin\phi - m\frac{3g}{4}\sin\phi = \frac{mg}{4}[4-3]\sin\phi = \frac{mg}{4}\sin\phi.$$

Thus the upper sphere will begin to slip when

$$\begin{aligned} \mu = \left|\frac{F}{R}\right| = \frac{\frac{\sin\phi}{4}}{\frac{5\cos\phi-3}{2}} &= \frac{\sin\phi}{2(5\cos\phi-3)}, \\ \text{or,} \quad \sin\phi &= 2\mu(5\cos\phi - 3). \end{aligned}$$

▶ **Problem 4.18.10** *A uniform solid cylinder rest on a smooth horizontal plane and on it placed a second equal cylinder touching it along its highest generator, if there is no slipping between the cylinders, and system moves from rest, show that the cylinder separate when the plane of either axes makes an angle ϕ with the vertical given by the equation*

$$2\cos^3\phi + 4\cos^2\phi - 35\cos\phi + 20 = 0.$$

Solution: Let C_1 and C_2 be the centres of the central sections of the two cylinder, let two fixed lines C_1A and C_2B in lower and upper cylinders, each of mass m and radius a, make angles ψ and θ with the fixed direction in space (vertical) at any time t. Let in this time, C_1C_2 makes an angle ϕ with the vertical and P is the point of contact of the cylinders. Let the vertical line through P meets the horizontal at O. We take x and y axes at O. Let $(-x_1, y_1)$ and (x_2, y_2) be the coordinates of C_1 and C_2 referred to rectangular axes Ox and Oy.

$$\therefore \left.\begin{aligned} x_1 &= a\sin\phi; & \dot{x}_1 &= a\cos\phi\cdot\dot{\phi} \\ y_1 &= a; & \dot{y}_1 &= 0 \\ x_2 &= a\sin\phi; & \dot{x}_2 &= a\cos\phi\cdot\dot{\phi} \\ y_2 &= a + 2a\cos\phi; & \dot{y}_2 &= -2a\sin\phi\cdot\dot{\phi} \end{aligned}\right\}. \tag{4.18.56}$$

Now

$$\begin{aligned} &\text{arc } AP = \text{arc } PB, \\ \text{or,}\quad & a(\phi-\psi) = a(\theta-\phi), \\ \text{or,}\quad & 2\phi = \theta+\psi, \quad \text{or,} \quad 2\dot{\phi} = \dot{\theta}+\dot{\psi}. \end{aligned} \tag{4.18.57}$$

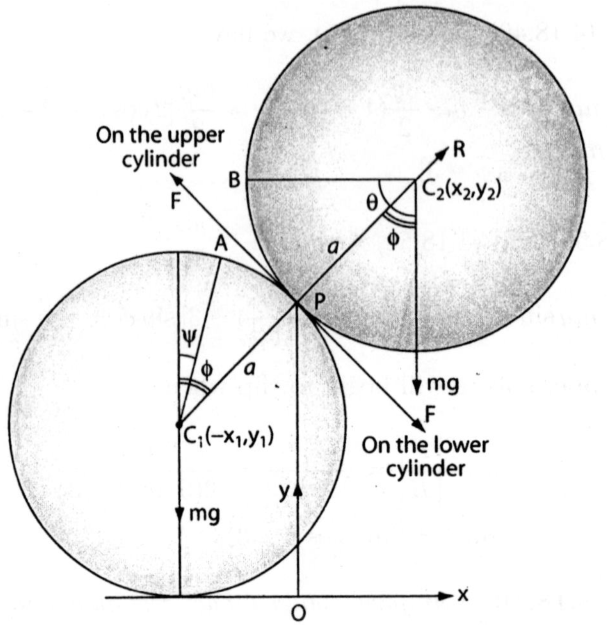

Fig 4.72

Considering the motion of the upper and lower cylinders, we get

$$m \cdot \frac{a^2}{2} \cdot \ddot{\psi} = F \cdot a \quad \left[\because K^2 = \frac{a^2}{2}\right] \tag{4.18.58}$$

$$m \cdot \frac{a^2}{2} \cdot \ddot{\theta} = F \cdot a \quad \left[\because K^2 = \frac{a^2}{2}\right]. \tag{4.18.59}$$

$$\therefore \ \ddot{\psi} = \ddot{\theta}.$$

Integrating

$$\left.\begin{aligned} \dot{\psi} &= \dot{\theta} \\ \psi &= \theta \end{aligned}\right\} \text{ as } \dot{\psi} = \dot{\theta} = \psi = \theta = 0; \text{ when } t = 0. \tag{4.18.60}$$

Thus from (4.18.57) and (4.18.60), $\theta = \phi = \psi$.
Considering the motion of the upper cylinder, we get

$$m\ddot{x}_2 = R\sin\phi - F\cos\phi$$

$$\text{or,} \quad m\left(-a\sin\phi \cdot \dot{\phi}^2 + a\cos\phi \cdot \ddot{\phi}\right) = -F \cdot \cos\phi. \tag{4.18.61}$$

[as $R = 0$, when the cylinders separate]

Again from (4.18.59), we get

$$m \cdot \frac{a^2}{2}\ddot{\theta} = m \cdot \frac{a^2}{2}\ddot{\phi} = F \cdot a, \quad \text{or,} \quad F = \frac{ma\ddot{\phi}}{2}. \tag{4.18.62}$$

From (4.18.62) and (4.18.61), we get

$$\frac{ma\ddot{\phi}}{2}\cos\phi + m(-a\sin\phi\cdot\dot{\phi}^2 + a\cos\phi\cdot\ddot{\phi}) = 0,$$

$$\text{or,}\quad \frac{\ddot{\phi}}{2}\cos\phi - \sin\phi\cdot\dot{\phi}^2 + \cos\phi\cdot\ddot{\phi} = 0,$$

$$\text{or,}\quad 3\cos\phi\cdot\ddot{\phi} - 2\sin\phi\cdot\dot{\phi}^2 = 0,$$

$$\text{or,}\quad \ddot{\phi} = \frac{2\sin\phi}{3\cos\phi}\cdot\dot{\phi}^2. \tag{4.18.63}$$

Now by conservation of energy, we get

$$\text{K.E.} + \text{P.E. of the system} = \text{constant}$$

$$\frac{1}{2}m\left[\dot{x}_1^2 + \dot{y}_1^2 + K^2\dot{\phi}^2\right] + \frac{1}{2}m\left[\dot{x}_2^2 + \dot{y}_2^2 + K^2\dot{\theta}^2\right]$$
$$+mg\cdot a + mg(a + 2a\cos\phi) = 4mga \quad [\text{as at } t = 0, \phi = 0],$$

$$\text{or,}\quad \frac{1}{2}\left[a^2\cos^2\phi\cdot\dot{\phi}^2 + \frac{a^2}{2}\dot{\phi}^2\right]$$
$$+\frac{1}{2}\left[a^2\cos^2\phi\cdot\dot{\phi}^2 + 4a^2\sin^2\phi\cdot\dot{\phi}^2 + \frac{a^2}{2}\cdot\dot{\phi}^2\right]$$
$$+2ga(1+\cos\phi) = 4ga \quad [\text{as } \dot{\theta} = \dot{\phi}],$$

$$\text{or,}\quad a^2\cos^2\phi\cdot\dot{\phi}^2 + 2a^2\sin^2\phi\cdot\dot{\phi}^2 + \frac{1}{2}a^2\dot{\phi}^2 = 2ga(1-\cos\phi),$$

$$\text{or,}\quad \left[\cos^2\phi + 2\sin^2\phi + \frac{1}{2}\right]a\dot{\phi}^2 = 2g(1-\cos\phi),$$

$$\text{or,}\quad (2\cos^2\phi + 4 - 4\cos^2\phi + 1)\,a\dot{\phi}^2 = 4g(1-\cos\phi),$$

$$\text{or,}\quad (5 - 2\cos^2\phi)\cdot a\dot{\phi}^2 = 4g(1-\cos\phi). \tag{4.18.64}$$

Differentiating with respect to t we get

$$4\cos\phi\sin\phi\cdot\dot{\phi}\cdot a\dot{\phi}^2 + 2a\dot{\phi}\ddot{\phi}\,(5 - 2\cos^2\phi) = 4g\sin\phi\cdot\dot{\phi},$$

$$\text{or,}\quad 2\sin\phi\cos\phi\cdot a\dot{\phi}^2 + a\ddot{\phi}\,(5-2\cos^2\phi) = 2g\sin\phi,$$

$$\text{or,}\quad 2\sin\phi\cos\phi\cdot a\dot{\phi}^2 + a\cdot\frac{2\sin\phi}{3\cos\phi}\,(5-2\cos^2\phi)\cdot\dot{\phi}^2 = 2g\sin\phi$$

[by (4.18.63)]

$$\text{or,}\quad (3\cos^2\phi + 5 - 2\cos^2\phi)\,a\dot{\phi}^2 = 3g\cos\phi \quad [\text{as } \sin\phi \neq 0],$$

$$\text{or,}\quad (\cos^2\phi + 5)\,a\dot{\phi}^2 = 3g\cos\phi. \tag{4.18.65}$$

Substituting (4.18.65) in (4.18.64), we get

$$(5 - 2\cos^2\phi)\,\frac{3g\cos\phi}{5+\cos^2\phi} = 4g(1-\cos\phi)$$

$$15\cos\phi - 6\cos^3\phi = 4\left(5 - 5\cos\phi + \cos^2\phi - \cos^3\phi\right)$$

$$2\cos^3\phi + 4\cos^2\phi - 35\cos\phi + 20 = 0.$$

► **Problem 4.18.11** *Two homogeneous spheres of equal radii and of masses M and M' rest on a smooth vertical plane with M' on the highest point of M. If the system be disturbed, show that the inclination θ of their common normal to the vertical is given by*

$$a\dot{\theta}^2\left(7M + 5M'\sin^2\theta\right) = 5g(M + M')(1 - \cos\theta).$$

Solution: Let C_1 and C_2 be the centres of the lower and upper spheres each of radius a and of masses M and M' respectively. Let at any time t, the fixed straight line C_1A and C_2B in the lower and upper spheres make angles ψ and ϕ with straight line fixed in space (vertical) and the line C_1C_2 makes an angle θ with the vertical. Let at time t, R, F are the normal reaction and friction at the point of contact P. We take x and y axes horizontally and vertically at O.

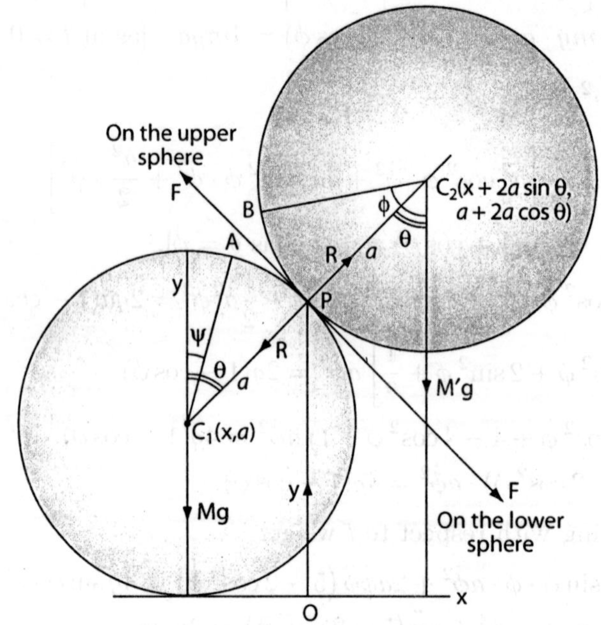

Fig 4.73

Thus

$$\begin{aligned}
&\text{arc } AP = \text{arc } PB,\\
\text{or,}\quad & a(\theta - \psi) = a(\phi - \theta),\\
\text{or,}\quad & \phi + \psi = 2\theta,\\
\text{or,}\quad & \dot{\phi} + \dot{\psi} = 2\dot{\theta} \quad \text{and} \quad \ddot{\phi} + \ddot{\psi} = 2\ddot{\theta}.
\end{aligned} \tag{4.18.66}$$

Again considering the motion of the spheres, taking moments separately, we get

$$M \cdot K^2 \ddot{\psi} = F \cdot a, \quad \text{or,} \quad M \cdot \frac{2a^2}{5} \ddot{\psi} = Fa \tag{4.18.67}$$

$$M' \cdot K^2 \ddot{\phi} = F \cdot a, \quad \text{or,} \quad M' \frac{2a^2}{5} \cdot \ddot{\phi} = Fa. \tag{4.18.68}$$

$$\therefore \quad \frac{M\ddot{\psi}}{M'\ddot{\phi}} = 1 \quad \text{or,} \quad \frac{\ddot{\psi}}{M'} = \frac{\ddot{\phi}}{M}.$$

Integrating we get

$$\frac{\dot{\psi}}{M'} = \frac{\dot{\phi}}{M} \quad [\text{as } \dot{\phi} = \dot{\psi} = 0 \text{ when } t = 0]$$

$$\therefore \quad \frac{\dot{\psi}}{M'} = \frac{\dot{\phi}}{M} = \frac{\dot{\psi} + \dot{\phi}}{M + M'} = \frac{2\dot{\theta}}{M + M'} \quad [\text{by (4.18.66)}].$$

$$\therefore \quad \dot{\psi} = \frac{2\dot{\theta} \cdot M'}{M + M'} \quad \text{and} \quad \dot{\phi} = \frac{2\dot{\theta} \cdot M}{M + M'}. \tag{4.18.69}$$

The equation of motion of the whole system along parallel to x-axis is

$$\frac{d^2}{dt^2}\left[Mx + M'(x + 2a\sin\theta)\right] = 0.$$

Integrating,

$$\frac{d}{dt}\left[Mx + M'(x + 2a\sin\theta)\right] = C,$$

$$\text{or,} \quad M\dot{\bar{x}} + M'\left(\dot{\bar{x}} + 2a\cos\theta \cdot \dot{\theta}\right) = C = 0 \quad \left[\text{as } \dot{x} = \dot{\theta} = 0 \text{ when } t = 0\right]$$

$$\therefore \quad \dot{x} = -\frac{2aM'\cos\theta \cdot \dot{\theta}}{M + M'}. \tag{4.18.70}$$

From conservation of energy, we get

$$\begin{aligned}
&\frac{1}{2}M\left[\dot{x}^2 + 0^2 + \frac{2a^2}{5}\dot{\psi}^2\right] \\
&\quad + \frac{1}{2}M'\left[(\dot{x} + 2a\cos\theta \cdot \dot{\theta})^2 + (-2a\sin\theta \cdot \dot{\theta})^2 + \frac{2a^2}{5}\dot{\phi}^2\right] \\
&\quad + Mga + M'g(a + 2a\cos\theta) \\
&\quad = \text{constant} = \text{initial value} = 3M'ga + Mga
\end{aligned}$$

$$\begin{aligned}
\text{or,} \quad & M\left[\dot{x}^2 + \frac{2a^2}{5}\dot{\phi}^2\right] + M'\left[\left(\dot{x} + 2a\cos\theta \cdot \dot{\theta}\right)^2 + 4a^2\sin^2\theta \cdot \dot{\theta}^2 + \frac{2a^2}{5}\dot{\phi}^2\right] \\
& = 4M'ga(1 - \cos\theta),
\end{aligned}$$

$$\text{or,}\quad M\left(\dot{x}^2+\frac{2a^2}{5}\dot{\psi}^2\right)+M'\left[\dot{x}^2+4a\cos\theta\cdot\dot{\theta}\cdot\dot{x}+4a^2\dot{\theta}^2+\frac{2a^2}{5}\dot{\phi}^2\right]$$
$$=4M'ga(1-\cos\theta),$$
$$\text{or,}\quad \dot{x}^2(M+M')+\frac{2Ma^2\dot{\psi}^2}{5}+\frac{2M'a^2\dot{\phi}^2}{5}+4M'a^2\dot{\theta}^2+4aM'\cos\theta\cdot\dot{\theta}\cdot\dot{x}$$
$$=4M'ga(1-\cos\theta).$$

Substituting $\dot{\psi}$, $\dot{\phi}$ and $\dot{x}$ from (4.18.69) and (4.18.70) we get

$$(M+M')\frac{4a^2M'^2\cos^2\theta\cdot\dot{\theta}^2}{(M+M')^2}+\frac{2M}{5}\cdot a^2\cdot\frac{4M'^2\dot{\theta}^2}{(M+M')^2}$$
$$+\frac{2M'a^2}{5}\cdot\frac{4M^2\dot{\theta}^2}{(M+M')^2}+4M'a^2\dot{\theta}^2-4aM'\cos\theta\cdot\dot{\theta}\cdot\frac{2aM'\cos\theta\cdot\dot{\theta}}{M+M'}$$
$$=4M'ga(1-\cos\theta)$$
$$a\dot{\theta}^2\left[\frac{4M'^2\cos^2\theta}{M+M'}+\frac{8MM'(M+M')}{5(M+M')^2}+4M'-\frac{8M'^2\cos^2\theta}{M+M'}\right]$$
$$=4M'g(1-\cos\theta)$$
$$\text{or,}\quad a\dot{\theta}^2\left[\frac{8MM'}{5(M+M')}+4M'-\frac{4M'^2}{M+M'}\cos^2\theta\right]$$
$$=4M'g(1-\cos\theta)$$
$$\text{or,}\quad a\dot{\theta}^2\left[\frac{2MM'}{5(M+M')}+M'-\frac{M'^2}{M+M'}\cos^2\theta\right]$$
$$=M'g(1-\cos\theta)$$
$$\text{or,}\quad a\dot{\theta}^2\left[2MM'+5M'(M+M')-5M'^2\cos^2\theta\right]$$
$$=5M'(M+M')g(1-\cos\theta)$$
$$\text{or,}\quad a\dot{\theta}^2\left[2MM'+5MM'+5M'^2(1-\cos^2\theta)\right]=5M'(M+M')g(1-\cos\theta)$$
$$\text{or,}\quad a\dot{\theta}^2\left[7M+5M'\sin^2\theta\right]=5(M+M')g(1-\cos\theta).$$

EXERCISES

1. A uniform rod is held at an inclination 45° to the vertical with one end in contact with a horizontal table whose angle of friction is λ. If it is then released, show that it will commence to slide if $\tan\lambda<\frac{3}{5}$.

2. If a uniform semi-circular wire be placed in a vertical plane with one extremity on a rough horizontal plane, and the diameter through that extremity vertical, show that the semi-circle will begin to roll or slide according as μ be greater than or less than $\frac{\pi}{\pi^2-2}$. If μ has this value, prove that wire will roll.

3. A homogeneous plane, of mass M, is placed on an imperfectly rough table and particle of mass m, is attached to the end of a horizontal diameter. Show that the sphere will begin to roll or slide according as μ is greater or less than
$$\frac{5(M+m)m}{7M^2+7Mm+5m^2}.$$
4. A rough cylinder of mass 2 nm, capable to motion about its horizontal axis, has a particle of mass m and coefficient of friction μ placed on it vertically above the axis. The system is then slightly disturbed. Show that the particle will slip on the cylinder after it has moved through an angle θ given by
$$(n+3)\cos\theta - 2 = \frac{n\sin\theta}{n}.$$
Hence find the value of θ, when the particle will leave the curve.
5. A homogeneous solid hemisphere, of mass M and radius a, rests with its vertex in contact with a rough horizontal plane, and a particle of mass m is placed on its base, which is smooth, at a distance c from the centre. Show that the hemisphere will commence to roll or slide according as the coefficient of friction μ is greater or less than
$$\frac{25mac}{26(M+m)a^2+40mc^2}.$$
6. A rough solid circular cylinder rolls down a second rough cylinder which is fixed with its axes horizontal. If the plane through which their axes makes an angle θ with the vertical when the first cylinder is at rest, show that the bodies will separate when this angle of inclination is
$$\cos^{-1}\left(\frac{4\cos\alpha}{7}\right).$$
7. A hollow cylinder, of radius a, is fixed with its axis horizontal inside it moves a solid cylinder, of radius b, whose angular velocity in its lowest position is Ω. If the friction between the cylinders be sufficient to prevent any sliding, prove that the small cylinder will be round the inner surface if
$$\Omega = \sqrt{\frac{11g}{3(a-b)}}.$$
8. A sphere of radius a is projected with a velocity V and under hand spin Ω up a slope of angle α. Show that, if V be large, the sphere will turn back after a time
$$\frac{5V+2a\Omega}{5g\sin\alpha}.$$

Chapter 5

Impulsive Forces

5.1 Definitions

The state of rest or motion of a body sometimes undergoes an apparently instantaneous change owing to the sudden application of a force which acts for a very short interval of time. If P be the variable force and τ be the very short interval of time during which it acts, then the impulse of the force is defined to be the time interval of the force for that time. Thus, the integral $\int_0^\tau P\,dt$ is the *impulse of the force.* Then P is called the *impulsive force.*

5.2 The Equations of Motion for Impulsive Forces (Three Dimension)

Let X, Y, Z be the components of the impulsive forces along parallel to the coordinate axes acting on a particle of mass m at (x, y, z) of a system for a very short interval of time τ. Then we have

$$\sum m\frac{d^2x}{dt^2} = \sum X, \quad \text{or,} \quad \frac{d}{dt}\left(\sum m\frac{dx}{dt}\right) = \sum X \tag{5.2.1}$$

$$\sum m\frac{d^2y}{dt^2} = \sum Y, \quad \text{or,} \quad \frac{d}{dt}\left(\sum m\frac{dy}{dt}\right) = \sum Y \tag{5.2.2}$$

$$\sum m\frac{d^2z}{dt^2} = \sum Z, \quad \text{or,} \quad \frac{d}{dt}\left(\sum m\frac{dz}{dt}\right) = \sum Z \tag{5.2.3}$$

and

$$\frac{d}{dt}\left[\sum m\left(y\frac{dz}{dt}-z\frac{dy}{dt}\right)\right]=\sum(yZ-zY) \tag{5.2.4}$$

$$\frac{d}{dt}\left[\sum m\left(z\frac{dx}{dt}-x\frac{dz}{dt}\right)\right]=\sum(zX-xZ) \tag{5.2.5}$$

$$\frac{d}{dt}\left[\sum m\left(x\frac{dy}{dt}-y\frac{dx}{dt}\right)\right]=\sum(xY-yX). \tag{5.2.6}$$

Let u_0, v_0, w_0 and u, v, w be the components of velocity of the particle of mass m along parallel to the axes, just before and just after the impulse. Then by integrating (5.2.1) in $(0,\tau)$, we get

$$\sum m\frac{dx}{dt}\bigg|_0^\tau = \int_0^\tau\left(\sum X\right)dt,$$

$$\text{or,}\quad \sum m\,(u-u_0) = \sum X', \tag{5.2.7}$$

where $X' = \int_0^\tau Xdt$ = impulse of the impulsive force X. Similarly, from (5.2.2) and (5.2.3) we get

$$\sum m(v-v_0)=\sum Y' \tag{5.2.8}$$

$$\sum m(w-w_0)=\sum Z' \tag{5.2.9}$$

where

$$Y'=\int_0^\tau Ydt \quad\text{and}\quad Z'=\int_0^\tau Zdt$$

are the impulse of the impulsive forces Y and Z respectively.

Again integrating (5.2.4) in $(0,\tau)$, we get

$$\sum m\left(y\frac{dz}{dt}-z\frac{dy}{dt}\right)\bigg|_0^\tau = \int_0^\tau\sum(yZ-zY)\,dt,$$

$$\text{or,}\quad \sum m(yw-zv)-\sum m(yw_0-zv_0)=\sum(yZ'-zY'). \tag{5.2.10}$$

Similarly from (5.2.5) and (5.2.6) we get, by integration in $(0,\tau)$

$$\sum m(zu-xw)-\sum m(zu_0-xw_0) = \sum(zX'-xZ') \tag{5.2.11}$$

$$\sum m(xv-yu)-\sum m(xv_0-yu_0) = \sum(xY'-yX'). \tag{5.2.12}$$

The above equations (5.2.7) to (5.2.12) are the required equation of the impulsive motion in three dimension.

5.3 Equations of Motion for Impulsive Forces in Two Dimension

[CUH 2003, '07; VUH 2001, '02, '04, '05; KUH 2004]

Let us take the plane of motion through centre of gravity of the today at $z = 0$-plane. Let $(\bar{x}, \bar{y})$ be the coordinates of c.g. and (x', y') be the coordinates of an elementary mass m at P with reference to parallel axes through the c.g. Let (X, Y) are the components of the impulsive forces parallel to the coordinate axes acting at P for a very short interval of time τ. Let u_0, v_0 and u, v are the components of velocity parallel of the axes just before and just after the impulse. The equations of motion are

$$M\frac{d^2\bar{x}}{dt^2} = \sum X \tag{5.3.1}$$

$$M\frac{d^2\bar{y}}{dt^2} = \sum Y \tag{5.3.2}$$

$$\text{and} \quad \frac{d}{dt}\left[MK^2\dot{\theta}\right] = \sum (x'Y - y'X), \tag{5.3.3}$$

where M is the mass of the system and MK^2 is the moment of inertia of the body about an axis through c.g. and perpendicular the axes considered.

Integrating (5.3.1) we get

$$M\frac{d\bar{x}}{dt}\bigg|_0^\tau = \int_0^\tau \left(\sum X\right) dt$$

$$\text{or,} \quad M(u - u_0) = \sum X' \tag{5.3.4}$$

where $X' = \displaystyle\int_0^\tau X dx$ = impulse of X.

Similarly, integrating (5.3.2) we get

$$M\frac{d\bar{y}}{dt}\bigg|_0^\tau = \int_0^\tau \left(\sum Y\right) dt, \quad \text{or,} \quad M(v - v_0) = \sum Y', \tag{5.3.5}$$

where

$$Y' = \int_0^\tau Y dt = \text{impulse of } Y.$$

Integrating (5.3.3) we get

$$MK^2\dot{\theta}\bigg|_0^\tau = \int_0^\tau \sum (x'Y - y'X)\, dt$$

$$\text{or,} \quad MK^2(w - w_0) = \sum (x'Y' - y'X'), \tag{5.3.6}$$

where w, w_0 are the angular velocities just after and just before the impulse.

5.4 Conservation of Linear Momentum

STATEMENT. If the sum of the impulse of the forces parallel to a fixed direction vanishes, the total linear momentum in that direction, before and after the application of impulses, will remain same.

Let $P(x, y, z)$ be the position of a particle of mass m at time t with respect to a set of axes Ox, Oy and Oz. Let (X, Y, Z) be the components of the external forces parallel to the axes. Then by D'Alembert's principle, we have

$$\left.\begin{aligned} \sum m\frac{d^2x}{dt^2} &= \sum X, \quad \text{or,} \quad \frac{d}{dt}\left(\sum m\frac{dx}{dt}\right) = \sum X \\ \sum m\frac{d^2y}{dt^2} &= \sum Y, \quad \text{or,} \quad \frac{d}{dt}\left(\sum m\frac{dy}{dt}\right) = \sum Y \\ \sum m\frac{d^2z}{dt^2} &= \sum Z, \quad \text{or,} \quad \frac{d}{dt}\left(\sum m\frac{dz}{dt}\right) = \sum Z \end{aligned}\right\}. \tag{5.4.1}$$

If τ be the time during which the impulsive forces act and u, v, w and u_0, v_0, w_0 are the components of the velocity just after and just before the impulse, then we have from (5.4.1), by integration

$$\left.\begin{aligned} \sum m\frac{dx}{dt}\bigg|_0^\tau &= \sum\int_0^\tau X dt, \quad \text{or,} \quad \sum m(u-u_0) = \sum X' \\ \sum m\frac{dy}{dt}\bigg|_0^\tau &= \sum\int_0^\tau Y dt, \quad \text{or,} \quad \sum m(v-v_0) = \sum Y' \\ \sum m\frac{dz}{dt}\bigg|_0^\tau &= \sum\int_0^\tau Z dt, \quad \text{or,} \quad \sum m(w-w_0) = \sum Z' \end{aligned}\right\}, \tag{5.4.2}$$

where

$$X' = \int_0^\tau X dt, \quad Y' = \int_0^\tau Y dt, \quad Z' = \int_0^\tau Z dt$$

are the impulse of the forces along parallel to x, y and z axes respectively. If

$$\sum X' = \sum Y' = \sum Z' = 0,$$

we have

$$\sum mu = \sum mu_0, \quad \sum mv = \sum mv_0, \quad \sum mw = \sum mw_0.$$

Thus the total momentum parallel to the Ox, Oy and Oz after and before the impulse will remain same.

5.5 Conservation of Angular Momentum or Moment of Momentum [CUH 2002]

STATEMENT. If the sum of the moments of the impulse of the forces about a fixed direction vanishes, the angular momentum in that direction, before and after the application of impulses, will remain same.

Let $P(x, y, z)$ be the position of a particle of mass m at time t, with respect to a set of axes Ox, Oy, Oz. Let X, Y, Z be the components of the external forces parallel to the axes and τ be the time during which the impulsive forces acts and u, v, w and u_0, v_0, w_0 are the components the velocities just after and just before the impulse. Then by D'Alembert's principle we have

$$\left.\begin{aligned} \sum m\left(y\frac{d^2z}{dt^2} - z\frac{d^2y}{dt^2}\right) &= \sum(yZ - zY) \\ \sum m\left(z\frac{d^2x}{dt^2} - x\frac{d^2z}{dt^2}\right) &= \sum(zX - xZ) \\ \sum m\left(x\frac{d^2y}{dt^2} - y\frac{d^2x}{dt^2}\right) &= \sum(xY - yX) \end{aligned}\right\}$$

or,

$$\left.\begin{aligned} \frac{d}{dt}\left[\sum m\left(y\frac{dz}{dt} - z\frac{dy}{dt}\right)\right] &= \sum(yZ - zY) \\ \frac{d}{dt}\left[\sum m\left(z\frac{dx}{dt} - x\frac{dz}{dt}\right)\right] &= \sum(zX - xZ) \\ \frac{d}{dt}\left[\sum m\left(x\frac{dy}{dt} - y\frac{dx}{dt}\right)\right] &= \sum(xY - yX) \end{aligned}\right\}. \tag{5.5.1}$$

Now integrating (5.5.1) in $(0, \tau)$ we get

$$\left.\begin{aligned} \sum m\left(y\frac{dz}{dt} - z\frac{dy}{dt}\right)\Big|_0^\tau &= \sum(yZ' - zY') \\ \sum m\left(z\frac{dx}{dt} - x\frac{dz}{dt}\right)\Big|_0^\tau &= \sum(zX' - xZ') \\ \sum m\left(x\frac{dy}{dt} - y\frac{dx}{dt}\right)\Big|_0^\tau &= \sum(xY' - yX') \end{aligned}\right\}, \tag{5.5.2}$$

where

$$X' = \int_0^\tau X\,dt, \quad Y' = \int_0^\tau Y\,dt, \quad Z' = \int_0^\tau Z\,dt$$

are the impulse of the forces along Ox, Oy and Oz.

Now if the sum of the moments of the impulse is zero, i. e., if

$$\begin{aligned} \sum (yZ' - zY') &= 0 \\ \sum (zX' - xZ') &= 0 \\ \text{and} \quad \sum (xY' - yX') &= 0 \end{aligned}$$

then from (5.5.2) we get

$$\sum m(yw - zv) - \sum m(yw_0 - zv_0) = 0,$$

$$\text{i.e.,} \quad \sum m(yw - zv) = \sum m(yw_0 - zv_0)$$

$$\sum m(zu - xw) - \sum m(zu_0 - xw_0) = 0,$$

$$\text{i.e.,} \quad \sum m(zu - xw) = \sum m(zu_0 - xw_0)$$

$$\sum m(xv - yu) - \sum m(xv_0 - yu_0) = 0,$$

$$\text{i.e.,} \quad \sum m(xv - yu) = \sum m(xv_0 - yu_0).$$

Hence the statement.

5.6 Centre of Percussion

If a rigid body rotating about a fixed axis and is so struck by an impulse that there is no pressure or reaction on the axis, then the point at which the line of action of the blow meets the plane through the centre of gravity of the body and the axis of rotation is called the *centre of percussion.* If the line of action of the blow be given, the axis about which the today begins to rotate is called the *axes of spontaneous rotation.*

5.7 Centre of Percussion of a Rod

Let OA be a rod of mass M and length $2a$. Let G be the centre of gravity of the rod. Let a horizontal blow P be given to the rod at C. Let $OC = x$. Now if due to the impulse, the angular velocity of the rod be w, then the velocity of the c.g. G of the rod is aw after impulse. Let X be the impulsive action at O.

Thus, the equation of motion are

$$M(aw - 0) = P + X \tag{5.7.1}$$

$$\text{and} \quad MK^2(w - 0) = P \cdot x. \tag{5.7.2}$$

If C be the centre of percussion, then $X = 0$.

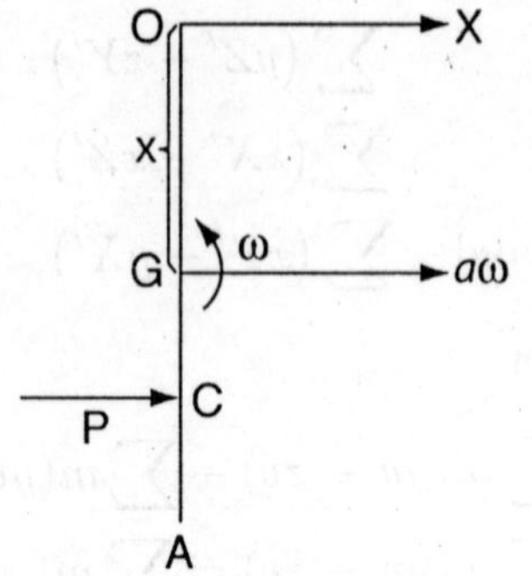

Fig 5.1

$\therefore$ From (5.7.1) we get

$$Maw = P. \tag{5.7.3}$$

From (5.7.2) and (5.7.3) we get

$$MK^2w = Maw \cdot x,$$

$$\text{or,} \quad x = \frac{K^2}{a} = \text{length of the simple equivalent pendulum of the rod.}$$

Thus the centre of percussion of the rod at a depth K^2/a from the point of suspension. So the centre of percussion coincides with its centre of oscillation.

5.8 Solved Problems

▶ **Problem 5.8.1** *A rod, of length 2a and mass m can turn freely about one of its end A which is fixed. If falls from a vertical position, when horizontal, it strikes a fixed inelastic obstacle at a distance b from the fixed end. Show that the impulse of the blow is* $m\frac{2a}{b}\sqrt{\frac{2ga}{3}}$ *and that the impulsive reaction at A is* $m\sqrt{\frac{3ga}{2}}\left(1 - \frac{4a}{3b}\right)$ *vertically upwards.* **[CUH 1983, 1997]**

Solution: Let G be the c.g. of the rod AB of mass m and length $2a$. Let P is the impulse at O when the rod is in horizontal position where $AO = b$. Let ω be the angular velocity of the rod just before the striking the obstacle at O. Thus from the energy equation we have (change in K.E. = work done)

$$\frac{1}{2} \cdot m \cdot \frac{4}{3}a^2\omega^2 - 0 = mga, \quad \text{or,} \quad \omega^2 = \frac{3g}{2a},$$

$$\text{or,} \quad \omega = \sqrt{\frac{3g}{2a}}. \tag{5.8.1}$$

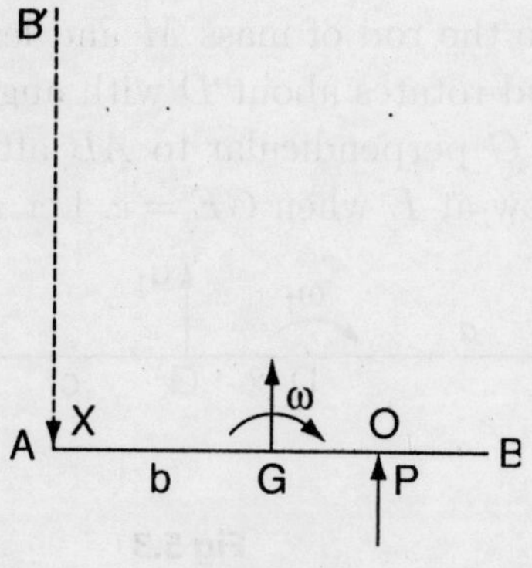

Fig 5.2

Let X be the impulsive reaction at A. As the rod comes to rest in horizontal position, we have

$$m \cdot \frac{4}{3}a^2(0-\omega) = -P \cdot b. \tag{5.8.2}$$

Again for the centre of gravity

$$m(0-a\omega) = X - P. \tag{5.8.3}$$

From (5.8.1) and (5.8.2), we get

$$P = \frac{4ma^2}{3b} \cdot \sqrt{\frac{3g}{2a}} = \frac{2ma}{b}\sqrt{\frac{4a^2 \cdot 3g}{9a \cdot 2}} = m \cdot \frac{2a}{b}\sqrt{\frac{2ag}{3}}.$$

Again from (5.8.3) we have,

$$\begin{aligned} -X &= ma\omega - P = ma\sqrt{\frac{3g}{2a}} - m\frac{2a}{b}\sqrt{\frac{2ag}{3}} \\ &= m\sqrt{\frac{3ga}{2}} - m \cdot \frac{2a}{b}\sqrt{\frac{2ag}{3}} \\ &= m\sqrt{\frac{3ga}{2}}\left[1 - \frac{2a}{b}\sqrt{\frac{2ag}{3}} \cdot \sqrt{\frac{2}{3ag}}\right] \\ &= m\sqrt{\frac{3ga}{2}}\left[1 - \frac{2a}{b} \cdot \frac{2}{3}\right] = m\sqrt{\frac{3ga}{2}}\left[1 - \frac{4a}{3b}\right]. \end{aligned}$$

► **Problem 5.8.2** *A uniform rod of mass M and length $2a$ at rest is struck by a blow J' at right angle to its length at a distance c from the centre. Find the point about which it will begin to turn when the ends are free. Find also the kinetic energy after the blow.* **[CUH 1984]**

Solution: Let AB be the rod of mass M and length $2a$ and G be the c.g. of the rod. Let the rod rotates about D with angular velocity ω_1 and u_1 be the linear velocity of G perpendicular to AB after the impulse. Let J' be the impulse of the blow at E when $GE = c$. Let $DG = x$.

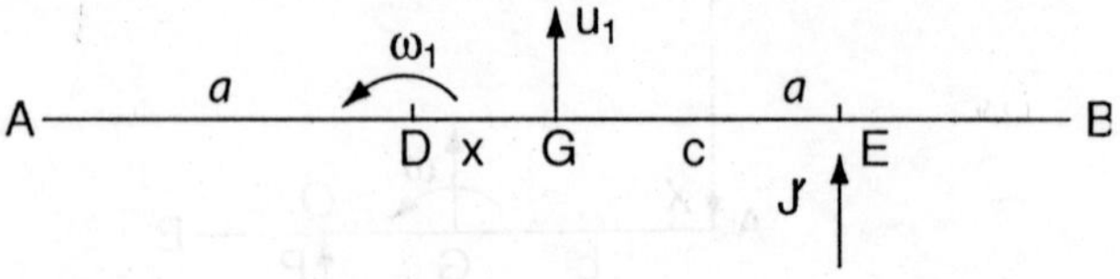

Fig 5.3

The equation of impulsive motion are

$$M(u_1 - 0) = J' \quad (5.8.4)$$

$$M\left(\frac{a^2}{3} + x^2\right)\omega_1 = J'(x + c). \quad (5.8.5)$$

The velocity of D is

$$u_1 - x\omega_1 = 0. \quad (5.8.6)$$

From (5.8.5) and (5.8.6), we get

$$M\left(\frac{a^2}{3} + x^2\right) \cdot \frac{u_1}{x} = J'(x + c),$$

$$\text{or,} \quad J'\left(\frac{a^2}{3} + x^2\right) = (x^2 + cx)\ J' \quad \text{[by (5.8.4)]},$$

$$\text{or,} \quad \frac{a^2}{3} + x^2 = x^2 + cx,$$

$$\text{or,} \quad x = \frac{a^2}{3c}. \quad (5.8.7)$$

Now,

$$\text{K.E.} = \frac{1}{2}M\left[\left(\frac{a^2}{3} + x^2\right)\right]\omega_1^2 = \frac{1}{2}M\left(\frac{a^2}{3} + \frac{a^4}{9c^2}\right)\omega_1^2$$

$$= \frac{1}{2}Ma^2\left(\frac{3c^2 + a^2}{9c^2}\right)\omega_1^2. \quad (5.8.8)$$

Dividing (5.8.5) by (5.8.4), we get

$$\frac{\left(\frac{a^2}{3} + x^2\right)\omega_1}{u_1} = x + c, \quad \text{or,} \quad \omega_1 = \frac{u_1(x + c)}{\frac{a^2}{3} + x^2} = \frac{\frac{a^2}{3c} + c}{\frac{a^2}{3} + \frac{a^4}{9c^2}} \cdot u_1$$

$$= \frac{a^2 + 3c^2}{3c} \times \frac{9c^2}{a^2\,(3c^2 + a^2)} \cdot u_1$$

$$= \frac{3c}{a^2}u_1 = \frac{3c}{a^2} \cdot \frac{J'}{M}.$$

$$\therefore \quad \text{K.E.} = \frac{1}{2}Ma^2 \cdot \frac{3c^2+a^2}{9c^2} \cdot \frac{9c^2}{a^4} \cdot \frac{J'^2}{M^2} = \frac{1}{2M} \cdot \frac{3c^2+a^2}{a^2} J'^2.$$

▶ **Problem 5.8.3** *AB, BC are two equal similar rods freely hinged at B and lie in a straight line on a table. The end A is struck by a blow perpendicular to AB; show that the resulting velocity of A is $3\frac{1}{2}$ times that of B.* **[CUH 2001]**

Solution: Let P be the impulse applied at A and G_1, G_2 be the centre of gravities of the rods AB and BC respectively. Let $2a$ be the length of each rod. Let after impulse u_1, ω_1 and u_2, ω_2 be the linear velocities and angular velocities of the rods AB and BC respectively after impulse. Let X be the impulsive action at B and M be the mass of the each rod.

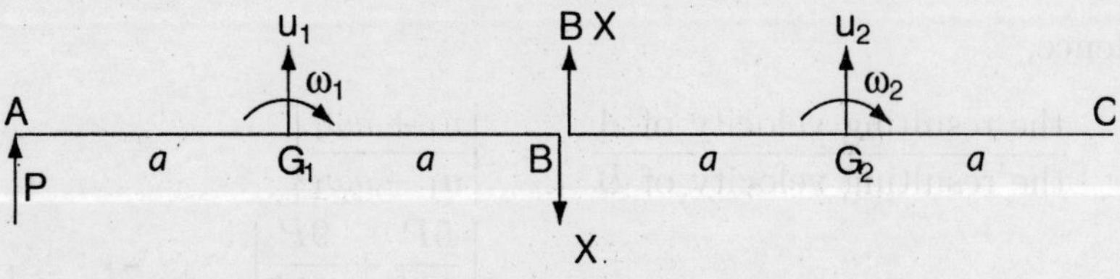

Fig 5.4

The equation of impulsive motion are, for rod AB,

$$M(u_1 - 0) = P - X \tag{5.8.9}$$

$$M \cdot \frac{a^2}{3}(\omega_1 - 0) = P \cdot a + X \cdot a \tag{5.8.10}$$

for the rod BC,

$$M(u_2 - 0) = X \tag{5.8.11}$$

$$M \cdot \frac{a^2}{3}(\omega_2 - 0) = X\ a. \tag{5.8.12}$$

The velocity of B of the rod $AB =$ the velocity of B of the rod BC

$$u_1 - a\omega_1 = u_2 + a\omega_2. \tag{5.8.13}$$

From (5.8.9), (5.8.10), (5.8.11) and (5.8.12), we get

$$\left.\begin{array}{ll} u_1 = \dfrac{P-X}{M}, & a\omega_1 = \dfrac{3(P+X)}{M} \\ u_2 = \dfrac{X}{M}, & a\omega_2 = \dfrac{3X}{M}. \end{array}\right\}. \tag{5.8.14}$$

Substituting (5.8.14) in (5.8.13), we get

$$\frac{P-X}{M} - \frac{3(P+X)}{M} = \frac{X}{M} + \frac{3X}{M}$$

$$\text{or,}\quad P - X - 3P - 3X = X + 3X,$$

$$\text{or,}\quad 8X = -2P, \quad \text{or,}\quad X = -P/4. \tag{5.8.15}$$

Thus from (5.8.14), we get

$$u_1 = \frac{P + \frac{P}{4}}{M} = \frac{5P}{4M}, \quad a\omega_1 = \frac{3\left(P - \frac{P}{4}\right)}{M} = \frac{9P}{4M}$$

$$u_2 = -\frac{P}{4M}, \quad a\omega_2 = -\frac{3P}{4M}$$

Hence,

$$\frac{\text{the resulting velocity of } A}{\text{the resulting velocity of } B} = \left|\frac{u_1 + a\omega_1}{u_1 - a\omega_1}\right| = \left|\frac{\frac{5P}{4M} + \frac{9P}{4M}}{\frac{5P}{4M} - \frac{9P}{4M}}\right| = \left|-\frac{7}{2}\right| = 3\frac{1}{2}.$$

► **Problem 5.8.4** *Two uniform rods AB, BC are freely jointed at B and laid on a horizontal table. AB is struck by a horizontal blow of impulsive force P in a direction perpendicular to AB at a distance c from its centre; length of AB, BC being 2a and 2b and their masses m and m_1, find the motion immediately after the blow.*

Solution: Let G_1, G_2 be the centre of the rod AB and BC of masses m and m_1 and lengths $2a$ and $2b$ respectively. Let u_1, ω_1 and u_2, ω_2 are the linear velocities and angular velocities of the rod AB and BC after the impulse respectively. Let the impulsive force P act at D, where $G_1D = c$. Let X be the impulsive action at B. Therefore the equation of impulsive motion are for the rod AB are

$$m(u_1 - 0) = P + X \tag{5.8.16}$$

$$m\frac{a^2}{3}(\omega_1 - 0) = P \cdot c + Xa \tag{5.8.17}$$

for the rod BC

$$m_1(u_2 - 0) = -X \tag{5.8.18}$$

$$m_1\frac{b^2}{3}(\omega_2 - 0) = -X \cdot b. \tag{5.8.19}$$

The velocity of B of the rod AB = the velocity of B of the rod BC

$$u_1 + a\omega_1 = u_2 + b\omega_2. \tag{5.8.20}$$

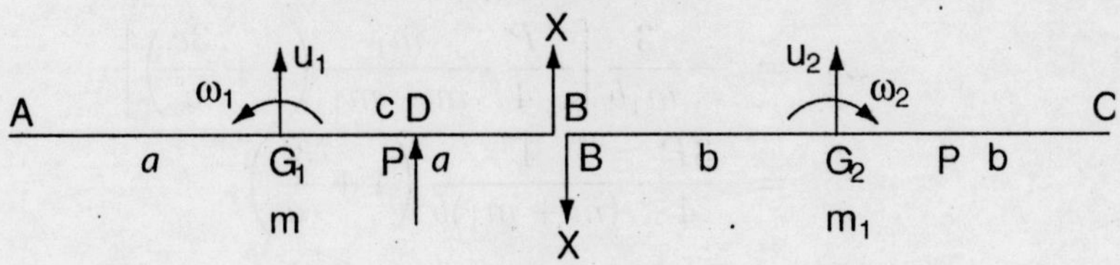

Fig 5.5

From (5.8.16), (5.8.17), (5.8.18) and (5.8.19), we get

$$\left.\begin{aligned} u_1 &= \frac{P+X}{M}, \quad a\omega_1 = \frac{3(P\cdot c + Xa)}{am} \\ u_2 &= -\frac{X}{m_1}, \quad b\omega_2 = -\frac{3X}{m_1} \end{aligned}\right\}. \tag{5.8.21}$$

Substituting (5.8.21) in (5.8.20), we get

$$\frac{P+X}{m} + \frac{3(P\cdot c + X\cdot a)}{am} = \frac{X}{m_1} - \frac{3X}{m_1},$$

$$\text{or,}\quad X\left[\frac{1}{m} + \frac{3}{m} + \frac{1}{m_1} + \frac{3}{m_1}\right] = -P\left[\frac{1}{m} + \frac{3c}{ma}\right],$$

$$\text{or,}\quad 4X\left(\frac{1}{m} + \frac{1}{m_1}\right) = -\frac{P}{m}\left[1 + \frac{3c}{a}\right],$$

$$\text{or,}\quad \frac{4(m+m_1)}{mm_1}X = -\frac{P}{m}\left(1 + \frac{3c}{a}\right),$$

$$\text{or,}\quad X = -\frac{P}{4}\frac{m_1}{m+m_1}\left(1 + \frac{3c}{a}\right).$$

Substituting in (5.8.21), we get

$$\begin{aligned} u_1 &= \frac{P}{m} - \frac{P}{4m}\cdot\frac{m_1}{m+m_1}\left(1 + \frac{3c}{a}\right) \\ &= \frac{P}{M}\left[1 - \frac{m_1}{4(m+m_1)}\left(1 + \frac{3c}{a}\right)\right] \\ a\omega_1 &= \frac{3}{ma}\left[Pc - \frac{Pa}{4}\cdot\frac{m_1}{m+m_1}\left(1 + \frac{3c}{a}\right)\right] \\ &= \frac{3P}{ma}\left[c - \frac{a}{4}\cdot\frac{m_1}{m+m_1}\left(1 + \frac{3c}{a}\right)\right] \end{aligned}$$

$$\text{or,}\quad \omega_1 = \frac{3P}{ma}\left[\frac{c}{a} - \frac{m_1}{4(m+m_1)}\cdot 1 + \left(1+\frac{3c}{a}\right)\right]$$

$$u_2 = -\frac{X}{m_1} = \frac{P}{4}\cdot\frac{1}{m+m_1}\left(1+\frac{3c}{a}\right)$$

$$\omega_2 = -\frac{3}{m_1 b}\left[-\frac{P}{4}\cdot\frac{m_1}{m+m_1}\left(1+\frac{3c}{a}\right)\right]$$

$$= \frac{3P}{4}\cdot\frac{1}{(m+m_1)b}\left(1+\frac{3c}{a}\right).$$

► **Problem 5.8.5** *Two equal uniform rods AB and BC, are freely joined at B and turn about a smooth joint at A. When the rods are in a straight line, ω being the angular velocity of AB and u be the linear velocity of the centre of gravity of BC; BC impinges on a fixed inelastic obstacle at a point D; show that the rods are instantly brought to rest if*

$$BD = 2a\cdot\frac{2u - a\omega}{3u + 2a\omega},$$

where 2a is the length of the either rod.

Solution: Let P be the impulse of the obstacle at D, where $BD = x$. Let m be the mass of the each rod and X is the impulsive reaction at B and ω_2 is the angular velocity of BC before the obstacle acts. Let G_1, G_2 the c.g.'s of the rod AB and BC.

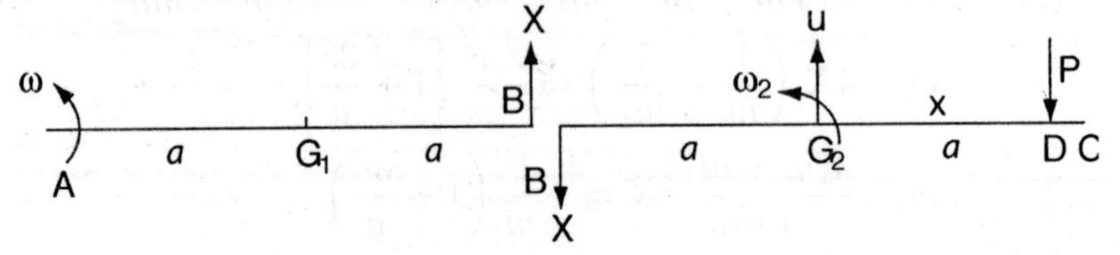

Fig 5.6

The velocity of B of the AB = the velocity of B of the BC

$$2a\omega = u - a\omega_2. \tag{5.8.22}$$

The equations of impulsive motion are:
for the rod AB,

$$m\frac{4a^2}{3}(0-\omega) = X\cdot 2a, \quad \text{or,} \quad \frac{2a\omega m}{3} = -X; \tag{5.8.23}$$

for the rod BC,

$$m(0-u) = -P - X, \quad \text{or,} \quad mu = P + X \tag{5.8.24}$$

and

$$m\frac{a^2}{3}(0-\omega_2) = -P(x-a)+Xa,$$

$$\text{or,}\quad -\frac{ma^2}{3}\omega_2 = -P(x-a)+X\cdot a. \qquad (5.8.25)$$

$$\left.\begin{aligned}
\text{From (5.8.22);}\quad a\omega_2 &= u-2a\omega\\
\text{From (5.8.23);}\quad X &= -\frac{2a\omega}{3}m\\
\text{From (5.8.24);}\quad P &= mu - X = mu+\frac{2a\omega m}{3}\\
&= m\left(u+\frac{2a\omega}{3}\right)
\end{aligned}\right\}. \qquad (5.8.26)$$

Using (5.8.26) in (5.8.25), we get

$$-\frac{ma}{3}(u-2a\omega) = -m\left(u+\frac{2a\omega}{3}\right)(x-a)-\frac{2a^2\omega m}{3}$$

$$\text{or,}\quad -\frac{a}{3}(u-2a\omega) = -\left(u+\frac{2a\omega}{3}\right)x+a\left(u+\frac{2a\omega}{3}\right)-\frac{2a^2\omega}{3}$$

$$= -\left(u+\frac{2a\omega}{3}\right)x+au+\frac{2a^2\omega}{3}-\frac{2a^2\omega}{3}$$

$$\text{or,}\quad \frac{3u+2a\omega}{3}x = au+\frac{au}{3}-\frac{2a^2\omega}{3}=\frac{4au-2a^2\omega}{3}$$

$$\text{or,}\quad x = \frac{2a(2u-a\omega)}{3u+2a\omega}.$$

► **Problem 5.8.6** *Two uniform rods AB and BC, are smoothly jointed at B and placed in a horizontal line; the rod BD is struck at G by a blow at right angle to it; find the position of G, so that the angular velocity of AB and BC may be equal in magnitude.*

Solution: Let $2a$ and $2b$ be the lengths of the rod AB and BC. Let G_1, G_2 be the c.g. of the rod AB and BC respectively. Let at G on the rod BC, the impulse P be applied where $G_2G = x$. Let u_1, ω_1 and u_2, ω_2 are the velocities and angular velocities of the rods after impulse respectively. Let X be the impulsive reaction at B. Let m be the line density of each rod.

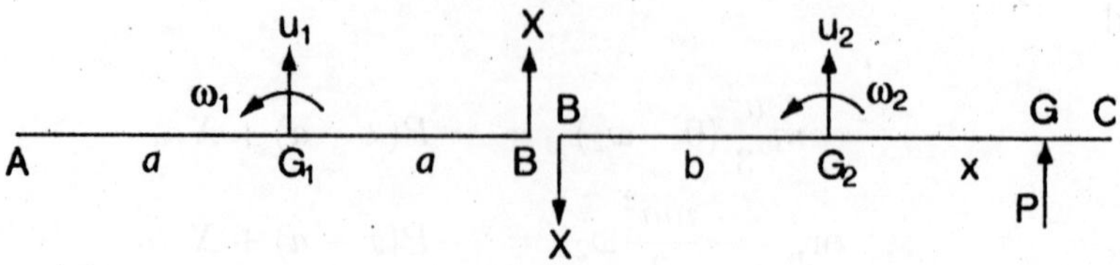

Fig 5.7

The equation of impulsive motions are:
for the rod AB

$$2am \cdot (u_1 - 0) = X \tag{5.8.27}$$

$$2am \cdot \frac{a^2}{3}(\omega_1 - 0) = X \cdot a. \tag{5.8.28}$$

for the rod BC

$$2bm \cdot (u_2 - 0) = P - X \tag{5.8.29}$$

$$2bm \cdot \frac{b^2}{3}(\omega_2 - 0) = P \cdot x + bX \tag{5.8.30}$$

the velocity of B of the rod AB = the velocity of B of the rod BC

$$u_1 + a\omega_1 = u_2 - b\omega_2. \tag{5.8.31}$$

Case I. *If* $\omega_1 = \omega_2$.

From (5.8.27) and (5.8.28), we get

$$2\frac{a^3m}{3}\omega_1 = 2a^2mu_1, \quad \text{or,} \quad u_1 = \frac{a\omega_1}{3} = \frac{a\omega_2}{3}. \tag{5.8.32}$$

Again from (5.8.31),

$$u_2 - b\omega_2 = \frac{a\omega_2}{3} + a\omega_1 = \frac{a\omega_2}{3} + a\omega_2 = \frac{4a\omega_2}{3} \quad [\text{as } \omega_1 = \omega_2]$$

$$\text{or,} \quad u_2 = \frac{4a + 3b}{3}\omega_2.$$

From (5.8.30),

$$Px = \frac{2b^3m}{3}\omega_2 - bX = \frac{2b^3m}{3}\omega_2 - 2mabu_1$$

$$= \frac{2b^3m}{3}\omega_2 - \frac{2ma^2b \cdot \omega_2}{3} = \frac{2mb}{3}\left(b^2 - a^2\right)\omega_2. \tag{5.8.33}$$

From (5.8.29),

$$P = 2bmu_2 + X = 2bm \cdot \frac{4a + 3b}{3}\omega_2 + 2am \cdot \frac{a\omega_2}{3} \quad [\text{by (5.8.32)}]$$

$$\therefore \quad P = \frac{2m\omega_2}{3}\left[4ab + 3b^2 + a^2\right] = \frac{2m}{3}(a+b)(a+3b)\omega_2. \tag{5.8.34}$$

From (5.8.33) and (5.8.34), we get

$$x = \frac{b(b-a)}{a+3b}.$$

Thus in this case the distance of G from the middle point of G_2

$$= \frac{b-a}{a+3b}.$$

Case II. *Where*

$$\omega_1 = -\omega_2. \tag{5.8.35}$$

From (5.8.27) and (5.8.28), we get

$$2a^2mu_1 = \frac{2a^3m}{3}\omega_1. \quad \therefore \quad u_1 = \frac{4}{3}\omega_1 = -\frac{a}{3}\omega_2. \tag{5.8.36}$$

From (5.8.31),

$$u_2 - b\omega_2 = -\frac{a\omega_2}{3} - a\omega_2 = -\frac{4a}{3}\omega_2. \quad \therefore \quad u_2 = \frac{3b-4a}{3}\omega_2.$$

From (5.8.30),

$$\begin{aligned} Px &= \frac{2mb^3}{3}\omega_2 - bX = 2mb^3\omega_2 - 2abmu_1 \\ &= \frac{2mb^3}{3}\omega_2 + \frac{2a^2b}{3}\omega_2 \quad \text{[by (5.8.36)]} \\ &= \frac{2mb}{3}\left(a^2+b^2\right)\omega_2. \end{aligned} \tag{5.8.37}$$

From (5.8.29),

$$\begin{aligned} P &= 2bmu_2 + X = 2bmu_2 + 2amu_1 \\ &= 2bm \cdot \frac{3b-4a}{3} - \frac{2a^2}{3}m \cdot \omega_2 \\ &= \frac{2m}{3}\left(3b^2 - 4ab - a^2\right)\omega_2. \end{aligned} \tag{5.8.38}$$

From (5.8.37) and (5.8.38), we get

$$x = \frac{b\left(a^2+b^2\right)}{3b^2-4ab-a^2}.$$

► **Problem 5.8.7** *Two rods AB and BC; of length $2a$ and $2b$ and of masses proportional to their lengths are freely joined at B and are being in a straight line. A blow is communicated to the end A; show that the kinetic energy when the system is free is to the energy when C is fixed is* $(4a + 3b)(3a + 4b) : 12(a + b)^2$.

Solution:

Case I. *When the system is free.*

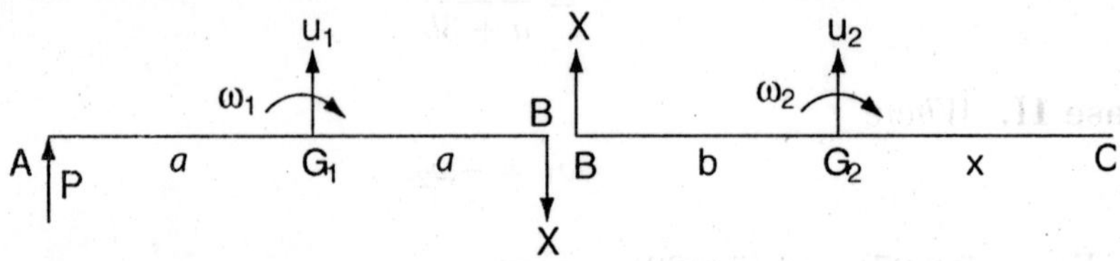

Fig 5.8

Let G_1 and G_2 be the centre of gravities of the rods AB and BC of length $2a$ and $2b$ respectively. Let the mass of the rod are $2am$ and $2bm$, where m is the proportion constant. Let the impulse P applied at the end A and u_1, ω_1 and u_2, ω_2 are the velocities and the angular velocities of the rods after impulse respectively. Let X be the impulsive action at B. The equations of the impulsive motion are:

for the rod AB

$$2am(u_1 - 0) = P - X \tag{5.8.39}$$

$$2am \cdot \frac{a^2}{3}(\omega_1 - 0) = (P + X)a \tag{5.8.40}$$

and for the rod BC

$$2bm(u_2 - 0) = X \tag{5.8.41}$$

$$\text{and} \quad 2bm \cdot \frac{b^2}{3}(\omega_2 - 0) = X \cdot b \tag{5.8.42}$$

the velocity of B of the rod AB = the velocity of B of the rod BC

$$u_1 - a\omega_1 = u_2 + b\omega_2. \tag{5.8.43}$$

$$\left.\begin{aligned}
\text{From (5.8.39)}, \quad u_1 &= \frac{P - X}{2am} \\
\text{From (5.8.40)}, \quad a\omega_1 &= \frac{3(P + X)}{2ma} \\
\text{From (5.8.41)}, \quad u_2 &= \frac{X}{2bm} \\
\text{From (5.8.42)}, \quad b\omega_2 &= \frac{3X}{2mb}
\end{aligned}\right\} \tag{5.8.44}$$

Substituting in (5.8.43), we get

$$\frac{P-X}{2am} - \frac{3(P+X)}{2am} = \frac{X}{2bm} + \frac{3X}{2bm}$$

$$(-2P-4X)\frac{1}{2am} = \frac{4X}{2bm}$$

$$\text{or,}\quad -2P\cdot\frac{1}{2am} = \frac{4X}{2m}\left(\frac{1}{a}+\frac{1}{b}\right) = \frac{2X(a+b)}{mab}$$

$$\text{or,}\quad X = -\frac{bP}{2(a+b)}$$

Thus from (5.8.44)

$$u_1 = \frac{P+\frac{bP}{2(a+b)}}{2am} = \frac{(2a+3b)P}{4ma(a+b)}$$

$$a\omega_1 = \frac{3\left[P-\frac{bP}{2(a+b)}\right]}{2am} = \frac{3P(2a+b)}{4am(a+b)}$$

$$u_2 = -\frac{\frac{bP}{2(a+b)}}{2bm} = -\frac{bP}{4bm(a+b)} = -\frac{P}{4m(a+b)}$$

$$b\omega_2 = \frac{3\left[-\frac{bP}{2(a+b)}\right]}{2mb} = -\frac{3P}{4m(a+b)}.$$

So, Kinetic energy in this case

$$\begin{aligned}
&= \text{K.E. of the rod } AB + \text{K.E. of the rod } BC \\
&= \frac{1}{2}\cdot 2am\left[u_1^2+\frac{a^2\omega_1^2}{3}\right] + \frac{1}{2}\cdot 2bm\left[u_2^2+\frac{b^2\omega_2^2}{3}\right] \\
&= am\left[\frac{(2a+3b)^2P^2}{16m^2a^2(a+b)^2} + \frac{9(b+2a)^2P^2}{3\cdot 16m^2a^2(a+b)^2}\right] \\
&\quad +bm\left[\frac{P^2}{16m^2(a+b)^2} + \frac{9P^2}{3\cdot 16m^2(a+b)^2}\right] \\
&= \frac{P^2(2a+3b)^2}{16am(a+b)^2} + \frac{3(b+2a)^2\cdot P^2}{16am(a+b)^2} + \frac{bP^2}{16(a+b)^2} + \frac{3P^2b}{16m(a+b)^2} \\
&= \frac{P^2}{16ma(a+b)^2}\left[(2a+3b)^2+3(b+2a)^2+4ab\right] \\
&= \frac{P^2}{16ma(a+b)^2}\left[4a^2+9b^2+12ab+3b^2+12a^2+12ab+4ab\right]
\end{aligned}$$

$$= \frac{P^2}{16ma(a+b)^2}\left[16a^2 + 28ab + 12b^2\right] = \frac{\left(4a^2 + 7ab + 3b^2\right)P^2}{4ma(a+b)^2}$$

$$= \frac{(4a+3b)(a+b)P^2}{4ma(a+b)^2} = \frac{(4a+3b)P^2}{4ma(a+b)}. \tag{5.8.45}$$

Case II. *When the end C fixed.*

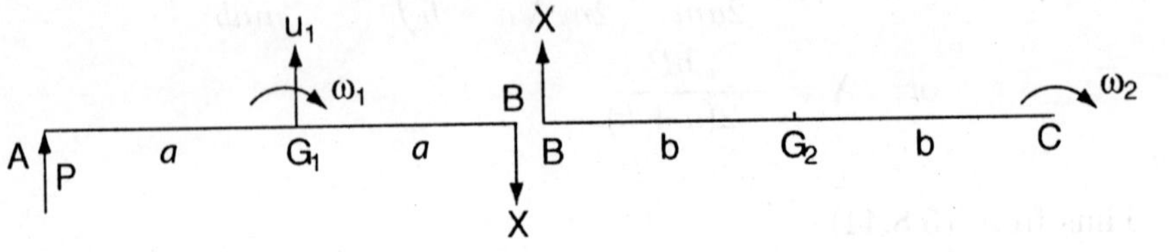

Fig 5.9

In this case,

for the rod AB

$$2am(u_1 - 0) = P - X \tag{5.8.46}$$

$$2am \cdot \frac{a^2}{3}(\omega_1 - 0) = (P + X)a \tag{5.8.47}$$

and for the rod BC,

$$2bm \cdot \frac{4b^2}{3}(\omega_2 - 0) = X \cdot 2b \tag{5.8.48}$$

the velocity of B of the rod AB = the velocity of B of the rod BC

$$u_1 - a\omega_1 = 2b\omega_2. \tag{5.8.49}$$

$$\left.\begin{aligned} &\text{From (5.8.46), we get} \quad u_1 = \frac{P-X}{2am} \\ &\text{From (5.8.47), we get} \quad a\omega_1 = \frac{3(P+X)}{2am} \\ &\text{From (5.8.48), we get} \quad 2b\omega_2 = \frac{3X}{2bm} \end{aligned}\right\}. \tag{5.8.50}$$

Substituting (5.8.50) in (5.8.49), we get

$$\frac{P-X}{2am} - \frac{3(P+X)}{2am} = \frac{3X}{2bm}$$

$$\text{or,} \quad -\frac{2P}{2am} - \frac{4X}{2am} = \frac{3X}{2bm}$$

$$\text{or,} \quad -\frac{P}{am} = \frac{X}{m}\left(\frac{3}{2b} + \frac{2}{a}\right) = \frac{X}{m}\left(\frac{3a+4b}{2ab}\right)$$

$$\text{or,} \quad X = -\frac{2Pb}{3a+4b}.$$

Now substituting in (5.8.50), we get

$$\begin{aligned}
u_1 &= \frac{P+\frac{2Pb}{3a+4b}}{2am} = \frac{P(3a+6b)}{2am(3a+4b)}\\
a\omega_1 &= \frac{3}{2am}\left[P-\frac{2Pb}{3a+4b}\right] = \frac{3P(3a+2b)}{2am(3a+4b)}\\
b\omega_2 &= \frac{3}{2bm}\left(\frac{-2Pb}{3a+4b}\right) = -\frac{3P}{m(3a+4b)}.
\end{aligned}$$

So, in this case,

$$\begin{aligned}
\text{K.E.} &= \text{K.E. of } AB + \text{K.E. of } BC\\
&= \frac{1}{2}\cdot 2am\left[u_1^2+\frac{1}{3}a^2\omega_1^2\right]+\frac{1}{2}\cdot 2bm\left[\frac{4}{3}b^2\omega_2^2\right]\\
&= am\left[\frac{P^2(3a+6b)^2}{4a^2m^2(3a+4b)^2}+\frac{1}{3}\cdot\frac{9P^2(3a+2b)^2}{4a^2m^2(3a+4b)^2}\right]\\
&\quad +mb\left[\frac{4}{3}\cdot\frac{9P^2}{m^2(3a+4b)^2}\right]\\
&= ma\left[\frac{P^2(3a+6b)^2}{4a^2m^2(3a+4b)^2}+\frac{3P^2(3a+2b)^2}{4a^2m^2(3a+4b)^2}\right]+b\cdot\frac{12P^2}{m(3a+4b)^2}\\
&= \frac{P^2}{4am(3a+4b)^2}\left[(3a+6b)^2+3(3a+2b)^2+12ab\right]\\
&= \frac{P^2}{4am(3a+4b)^2}\left[9a^2+36b^2+36ab+27a^2+12b^2+36ab+12ab\right]\\
&= \frac{P^2}{4am(3a+4b)^2}\left[36a^2+84ab+48b^2\right]\\
&= \frac{12P^2}{4am(3a+4b)^2}\left[3a^2+7ab+4b^2\right]\\
&= \frac{3P^2(a+b)(3a+4b)}{am(3a+4b)^2}\\
&= \frac{3P^2(a+b)}{am(3a+4b)}.
\end{aligned}$$

Thus the ratio of K.E. is

$$\begin{aligned}
&\frac{(4a+3b)P^2}{4ma(a+b)} : \frac{3P^2(a+b)}{am(3a+4b)}\\
\text{or,}\quad & 12(a+b)^2 : (4a+3b)(3a+4b).
\end{aligned}$$

► **Problem 5.8.8** *Two equal uniform rod AB and AC, are freely hinged at A and rest in a straight line on a smooth table. A blow is struck at B perpendicular to the rods; show that the kinetic energy generated is $\frac{7}{4}$ times what it would be if the rods were rigidly fastend together at A.* **[CUH 1988]**

Solution:

Case I. *When the rods are freely hinged at A.*

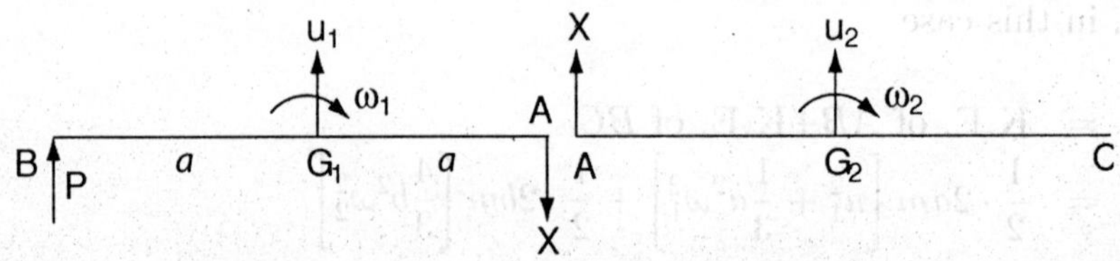

Fig 5.10

Let G_1, G_2 be the centre of gravities of the rod AB and AC each of length $2a$ and of mass m. Let u_1, u_2 are the linear velocities of the c.g.'s G_1 and G_2 and ω_1 and ω_2 are the angular velocities of the rods AB and AC respectively after the impulse P at B. Let X be the impulsive action at A.

The equation of impulsive motion are:
for the rod AB,

$$m(u_1 - 0) = P - X \quad \text{or,} \quad u_1 = \frac{P - X}{m} \tag{5.8.51}$$

$$m\frac{a^2}{3}(\omega_1 - 0) = P \cdot a + X \cdot a \quad \text{or,} \quad a\omega_1 = 3\frac{(P + X)}{m} \tag{5.8.52}$$

for the rod AC,

$$m(u_2 - 0) = X \quad \text{or,} \quad u_2 = \frac{X}{m} \tag{5.8.53}$$

$$m\frac{a^2}{3}(\omega_2 - 0) = X \cdot a \quad \text{or,} \quad a\omega_2 = \frac{3X}{m} \tag{5.8.54}$$

the velocity of A of the rod AB = the velocity of A of the rod AC

$$u_1 - a\omega_1 = u_2 + a\omega_2. \tag{5.8.55}$$

Using (5.8.51) to (5.8.54) in (5.8.55), we get

$$\frac{P - X}{m} - \frac{3(P + X)}{m} = \frac{X}{m} + \frac{3X}{m}$$

$$\text{or,} \quad -2P = 8X \quad \text{or,} \quad X = -\frac{P}{4}$$

$$\therefore \quad u_1 = \frac{P + \frac{P}{4}}{m} = \frac{5P}{4m}, \quad a\omega_1 = \frac{3}{m}\left(P - \frac{P}{4}\right) = \frac{9P}{4m}$$

$$u_2 = -\frac{P}{4m}, \quad a\omega_2 = -\frac{3P}{4m}.$$

Thus, the

$$\begin{aligned}
\text{K.E.} &= \text{K.E. of } AB + \text{K.E. of } AC \\
&= \frac{1}{2}m\left[u_1^2 + \frac{a^2}{3}\omega_1^2\right] + \frac{1}{2}m\left[u_2^2 + \frac{a^2}{3}\omega_2^2\right] \\
&= \frac{1}{2}m\left[\frac{25P^2}{16m^2} + \frac{1}{3}\cdot\frac{81P^2}{16m^2}\right] + \frac{1}{2}m\left[\frac{P^2}{16m^2} + \frac{1}{3}\cdot\frac{9P^2}{16m^2}\right] \\
&= \frac{1}{32m}\left[25P^2 + 27P^2 + P^2 + 3P^2\right] = \frac{56P^2}{32m} = \frac{7P^2}{4m}.
\end{aligned}$$

Case II. *When the rods are rigidly fastened at A.*

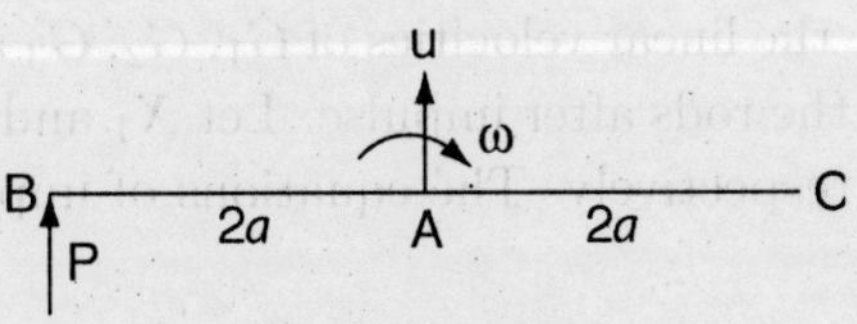

Fig 5.11

In this case, the rods may consider as one rod BC of mass $2m$ and length $4a$ and the point A as c.g. of the rod. Let u be the linear velocity of c.g. A and ω be the angular velocity. In this case, the equation of the impulsive motion are

$$2m(u - 0) = P, \quad \text{or,} \quad u = \frac{P}{2m} \tag{5.8.56}$$

$$2m \cdot \frac{(2a)^2}{3}(\omega - 0) = P \cdot 2a, \quad \text{or,} \quad a\omega = \frac{3P}{4m}. \tag{5.8.57}$$

So, in this case,

$$\begin{aligned}
\text{K.E.} &= \frac{1}{2}\cdot 2m\left[u^2 + \frac{4a^2}{3}\omega^2\right] = m\left[\frac{P^2}{4m^2} + \frac{4}{3}\cdot\frac{9P^2}{4m^2}\right] \\
&= \frac{4P^2}{4m} = \frac{P^2}{m}.
\end{aligned}$$

Now the ratio of K.E. of the two cases

$$\frac{7P^2}{4m} : \frac{P^2}{m} = \frac{7}{4}.$$

► **Problem 5.8.9** *Three equal uniform rods AB, BC, CD are freely jointed and placed in a straight line on a smooth table. The rod AB is struck at its end A by a blow which is perpendicular to its length; find the resulting motion, and show that the velocity of the centre of AB is 19 times that of CD and its angular velocity 11 times that of CD.*

[CUH 1990, 1999, 2003; VUH 2003]

Solution: Let G_1, G_2, G_3 are the c.g.'s of the three uniform rods each of length $2a$ and each of mass m. Let P is the impulse acted at A.

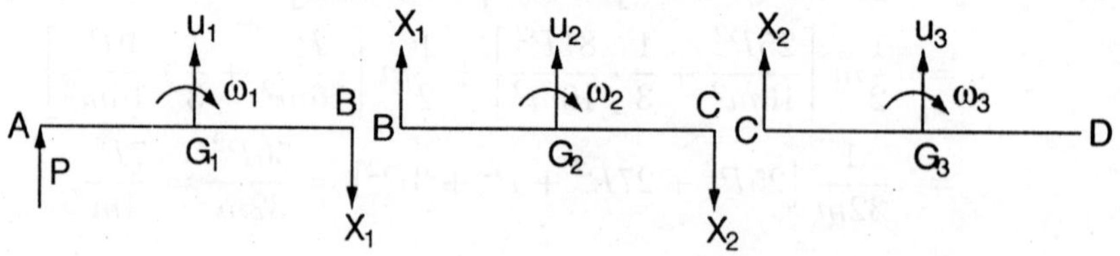

Fig 5.12

Let u_1, u_2, u_3 are the linear velocities of G_1, G_2, G_3 and ω_1, ω_2, ω_3 are the angular velocities of the rods after impulse. Let X_1 and X_2 are the impulsive actions at B and C respectively. The equations of impulsive motion are for rod AB,

$$m(u_1 - 0) = P - X_1, \quad \text{or,} \quad u_1 = \frac{P - X_1}{m} \tag{5.8.58}$$

$$m \cdot \frac{a^2}{3}(\omega_1 - 0) = P \cdot a + X_1 \cdot a, \quad \text{or,} \quad a\omega_1 = \frac{3(P + X_1)}{m} \tag{5.8.59}$$

for rod BC,

$$m(u_2 - 0) = X_1 - X_2, \quad \text{or,} \quad u_2 = \frac{X_1 - X_2}{m} \tag{5.8.60}$$

$$m \cdot \frac{a^2}{3}(\omega_2 - 0) = X_1 \cdot a + X_2 \cdot a, \quad \text{or,} \quad a\omega_2 = \frac{3(X_1 + X_2)}{m} \tag{5.8.61}$$

for rod CD,

$$m(u_3 - 0) = X_2, \quad \text{or,} \quad u_3 = \frac{X_2}{m} \tag{5.8.62}$$

$$m \cdot \frac{a^2}{3}(\omega_3 - 0) = X_2 a, \quad \text{or,} \quad a\omega_3 = \frac{3X_2}{m}. \tag{5.8.63}$$

Again considering the velocities of B and C we get,
the velocity of B of the rod AB = the velocity of B of the rod BC

$$u_1 - a\omega_1 = u_2 + a\omega_2 \tag{5.8.64}$$

the velocity of C of the rod BC = the velocity of C of the rod CD

$$u_2 - a\omega_2 = u_3 + a\omega_3. \tag{5.8.65}$$

Using (5.8.58) to (5.8.61) in (5.8.64), we get

$$\frac{P - X_1}{m} - \frac{3(P + X_1)}{m} = \frac{X_1 - X_2}{m} + \frac{3(X_1 + X_2)}{m},$$

$$\text{or,} \quad -2P - 4X_1 = 4X_1 + 2X_2,$$

$$\text{or,} \quad -2P = 8X_1 + 2X_2, \quad \text{or,} \quad -P = 4X_1 + X_2. \tag{5.8.66}$$

Using (5.8.60), (5.8.61), (5.8.62) and (5.8.63) in (5.8.65), we get

$$\frac{X_1 - X_2}{m} - \frac{3(X_1 + X_2)}{m} = \frac{X_2}{m} + \frac{3X_2}{m},$$

$$\text{or,} \quad -2X_1 = 8X_2, \quad \text{or,} \quad X_1 = -4X_2. \tag{5.8.67}$$

From (5.8.67) and (5.8.66), we get

$$-P = -16X_2 + X_2. \quad \therefore \quad X_2 = \frac{P}{15} \quad \text{and} \quad X_1 = -\frac{4P}{15}.$$

$$\therefore \quad \frac{u_1}{u_3} = \frac{P - X_1}{X_2} = \frac{P + \frac{4P}{15}}{\frac{P}{15}} = \frac{19}{1}, \quad \text{or,} \quad u_1 = 19u_3$$

$$\text{and} \quad \frac{\omega_1}{\omega_3} = \frac{P + X_1}{X_2} = \frac{P - \frac{4P}{15}}{\frac{P}{15}} = \frac{11}{1}, \quad \text{or,} \quad \omega_1 = 11\omega_3$$

▶ **Problem 5.8.10** *Two equal rods, AB and AC, are freely jointed at A, and are placed on a smooth table so as to be at right angles. The rod AC is struck by a blow at C in a direction perpendicular to itself; show that the resulting velocities of the middle points of AB and AC are in the ratio* 2 : 7.

[CUH 1989, 2007; VUH 1993]

Solution: Let G_1, G_2 are the c.g.'s of the rod AC and AB each of length $2a$ and each of mass m. Let P be the impulse acts at C and u_1, u_2 are the linear velocities of the c.g.'s G_1 and G_2 after impulse and ω_1 is the angular velocity of the rod AC. Thus the equations of impulsive motion are:

for rod AC,

$$m(u_1 - 0) = P - X, \quad \text{or,} \quad u_1 = \frac{P - X}{m} \tag{5.8.68}$$

$$m\frac{a^2}{3}(\omega_1 - 0) = (P + X)a, \quad \text{or,} \quad a\omega_1 = \frac{3(P + X)}{m}. \tag{5.8.69}$$

for rod AB

$$mu_2 = -X, \quad \text{or,} \quad u_2 = -\frac{X}{m}, \tag{5.8.70}$$

where X is the impulsive action at A.

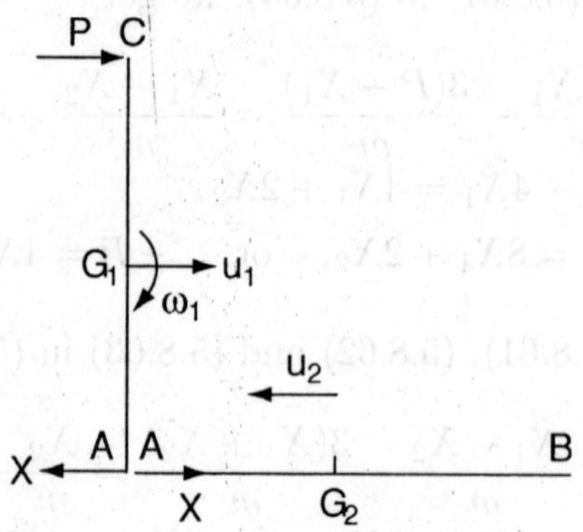

Fig 5.13

Now the velocity of A of rod AC = the velocity of A of rod AB

$$u_1 - a\omega_1 = u_2. \tag{5.8.71}$$

Substituting (5.8.68) to (5.8.70) in (5.8.71), we get

$$\frac{P - X}{m} - \frac{3(P+X)}{m} = \frac{X}{m}, \quad \text{or,} \quad -2P = 5X, \quad \text{or,} \quad X = -\frac{2}{5}P.$$

$$\therefore \quad u_1 = \frac{P + \frac{2}{5}P}{m} = \frac{7P}{5m} \quad \text{and} \quad u_2 = \frac{\frac{2}{5}P}{m} = \frac{2P}{5m}.$$

$$\therefore \quad \frac{\text{Velocity of } G_2}{\text{Velocity of } G_1} = \frac{u_2}{u_1} = \frac{2}{7}.$$

▶ **Problem 5.8.11** *Two uniform rods AB and AC of mass m and m' respectively are freely jointed at A and laid on a smooth horizontal table in such a way that $\angle BAC = \pi/2$. The rod AB is suddenly struck by a blow P at B in a direction perpendicular to AB. Show that the initial velocity of A is $\frac{2P}{4m'+m}$.* **[CUH 1996, 2004; VUH 2002]**

Solution: Let G_1, G_2 are the centres of gravity of the rods AB and AC of mass m and m' and length $2a$ and $2b$ respectively. Let P be the impulse acts at B. Let u_1 and u_2 are the linear velocities of the c.g.'s G_1 and G_2 after impulse and ω_1 is the angular velocity of the rod AB. Let X_1 the impulsive action at A.

The equations of the impulsive motion are:
for rod AB,

$$m(u_1 - 0) = P + X_1, \quad \text{or,} \quad u_1 = \frac{P + X_1}{m}. \tag{5.8.72}$$

$$m\frac{a^2}{3}(\omega_1 - 0) = P \cdot a - X_1 a, \quad \text{or,} \quad a\omega_1 = \frac{3(P - X_1)}{m} \tag{5.8.73}$$

for rod AC,

$$m'(u_2 - 0) = X_1, \quad u_2 = \frac{X_1}{m'}. \tag{5.8.74}$$

Considering the velocity of A

$$u_1 - a\omega_1 = -u_2. \tag{5.8.75}$$

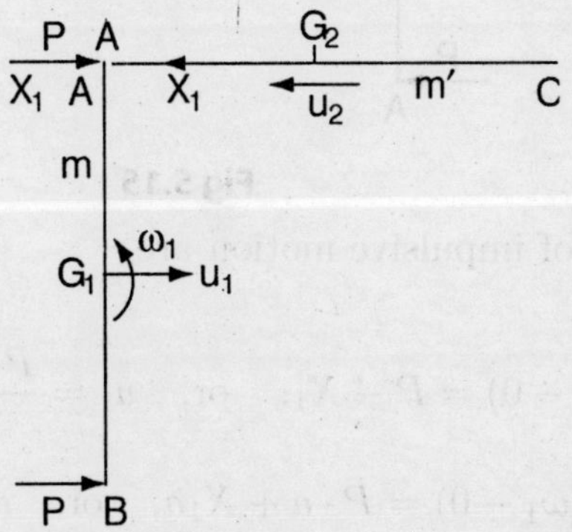

Fig 5.14

Substituting (5.8.72) to (5.8.74) in (5.8.75), we get

$$\frac{P + X_1}{m} - \frac{3(P - X_1)}{m} = -\frac{X_1}{m'}, \quad \text{or,} \quad -\frac{2P}{m} + \frac{4X_1}{m} = -\frac{X_1}{m'},$$

$$\text{or,} \quad -\frac{2P}{m} = -X_1\left(\frac{4m' + m}{mm'}\right),$$

$$\text{or,} \quad X_1 = \frac{2m'P}{m + 4m'}.$$

Thus, from (5.8.74)

$$u_2 = \frac{X_1}{m'} = \frac{2P}{m + 4m'}.$$

► **Problem 5.8.12** *Three equal uniform rods AB, BC, CD are hinged freely at their ends B and C, so as to form three sides of a square and are laid on a smooth table; the end A is struck by horizontal blow P at right angle to AB. Show that the initial velocity of A is nineteen times that of D, and that the impulsive actions at B and C are respectively* $5P/12$ *and* $P/12$.

[CUH 2005, 1998]

Solution: Let G_1, G_2, G_3 are the centre of gravities of the rods AB, BC and CD each of mass m and length $2a$. Let the impulse P is applied at A and X_1 and X_2 are the impulsive actions at B and C. Let u_1, u_2, u_3 are the linear velocities of G_1, G_2, G_3 and ω_1, ω_3 are the angular velocities of the rod AB and CD respectively.

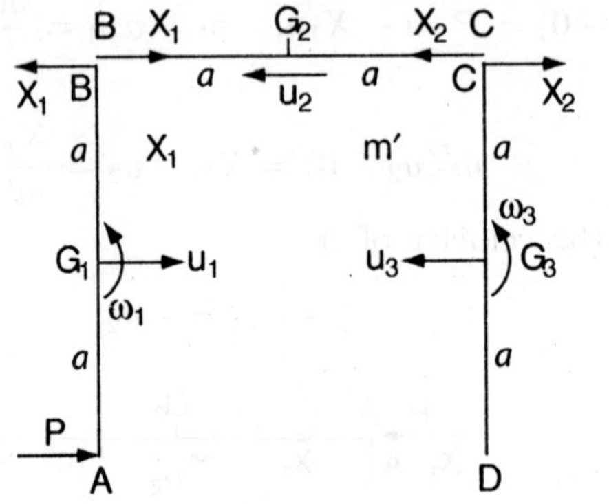

Fig 5.15

The equations of impulsive motion are
for the rod AB

$$m(u_1 - 0) = P - X_1, \quad \text{or,} \quad u_1 = \frac{P - X_1}{m} \tag{5.8.76}$$

$$m\frac{a^2}{3}(\omega_1 - 0) = P \cdot a + X_1 a, \quad \text{or,} \quad a\omega_1 = \frac{3(P + X_1)}{m} \tag{5.8.77}$$

for the rod BC,

$$m(u_2 - 0) = X_2 - X_1, \quad \text{or,} \quad u_2 = \frac{X_2 - X_1}{m} \tag{5.8.78}$$

for the rod CD,

$$m(u_3 - 0) = -X_2, \quad \text{or,} \quad u_3 = -\frac{X_2}{m} \tag{5.8.79}$$

$$m\frac{a^2}{3}(\omega_3 - 0) = -X_2 a, \quad \text{or,} \quad a\omega_3 = -\frac{3X_2}{m} \tag{5.8.80}$$

the velocity of B of the rod AB = the velocity of B of the rod BC

$$u_1 - a\omega_1 = -u_2 \tag{5.8.81}$$

$$u_3 + a\omega_3 = u_2. \tag{5.8.82}$$

Using (5.8.76) to (5.8.78) in (5.8.81), we get

$$\frac{P - X_1}{m} - \frac{3(P + X_1)}{m} = -\frac{X_2 - X_1}{m},$$

$$\text{or,} \quad P - X_1 - 3P - 3X_1 = -X_2 + X_1,$$

$$\text{or,} \quad -2P = 5X_1 - X_2. \tag{5.8.83}$$

Using (5.8.78) to (5.8.80) in (5.8.81), we get

$$-\frac{X_2}{m}-\frac{3X_2}{m}=\frac{X_2-X_1}{m}, \quad \text{or,} \quad X_1=5X_2. \tag{5.8.84}$$

From (5.8.83) and (5.8.84), we get

$$-2P=25X_2-X_2=24X_2. \quad \therefore \quad X_2=-\frac{P}{12}.$$

$$\therefore \quad X_1=-\frac{5P}{12}.$$

$$\therefore \quad u_1=\frac{P+\frac{5P}{12}}{m}=\frac{17P}{12m}, \quad a\omega_1=\frac{3}{m}\left(P-\frac{5P}{12}\right)=\frac{7P}{4m}$$

$$u_3=\frac{P}{12m}, \quad a\omega_3=\frac{P}{4m}.$$

Now

$$\frac{\text{the velocity of } A}{\text{velocity of } D}=\frac{u_1+a\omega_1}{-u_3+a\omega_3}=\frac{\frac{17P}{12m}+\frac{7P}{4m}}{-\frac{P}{12m}+\frac{P}{4m}}=\frac{17+21}{3-1}=\frac{38}{2}=19.$$

Thus, velocity of $A=19\times$ velocity of D.

▶ **Problem 5.8.13** *A lamina in the from of an ellipse is rotating in its own plane about one of its foci with angular velocities ω. This, focus is set free and the other, at the same instant is fixed, show that the ellipse now rotate about it with angular velocity $\omega\frac{2-5e^2}{2+3e^2}$.*

Solution: Let C be the centre of the elliptic lamina of mass m and $2a$ and $2b$ be the lengths of the major and minor axes. Let the elliptic lamina rotating with angular velocity ω about the focus S. Let the focus S is freed and the other focus S' is fixed, then ω' be the angular velocity of the elliptic lamina with which it is rotating about S'. Thus the angular momentum of the lamina about S' just after the fixture = angular momentum just before the fixture.

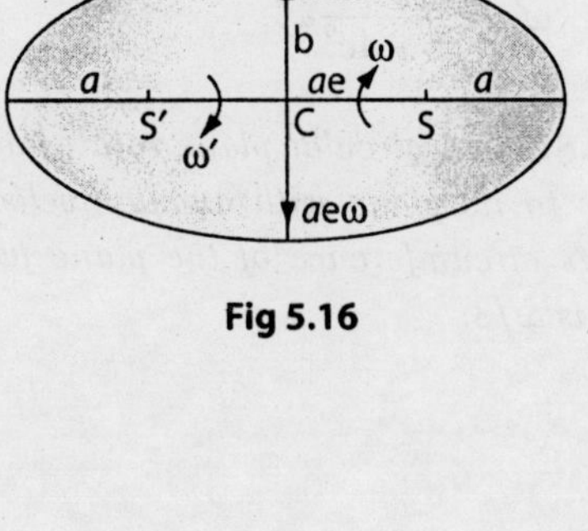

Fig 5.16

Thus

$$mK^2\omega' + mae\omega' \cdot ae = mK^2\omega - mae\omega \cdot ae,$$

$$\text{or,} \quad \frac{a^2+b^2}{4}\omega' + a^2e^2\omega' = \frac{a^2+b^2}{4}\omega - a^2e^2\omega,$$

$$\text{or,} \quad \omega' = \frac{a^2+b^2-4e^2a^2}{a^2+b^2+4a^2e^2}\cdot\omega = \frac{a^2+a^2-a^2e^2-4a^2e^2}{a^2+a^2-a^2e^2+4a^2e^2}\cdot\omega = \frac{2-5e^2}{2+3e^2}\cdot\omega.$$

► **Problem 5.8.14** *An elliptic area,of eccentricity e, is rotating with angular velocity ω about one latus rectum; suddenly this latus rectum is freed and the other fixed. Show that the new angular velocity is* $\omega\frac{1-4e^2}{1+4e^2}$

[CUH 1996, 2002]

Solution: Let C be the centre of the elliptic area of mass m and $2a$ and $2b$ be the length of major and minor axes. Let ω be the angular velocity of the area with which the area rotates about latus rectum LSL' through the focus S. Let when suddenly LSL', is loosed, the area rotates with angular velocity ω', about the latus rectum $MS'M'$ through focus S'.

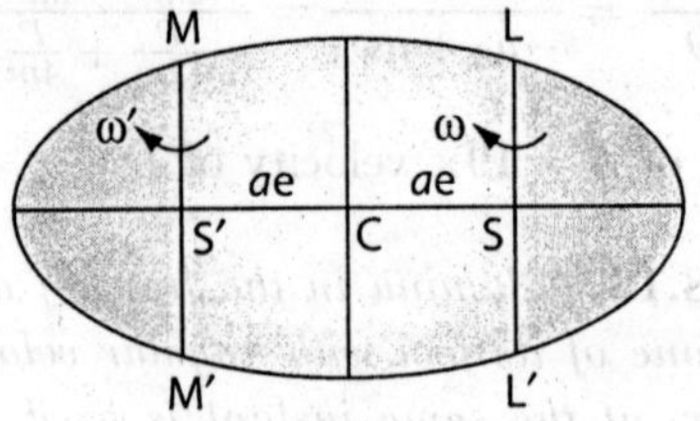

Fig 5.17

Thus, the moment of momentum about MSM' will remain same before and after the fixture.

$$\therefore \quad mK^2\omega' + m\cdot ae\omega'\cdot ae = mK^2\omega - mae\omega\cdot ae,$$

$$\text{or,} \quad k^2\omega' + a^2e^2\omega' = k^2\omega - a^2e^2\omega,$$

$$\text{or,} \quad \left(\frac{a^2}{4} + a^2e^2\right)\omega' = \left(\frac{a^2}{4} - a^2e^2\right)\omega,$$

$$\text{or,} \quad \omega' = \frac{1-4e^2}{1+4e^2}\omega.$$

► **Problem 5.8.15** *A circular plane rotates about an axis through its centre perpendicular to its plane with angular velocity ω. This axis is set free and a point in its circumference of the plane fixed, show that the resulting angular velocity is ω/3.*

Solution: Let C be the centre of the circular plane of mass m and radius a. Let the circular plane rotating with angular velocity ω about an axis through C and suddenly C is freed and a point A on the circumference is fixed. Let ω' be the angular velocity with which the plane rotates.

Thus the moment of momentum about the axis through A remains same before and after the fixture.

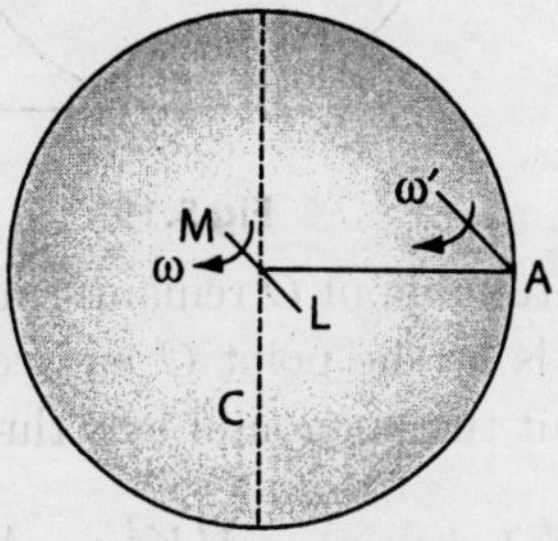

Fig 5.18

Thus,

$$mk^2\omega' + ma\omega' \cdot a = mk^2\omega, \quad \text{or,} \quad \frac{a^2}{2}\omega' + a^2\omega' = \frac{1}{2}a^2\omega,$$

$$\text{or,} \quad \frac{3}{2}a^2\omega' = \frac{a^2\omega}{2},$$

$$\text{or,} \quad \omega' = \frac{\omega}{3}.$$

► **Problem 5.8.16** *A uniform disc of radius a is rolling without slipping along a smooth horizontal table with velocity V, when the highest point becomes suddenly fixed. Prove that the disc will make a complete revolution round the point if $V^2 > 24ag$, where g is the acceleration due to gravity.*

[VUH 2005]

Solution: Let M be the mass of the disc. Let ω be the angular velocity of the disc and V be the velocity of the centre of the disc, just before the fixture. As the disc is rolling only, the velocity of the point of contact

$$V - a\omega = 0. \quad \therefore \quad \omega = \frac{V}{a}. \tag{5.8.85}$$

Let ω' be the angular velocity of the disc after fixture at the highest point O then the velocity of the centre of the disc is $a\omega'$.

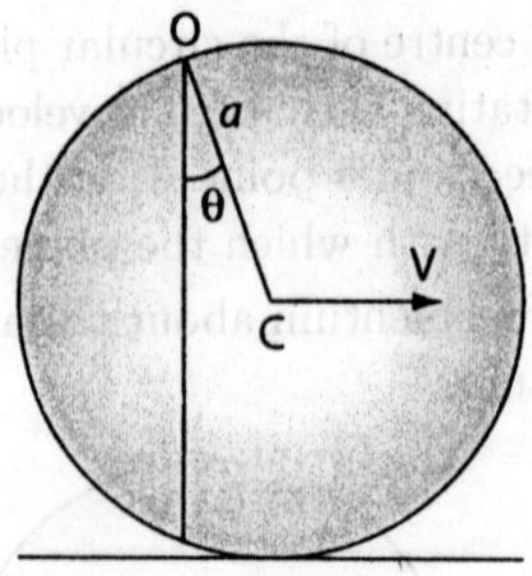

Fig 5.19

The angular momentum about O remains same just before and after the fixture, as the impulse is at the point O and consequently the moment of the impulse is zero about the horizontal axis through O.

$$\therefore \quad M\left(K^2\omega' + a\omega' \cdot a\right) = MK^2\omega - MVa$$

$$\text{or,} \quad \left(\frac{a^2}{2} + a^2\right)\omega' = \frac{a^2}{2}\omega - aV \qquad \left[\because K^2 = \frac{a^2}{2} \text{ for disc}\right]$$

$$\text{or,} \quad \frac{3a^2}{2}\omega' = \frac{a^2}{2}\frac{V}{a} - aV = -\frac{aV}{2} \qquad \text{[by (5.8.85)]}$$

$$\text{or,} \quad \omega' = -\frac{V}{3a}. \tag{5.8.86}$$

Let when the disc is turning about O, the radius through O makes an angle θ with the vertical. Therefore the energy equation is

$$\frac{1}{2}M\left(K^2\dot{\theta}^2 + a^2\dot{\theta}^2\right) - \frac{1}{2}M\left(K^2\omega'^2 + a^2\omega'^2\right) = -Mg(a - a\cos\theta)$$

$$\text{or,} \quad \left(\frac{a^2}{2} + a^2\right)\dot{\theta}^2 - \left(\frac{a^2}{2} + a^2\right)\omega'^2 = -2g(a - a\cos\theta)$$

$$\text{or,} \quad \dot{\theta}^2 - \omega'^2 = -\frac{4ga(1-\cos\theta)}{3a^2} = -\frac{4g(1-\cos\theta)}{3a}$$

$$\text{or,} \quad \dot{\theta}^2 = \omega'^2 - \frac{4g}{3a}(1-\cos\theta) = \frac{V^2}{9a^2} - \frac{4g}{3a}(1-\cos\theta) \quad \text{[by (5.8.86)]}$$

$$\text{or,} \quad \dot{\theta} = \sqrt{\frac{V^2}{9a^2} - \frac{4g}{3a}(1-\cos\theta)}.$$

In order that the disc may make complete revolution, $\dot{\theta} > 0$ when $\theta = \pi$.

$$\therefore \quad \sqrt{\frac{V^2}{9a^2} - \frac{4g}{3a}(1+1)} > 0$$

$$\frac{V^2}{9a^2} - \frac{8g}{3a} > 0, \quad \text{or,} \quad V^2 > 24ga.$$

► **Problem 5.8.17** *A circular ring of mass M and radius a lies on a smooth horizontal plane and an insect of mass m resting on it starts walking round it with uniform velocity v relative to the ring. Show that centre of the ring describes a circle with angular velocity* $\frac{mv}{(M+2m)a}$. **[VUH 2006]**

Solution: Let O be the centre of the ring of radius a and mass M and P be the position of the insect of mass m. Let G be the common centre of gravity of the ring and the insect. Let $OG = x \quad \therefore PG = a - x$.

$$\therefore \quad \frac{x}{m} = \frac{a-x}{M} = \frac{a}{M+m}. \tag{5.8.87}$$

$$\therefore \quad x = \frac{am}{M+m}.$$

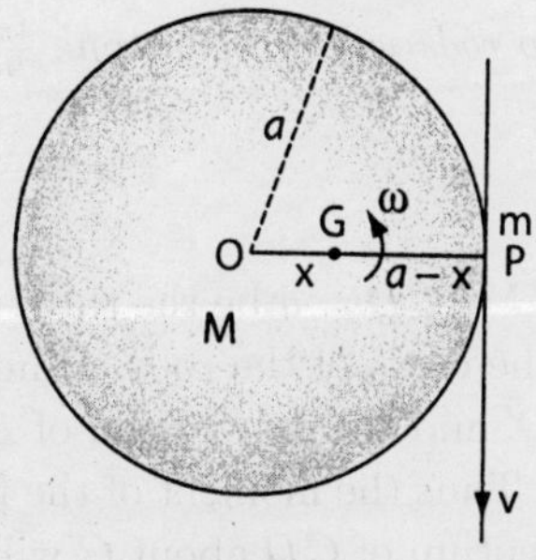

Fig 5.20

The total horizontal momentum remains unaltered through out the motion, as there is no horizontal forces on the system. So the velocity of the common centre of gravity G is zero. Throughout the motion as the initial velocity of G is zero, so G remains fixed. Thus O, the centrie of the ring describes as circle of radius $x = \frac{am}{M+m}$ with G as centre.

Let ω be the angular velocity of the ring about G. As all the forces acting on the system are vertical, their moments vanish about the vertical through G. So moment of momentum about the vertical through G remains unaltered.

So,

$$M\left[K^2\omega + x\omega \cdot x\right] + m(a-x)^2\omega - mv(a-x) = 0$$

$$\text{or,} \quad M\left(a^2 + x^2\right)\omega + m(a-x)^2\omega = mv(a-x)$$

$$\text{or,} \quad M\frac{a^2+x^2}{a-x}\omega + m(a-x)\omega = mv$$

$$\text{or,} \quad M\frac{(a-x)^2+2ax}{a-x}\omega + m(a-x)\omega = mv$$

$$\text{or,}\quad M\left(a-x+\frac{2ax}{a-x}\right)\omega+m(a-x)\omega=mv$$

$$\text{or,}\quad [M+m](a-x)\omega+2Ma\omega\cdot\frac{x}{a-x}=mv$$

$$\text{or,}\quad [M+m]\frac{Ma}{M+m}\omega+2Ma\omega\cdot\frac{m}{M}=mv\quad \text{[by (5.8.84)]}$$

$$\text{or,}\quad a\omega(M+2m)=mv$$

$$\text{or,}\quad \omega=\frac{mv}{(M+2m)a}.$$

► **Problem 5.8.18** *Three equal uniform rods placed in a straight line are freely jointed at junctions and move with a velocity v perpendicular to their lengths. If the middle point of the middle rod be suddenly fixed, show that the ends of the other two rods will meet in time $\frac{4\pi a}{9v}$, where a is the length of the each rod.* **[CUH 1985,1987]**

Solution: Let $AB = BC = CD = a$ be the three uniform rods each of mass m and G_1, G_2, G_3 are the c.g.'s of the rods. When the rod BC is fixed the ends A and D meet at D' and the only action of BC on CD at C and there is no other force on CD. Thus the moment of the force at C about C is zero. Hence the angular momentum of CD about C will be same before and after the fixture. If ω be the angular velocity of CD just after BC is fixed, then

$$m\cdot K^2\cdot\omega=mv\cdot\frac{a}{2}$$

$$\text{or,}\quad m\cdot\frac{a^2}{3}\omega=mv\frac{a}{2},\quad \text{or,}\quad \omega=\frac{3v}{2a}$$

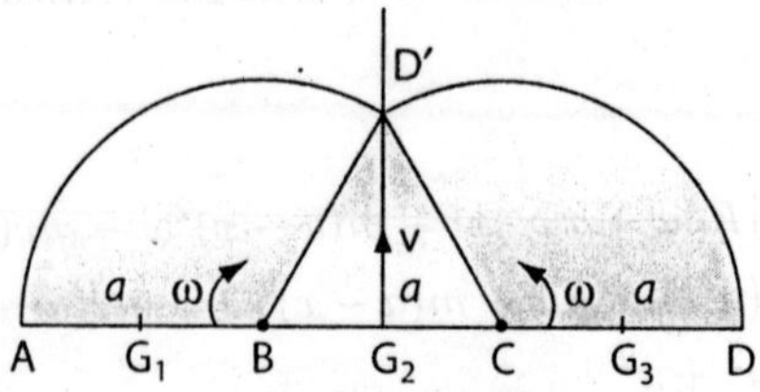

Fig 5.21

$$\therefore \text{ Time required by } D \text{ to reach at } D'=\frac{2\pi}{3}\Big/\frac{3v}{2a}=\frac{4\pi a}{9v}.$$

EXERCISES

1. A uniform rod at rest is struck at one end by a heavy blow in a direction perpendicular to the rod. Prove that the kinetic energy generated will be greater than if the other end had been fixed, in the ratio 4 : 3.

2. An elliptic lamina is rotating in its own plane about one of its foci with angular velocity ω. This focus is set free and the other focus is fixed at the same instant. Show that the resulting angular velocity is $\frac{\omega}{3}$, if the eccentricity of the ellipse is $\frac{\sqrt{2}}{3}$. **[CUH 2005]**

   [*Hints*: Put $e = \frac{\sqrt{2}}{3}$ in Problem 5.8.13.]

3. A rod OA is spinning in a horizontal plane with angular velocity ω about the fixed end O. Suddenly this end A is released and the other end fixed. Find the resulting motion.

4. Four equal uniform rods AB, BC, CD and DE are freely joined at B, C and D and lie on a smooth table in the form of a square. The rod AB is struck by a blow at right angles to AB from the inside of the square; show that the initial velocity of A is 79 times that of E.

5. Three particles of equal masses are attached to the ends A and C and the middle point B of a light rod ABC, and the system is at rest on a smooth table. The particle C is struck a blow at right angles to the rod; show that the energy communicated to the system when A is fixed, is to the energy communicated when the system is free as 24 : 25.

6. Find the centre of percussion of a circular lamina when a horizontal tangent is the axis of suspension.

7. A uniform rod AB can turn about its end A and is in equilibrium. Find the points of its length where a blow must be applied to it so that the impulses at A may be in each case $\frac{1}{n}$th of that of the blow.

8. A uniform rod AB of mass m and length 2a lies at rest on a smooth horizontal table. The end A of the rod is smoothly hinged to a point of the tables. It is struck by a horizontal blow of impulse J perpendicularly to its length at a distance c from the centre. Show that the angular velocity of the initial motion is $\frac{3J}{4M} \cdot \frac{a+c}{a^2}$ and the kinetic energy generated is $\frac{3J^2}{8M} \cdot \left(\frac{a+c}{a}\right)^2$.

9. A rod OA is spinning in a horizontal plane with angular velocity ω about a fixed end O. Suddenly this end is released and other end fixed. Find the resulting motion.

Appendices

Appendix

2.2.1 Vector Method (D'Alembert's Principle)

Let $\vec{r}$ be position vector of a particle of mass m within the rigid body in motion at any time t. Let $\vec{F}$ and $\vec{R}$ be external and internal forces acting respectively on it.

Therefore by Newton's law, we have

$$m\frac{d^2\vec{r}}{dt^2} = \vec{F} + \vec{R}$$

$$\text{or} \quad \vec{F} + \vec{R} - m\frac{d^2\vec{r}}{dt^2} = 0$$

Thus external force $\vec{F}$, internal force $\vec{R}$ and reversed effective force $-m\frac{d^2\vec{r}}{dt^2}$ acting on the particle m are in equilibrium.

Thus using same principles to each and every particle by which the rigid body constituted, the total external forces $\sum \vec{F}$, internal forces $\sum \vec{R}$ and the reversed effective forces $-\sum m\frac{d^2\vec{r}}{dt^2}$ are in equilibrium, where the summation is extended over the rigid body.

As by the assumption of D'Alembert's principle that internal forces on the body in motion are in equilibrium, $\sum \vec{R} = 0$.

Thus $\sum \vec{F}$, $-\sum m\frac{d^2\vec{r}}{dt^2}$ are in equilibrium.

So,

$$\sum \vec{F} - \sum m\frac{d^2\vec{r}}{dt^2} = \vec{0} \tag{2.2.1}$$

Hence the D'Alembert's principle.

Deduction in Scalar form

Let (x, y, z) be the cartesian coordinates of the particle of mass m w.r.t the rectangular axes and X, Y, Z are the components to the external force $\vec{F}$ parallel to the axes, then

$$\vec{r} = x\vec{i} + y\vec{j} + z\vec{k} \quad \text{and} \quad \vec{F} = X\vec{i} + Y\vec{j} + Z\vec{k}$$

where $\vec{i}, \vec{j}, \vec{k}$ are the unit vector along parallel to axes.

Thus

$$\frac{d^2\vec{r}}{dt^2} = \frac{d^2x}{dt^2}\vec{i} + \frac{d^2y}{dt^2}\vec{j} + \frac{d^2z}{dt^2}\vec{k}$$

Now by (2.2.1) we get,

$$\sum X\cdot\vec{i} + \sum Y\cdot\vec{j} + \sum Z\cdot\vec{k} - \left(\sum \frac{d^2x}{dt^2}\cdot\vec{i} + \sum \frac{d^2y}{dt^2}\cdot\vec{j} + \sum \frac{d^2z}{dt^2}\cdot\vec{k}\right) = \vec{0}$$

$$\text{Thus}\quad \sum \frac{d^2x}{dt^2} = \sum X$$

$$\sum \frac{d^2y}{dt^2} = \sum Y$$

$$\sum \frac{d^2z}{dt^2} = \sum Z.$$

2.3.2 Vector Method (Equation of Motion of a Rigid Eody due to D'Alembert's Principle)

Let $\vec{r}$ be the position vector of a particle of mass m within the rigid body in motion at any time t. Let $\vec{F}$ be the external force act on it. Then by D'Alembert's principle,

$$\sum \vec{F} - \sum m\frac{d^2\vec{r}}{dt^2} = 0$$

$$\text{i.e.,}\quad \sum m\frac{d^2\vec{r}}{dt^2} = \sum \vec{F} \tag{2.3.7}$$

Again, taking moment of external forces i.e., taking cross product by $\vec{r}$, we get

$$\sum m\left(\vec{r}\times\frac{d^2\vec{r}}{dt^2}\right) = \sum(\vec{r}\times\vec{F}) \tag{2.3.8}$$

The equations (2.3.7) and (2.3.8) are the vector equations of a rigid body in motion.

Deduction in Scalar form

Let (x, y, z) be the cartesian co-ordinates of the particle of mass m, at any time t, w.r.t the rectangular axes and X, Y, Z be the components of the external forces $\vec{F}$ parallel to the axes, then

$$\vec{r} = x\cdot\vec{i} + y\cdot\vec{j} + z\cdot\vec{k} \tag{2.3.9}$$

$$\text{and}\quad \vec{F} = X\cdot\vec{i} + Y\cdot\vec{j} + Z\cdot\vec{k} \tag{2.3.10}$$

Thus

$$\frac{d^2\vec{r}}{dt^2} = \frac{d^2x}{dt^2}\cdot\vec{i} + \frac{d^2y}{dt^2}\cdot\vec{j} + \frac{d^2z}{dt^2}\cdot\vec{k} \qquad (2.3.11)$$

where $\vec{i}, \vec{j}, \vec{k}$ are the unit vectors along parallel to the axes.

Now,

$$\vec{r}\times\vec{F} = \begin{vmatrix} \vec{i} & \vec{j} & \vec{k} \\ x & y & z \\ X & Y & Z \end{vmatrix}$$

$$= (yZ - zY)\cdot\vec{i} + (zX - xZ)\cdot\vec{j} + (xY - yX)\cdot\vec{k} \qquad (2.3.12)$$

and $\vec{r}\times\dfrac{d^2\vec{r}}{dt^2} = \begin{vmatrix} \vec{i} & \vec{j} & \vec{k} \\ x & y & z \\ \frac{d^2x}{dt^2} & \frac{d^2y}{dt^2} & \frac{d^2z}{dt^2} \end{vmatrix}$

$$= \left(y\frac{d^2z}{dt^2} - z\frac{d^2y}{dt^2}\right)\cdot\vec{i} + \left(z\frac{d^2x}{dt^2} - x\frac{d^2z}{dt^2}\right)\cdot\vec{j} + \left(x\frac{d^2y}{dt^2} - y\frac{d^2x}{dt^2}\right)\cdot\vec{k} \qquad (2.3.13)$$

Using (2.3.10) and (2.3.11) in (2.3.7) we get

$$\sum m\left(\frac{d^2x}{dt^2}\cdot\vec{i} + \frac{d^2y}{dt^2}\cdot\vec{j} + \frac{d^2z}{dt^2}\cdot\vec{k}\right) = \sum(X\cdot\vec{i} + Y\cdot\vec{j} + Z\cdot\vec{k})$$

Equating the co-efficients of $\vec{i}, \vec{j}, \vec{k}$ on both sides, we get

$$\left.\begin{aligned} \sum m\frac{d^2x}{dt^2} &= \sum X \\ \sum m\frac{d^2y}{dt^2} &= \sum Y \\ \sum m\frac{d^2z}{dt^2} &= \sum Z \end{aligned}\right\} \qquad (2.3.14)$$

Using (2.3.13) and (2.3.12) in (2.3.8) we get,

$$\sum m\left[\left(y\frac{d^2z}{dt^2} - z\frac{d^2y}{dt^2}\right)\cdot\vec{i} + \left(z\frac{d^2x}{dt^2} - x\frac{d^2z}{dt^2}\right)\cdot\vec{j} + \left(x\frac{d^2y}{dt^2} - y\frac{d^2x}{dt^2}\right)\cdot\vec{k}\right]$$

$$= \sum\left[(yZ - zY)\cdot\vec{i} + (zX - xZ)\cdot\vec{j} + (xY - yX)\cdot\vec{k}\right]$$

Equating the co-efficients of $\vec{i}, \vec{j}, \vec{k}$ from both sides we get,

$$\left.\begin{aligned}\sum m\left(y\frac{d^2z}{dt^2} - z\frac{d^2y}{dt^2}\right) &= \sum(yZ - zY)\\ \sum m\left(z\frac{d^2x}{dt^2} - x\frac{d^2z}{dt^2}\right) &= \sum(zX - xZ)\\ \sum m\left(x\frac{d^2y}{dt^2} - y\frac{d^2x}{dt^2}\right) &= \sum(xY - yX)\end{aligned}\right\} \tag{2.3.15}$$

The equation (2.3.14) and (2.3.15) are the general equations of motions of a rigid body.

2.4.1 Vector Method (Motion of Centre of Inertia and Equations of Motion Relative to the Centre of Inertia)

Let $\vec{r}$ be the position vector of a particle of mass m at P of a rigid body of mass M in motion in any time t, w.r.t a fixed origin O.

Let $\vec{F}$ be the external force acting on the particle

Let G be the centre of intertia of the body and $\vec{\bar{r}}$ be the position vector of G w.r.t. a fixed origin O. Let $\vec{r'}$ be the position vector of the particle at P w.r.t the centre of inertia G.

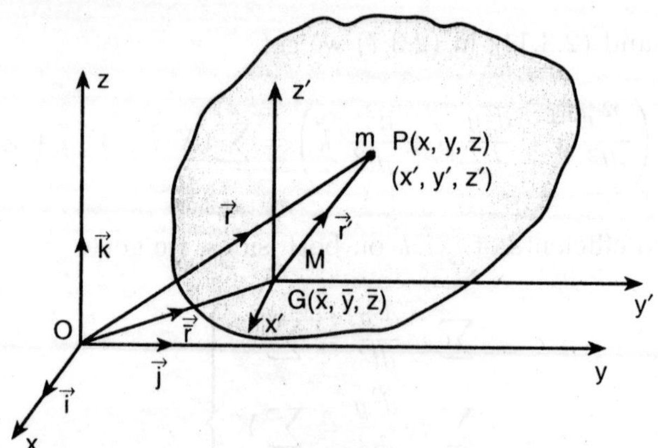

Fig 2.1(a)

Then,

$$\vec{r} = \vec{\bar{r}} + \vec{r'} \tag{2.4.14}$$

$$\therefore \quad \frac{d^2\vec{r}}{dt^2} = \frac{d^2\vec{\bar{r}}}{dt^2} + \frac{d^2\vec{r'}}{dt^2} \tag{2.4.15}$$

$$\text{and } \sum m\cdot\vec{\bar{r}} = \sum m\vec{r}$$

$$M\cdot\vec{\bar{r}} = \sum m\vec{r}$$

$$\text{or}\quad M\cdot\frac{d^2\bar{\vec{r}}}{dt^2}=\sum m\frac{d^2\vec{r}}{dt^2} \tag{2.4.16}$$

$$\text{and}\quad \sum m\vec{r'}=\bar{0}\quad\text{or}\quad \sum m\frac{d^2\vec{r'}}{dt^2}=0 \tag{2.4.17}$$

The equation of motion of the rigid body, by D'Alembert's principles is

$$\sum m\frac{d^2\vec{r}}{dt^2}=\sum\vec{F} \tag{2.4.18}$$

and the moment vector equation of motion for the rigid body is

$$\sum m\left(\vec{r}\times\frac{d^2\vec{r}}{dt^2}\right)=\sum(\vec{r}\times\vec{F}) \tag{2.4.19}$$

Now, using (2.4.16) in (2.4.18), we have

$$M\frac{d^2\bar{\vec{r}}}{dt^2}=\sum F \tag{2.4.20}$$

Thus the equation (2.4.20) shows that the equation of motion of a rigid body is invariant to the motion of centre of inertia of the body moves as if all the mass of the body were collected at it and as if all the external forces acting on the body were acting on it in directions parallel to those in which they act.

Again using (2.4.14) in (2.4.19), we get

$$\sum m\left[(\bar{\vec{r}}+\vec{r}')\times\frac{d^2}{dt^2}(\bar{\vec{r}}+\vec{r'})\right]=\sum(\bar{\vec{r}}+\vec{r}')\times\vec{F}$$

$$\text{or,}\quad \sum m\left[\bar{\vec{r}}\times\frac{d^2\bar{\vec{r}}}{dt^2}+\bar{\vec{r}}\times\frac{d^2\vec{r'}}{dt^2}+\vec{r'}\times\frac{d^2\bar{\vec{r}}}{dt^2}+\vec{r'}\times\frac{d^2\vec{r'}}{dt^2}\right]=\sum(\bar{\vec{r}}+\vec{r'})\times\vec{F}$$

$$\text{or,}\quad \left(\bar{\vec{r}}\times\frac{d^2\bar{\vec{r}}}{dt^2}\right)\sum m+\sum m\left(\vec{r'}\times\frac{d^2\vec{r'}}{dt^2}\right)+\bar{\vec{r}}\sum m\frac{d^2\vec{r'}}{dt^2}+\frac{d^2\bar{\vec{r}}}{dt^2}\sum m\vec{r'}$$
$$=\sum(\bar{\vec{r}}+\vec{r'})\times\vec{F}$$

$$\text{or,}\quad \bar{\vec{r}}\times\frac{d^2\bar{\vec{r}}}{dt^2}\cdot M+\sum m\left(\vec{r'}\times\frac{d^2\vec{r'}}{dt^2}\right)=\sum(\bar{\vec{r}}\times\vec{F})+\sum(\vec{r'}\times\vec{F})$$

[by (2.4.17)]

$$\text{or,}\quad \bar{\vec{r}}\times\sum\vec{F}+\sum m\left(\vec{r'}\times\frac{d^2\vec{r'}}{dt^2}\right)=\bar{\vec{r}}\times\sum\vec{F}+\sum(\vec{r'}\times\vec{F})\quad\text{[by (2.4.20)]}$$

$$\text{or,}\quad \sum m\left(\vec{r'}\times\frac{d^2\vec{r'}}{dt^2}\right)=\sum(\vec{r'}\times\vec{F}) \tag{2.4.21}$$

The above equation is the vector equation of motion of a rigid body relative to centre of inertia.

The equation (2.4.21) shows that the motion of a rigid body about its centre of inertia is the same as it would be if the centre of inertia is fixed and the same forces act on the body.

Deduction in Scalar form

Let $(\bar{x}, \bar{y}, \bar{z})$ be the co-ordinates of the centre of inertia G w.r.t Ox, Oy and Oz as rectangular axes at a fixed point O. Let (x, y, z) be the co-ordinates of particle of mass m at P w.r.t. ox, oy and oz as axis at O and (x', y', z') be the co-ordinates of P w.r.t parallel rectangular axes Gx', Gy' and Gz' at G.

Then,

$$\left.\begin{aligned}
\vec{r} = x\vec{i} + y\vec{j} + z\vec{k} \quad &\Rightarrow \frac{d^2\vec{r}}{dt^2} = \frac{d^2x}{dt^2}\vec{i} + \frac{d^2y}{dt^2}\vec{j} + \frac{d^2z}{dt^2}\vec{k} \\
\vec{\bar{r}} = \bar{x}\vec{i} + \bar{y}\vec{j} + \bar{z}\vec{k} \quad &\Rightarrow \frac{d^2\vec{\bar{r}}}{dt^2} = \frac{d^2\bar{x}}{dt^2}\vec{i} + \frac{d^2\bar{y}}{dt^2}\vec{j} + \frac{d^2\bar{z}}{dt^2}\vec{k} \\
\vec{r'} = x'\vec{i} + y'\vec{j} + z'\vec{k} \Rightarrow &\frac{d^2\vec{r'}}{dt^2} = \frac{d^2x'}{dt^2}\vec{i} + \frac{d^2y'}{dt^2}\vec{j} + \frac{d^2z'}{dt^2}\vec{k}
\end{aligned}\right\} \tag{2.4.22}$$

where $\vec{i}, \vec{j}, \vec{k}$ are the unit vectors along parallel to the axes.

If (X, Y, Z) be the components of the force $\vec{F}$ along parallel to the axes, we have

$$\vec{F} = X\vec{i} + Y\vec{j} + Z\vec{k} \tag{2.4.23}$$

where $\vec{i}, \vec{j}, \vec{k}$ are unit vectors along parallel to the axes.

Thus from equation (2.4.20) we get

$$M\left(\frac{d^2\bar{x}}{dt^2}\vec{i} + \frac{d^2\bar{y}}{dt^2}\vec{j} + \frac{d^2\bar{z}}{dt^2}\vec{k}\right) = \sum(X\vec{i} + Y\vec{j} + Z\vec{k})$$

Equating the co-efficients of $\vec{i}, \vec{j}, \vec{k}$ from both sides we get

$$M\frac{d^2\bar{x}}{dt^2} = \sum X, \; M\frac{d^2\bar{y}}{dt^2} = \sum Y, \; M\frac{d^2\bar{z}}{dt^2} = \sum Z \tag{2.4.24}$$

Using (2.4.22) and (2.4.23) we have

$$\vec{r'} \times \frac{d^2\vec{r'}}{dt^2} = \begin{vmatrix} \vec{i} & \vec{j} & \vec{k} \\ x' & y' & z' \\ \frac{d^2x'}{dt^2} & \frac{d^2y'}{dt^2} & \frac{d^2z'}{dt^2} \end{vmatrix}$$

$$= \left(y'\frac{d^2z'}{dt^2} - z'\frac{d^2y'}{dt^2}\right)\vec{i} + \left(z'\frac{d^2x'}{dt^2} - x'\frac{d^2z'}{dt^2}\right)\vec{j} + \left(x'\frac{d^2y'}{dt^2} - y'\frac{d^2x'}{dt^2}\right)\vec{k}$$

and $\vec{r'} \times \vec{F} = \begin{vmatrix} \vec{i} & \vec{j} & \vec{k} \\ x' & y' & z' \\ X & Y & Z \end{vmatrix}$

$$= (y'Z - z'Y)\vec{i} + (z'X - x'Z)\vec{j} + (x'Y - y'X)\vec{k}$$

Thus from equation (2.4.21) we get

$$\sum m \left[\left(y'\frac{d^2z'}{dt^2} - z'\frac{d^2y'}{dt^2} \right) \vec{i} + \left(z'\frac{d^2x'}{dt^2} - x'\frac{d^2z'}{dt^2} \right) \vec{j} + \left(x'\frac{d^2y'}{dt^2} - y'\frac{d^2x'}{dt^2} \right) \vec{k} \right]$$

$$= \sum \left[(y'Z - z'Y)\vec{i} + (z'X - x'Z)\vec{j} + (x'Y - y'X)\vec{k} \right]$$

Equating the co-efficient $\vec{i}, \vec{j}, \vec{k}$ from both sides we have

$$\left.\begin{aligned} \sum m \left(y'\frac{d^2z'}{dt^2} - z'\frac{d^2y'}{dt^2} \right) &= \sum (y'Z - z'Y) \\ \sum m \left(z'\frac{d^2x'}{dt^2} - x'\frac{d^2z'}{dt^2} \right) &= \sum (z'X - x'Z) \\ \sum m \left(x'\frac{d^2y'}{dt^2} - y'\frac{d^2x'}{dt^2} \right) &= \sum (x'Y - y'X) \end{aligned}\right\} \quad (2.4.25)$$

4.2.1 Vector Method (Equations of Motion)

The motion of a rigid body consists of two different independent motions

(i) *the motion of centre of inertia i.e., the motion of centre of inertia means that the total mass of the rigid body M be placed at the c.g. and all the external forces are transferred parallel to themselves to act at the c.g. G.*

(ii) *the motion about the centre of inertia i.e., the algebraic sums of the moment of the effective forces about the centre of inertia is equal to the algebraic sums of the moment of the external forces about the centre of inertia G.*

(i) Let $\vec{r}$ be the position vector of the centre of inertia G at any time t w.r.t certain origin O. Let $\vec{F}$ be the external force acting at any particle of mass m at P of the rigid body and M be the mass of the rigid body. Therefore

$$M\frac{d^2\vec{r}}{dt^2} = \sum \vec{F} \tag{4.2.6}$$

where the summation is extended to cover the whole body.

(ii) Let $\vec{r'}$ be the position of a particle of mass m at P relative to the centre of inertia G, then we have

$$\sum\left(\vec{r'} \times m\frac{d^2\vec{r'}}{dt^2}\right) = \sum(\vec{r'} \times \vec{F})$$

$$\text{or,} \quad \frac{d}{dt}\left[\sum m\vec{r'} \times \frac{d\vec{r'}}{dt}\right] = \sum(\vec{r} \times \vec{F}) \tag{4.2.7}$$

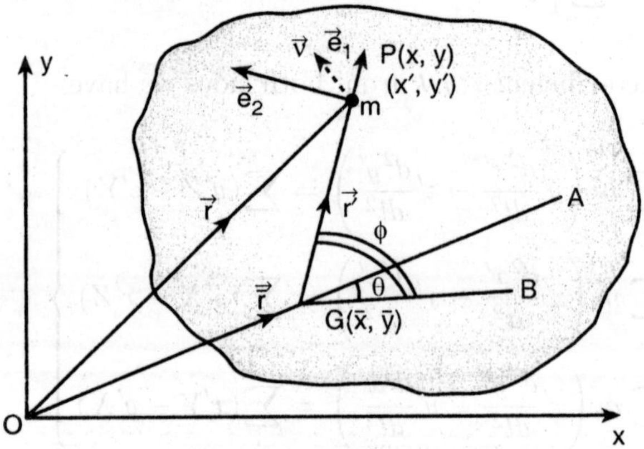

Fig 4.1(a)

Let the line GB fixed in space and the line GA fixed in the body. Let $\angle AGB = \theta$ and $\angle PGB = \phi$.

Thus $\angle PGB - \angle AGB =$ constant

$$\therefore \ \phi - \theta = \text{constant}$$

$$\text{or,} \quad \frac{d\phi}{dt} = \frac{d\theta}{dt} \tag{4.2.8}$$

As $GP = r'$, then the velocity of mass m at P relative to centre of inertia G

$$= r' \cdot \frac{d\phi}{dt} = r' \cdot \frac{d\theta}{dt} \tag{4.2.9}$$

in the direction perpendicular r' in the plane PGA.

Let $\vec{e_1}, \vec{e_2}$ be the unit vecors along and perpendicular to $\vec{r'}$ respectively, in the plane PGA. Let $\vec{\nu}$ be the unit vector normal to the plane PGA.

Thus

$$\vec{r'} = r'\vec{e}_1 \quad \text{and} \quad \frac{d\vec{r'}}{dt} = r'\frac{d\phi}{dt}\vec{e}_2 = r'\frac{d\theta}{dt}\vec{e}_2 \quad .$$

Now,

$$\sum\left(m\vec{r'} \times \frac{d\vec{r'}}{dt}\right) = \sum\left(mr'\vec{e}_1 \times r'\frac{d\theta}{dt}\vec{e}_2\right)$$

$$= \sum m\,r'^2 \cdot \frac{d\theta}{dt}(\vec{e}_1 \times \vec{e}_2) = \sum m\,r'^2\frac{d\theta}{dt}\cdot\vec{\nu}$$

$$= \frac{d\theta}{dt}\sum m\,r'^2 \cdot \hat{\nu} = MK^2\frac{d\theta}{dt}\cdot\hat{\nu} \qquad (4.2.10)$$

where K is the radius of gyration of the rigid body about G.

Again, by the definition of moment, we have

$$\sum \vec{r'} \times \vec{F} = \sum p' \cdot \vec{F}\vec{\nu} \qquad (4.2.11)$$

where p' is the perpendicular distance of G from the direction of the force $\vec{F}$.

Thus using (4.2.10) and (4.2.11) in (4.2.7) we get

$$\frac{d}{dt}\left[MK^2\frac{d\theta}{dt}\right]\cdot\vec{\nu} = \sum p' \cdot \vec{F}\cdot\vec{\nu}$$

$$\therefore \quad MK^2\frac{d^2\theta}{dt^2} = \sum r' \cdot \vec{F} \qquad (4.2.12)$$

The equations (4.2.6) and (4.2.12) are the vector equations of motion of a rigid body moving in two dimension.

Deduction in Scalar form

Let the cartesian co-ordinate of the particle of mass m at P is (x, y) and that of the centre of inertia G is $(\bar{x}, \bar{y})$ (w.r.t ox and oy as axes) and the co-ordinates of P relative to G is (x', y') at any time t. Let X, Y be the components of the force $\vec{F}$ along parallel to the axes.

Thus

$$\left.\begin{aligned} \vec{r} &= x\vec{i} + y\vec{j}, \\ \vec{\bar{r}} &= \bar{x}\vec{i} + \bar{y}\vec{j}, \qquad \therefore \quad \frac{d^2\vec{\bar{r}}}{dt^2} = \frac{d^2\bar{x}}{dt^2}\vec{i} + \frac{d^2\bar{y}}{dt^2}\vec{j} \\ \vec{r'} &= x'\vec{i} + y'\vec{j}, \\ \text{and} \quad \vec{F} &= X\vec{i} + Y\vec{j} \end{aligned}\right\} \qquad (4.2.13)$$

where $\vec{i}, \vec{j}$ are the unit vectors parallel to axes.

Therefore

$$p'\vec{F} = (x'Y - y'X). \qquad (4.2.14)$$

Now using (4.2.13) in (4.2.6), we get

$$M\left(\frac{d^2\bar{x}}{dt^2}\vec{i}+\frac{d^2\bar{y}}{dt^2}\vec{j}\right)=\sum(X\vec{i}+Y\vec{j})$$

equating the co-efficients of $\vec{i}$ and $\vec{j}$ from both sides, we get

$$\left.\begin{aligned} M\frac{d^2\bar{x}}{dt^2}&=\sum X\\ M\frac{d^2\bar{y}}{dt^2}&=\sum Y\end{aligned}\right\} \tag{4.2.15}$$

By (4.2.14) we get from (4.2.12)

$$MK^2\frac{d^2\theta}{dt^2}=\sum(x'Y-y'X) \tag{4.2.16}$$

4.3.1 Vector Method (Kinetic Energy of a Rigid Body Moving in Two Dimensions)

Let m be the mass of a particle of the rigid body of mass M at P. Let at any time t, $\vec{\bar{r}}$ be the position vector of the centre of inertia G of the rigid body relative to the origin O, and $\vec{r}$ be the position vector of the mass at P relative of the origin O. Let $\vec{r'}$ be the position vector of the point P relative to the centre of inertia G.

Thus

$$\left.\begin{aligned} &\vec{r}=\vec{\bar{r}}+\vec{r'} \qquad\qquad \therefore\ \frac{d\vec{r}}{dt}=\frac{d\vec{\bar{r}}}{dt}+\frac{d\vec{r'}}{dt}\\ &\therefore\ \sum m\vec{r'}=0,\ \sum m\frac{d\vec{r'}}{dt}=0 \quad\text{and}\quad \sum m=M\end{aligned}\right\} \tag{4.3.3}$$

Therefore, the kinetic energy of the rigid body

$$\begin{aligned} \text{K.E.}&=\tfrac{1}{2}\sum m\left(\frac{d\vec{r}}{dt}\right)^2=\tfrac{1}{2}\sum m\left[\frac{d\vec{\bar{r}}}{dt}+\frac{d\vec{r'}}{dt}\right]^2\\ &=\tfrac{1}{2}\sum m\left(\frac{d\vec{\bar{r}}}{dt}\right)^2+\tfrac{1}{2}\sum m\left(\frac{d\vec{r'}}{dt}\right)^2+\sum m\frac{d\vec{\bar{r}}}{dt}\cdot\frac{d\vec{r'}}{dt}\\ &=\tfrac{1}{2}\left(\frac{d\vec{\bar{r}}}{dt}\right)^2\sum m+\tfrac{1}{2}\sum m\left(\frac{d\vec{r'}}{dt}\right)^2+\frac{d\vec{\bar{r}}}{dt}\sum m\frac{d\vec{r'}}{dt}\\ &=\tfrac{1}{2}\left(\frac{d\vec{\bar{r}}}{dt}\right)^2+\tfrac{1}{2}\sum m\left(\frac{d\vec{r'}}{dt}\right)^2 \qquad [\text{by (4.3.3)}]\end{aligned} \tag{4.3.4}$$

Thus the kinetic energy of a rigid body moving in two dimensions is equal to the sum of the kinetic energy of particle of mass M placed at the centre of inertia and the kinetic energy of the body relative to the centre of inertia.

Deduction in Scalar form

Let the line GA fixed in the body makes an angle θ with line GB fixed in space and line PG makes an angle ϕ with line GB.

Thus $\phi - \theta =$ constant

$$\therefore \quad \frac{d\phi}{dt} = \frac{d\theta}{dt} \tag{4.3.5}$$

Let v be the velocity of the centre of inertia and $\vec{e}_1, \vec{e}_2$ be the unit vectors along and perependicular to GP in the plane PGA,

Thus,

$$v = \frac{d\vec{\bar{r}}}{dt} \quad \text{and} \quad \left(\frac{d\vec{r'}}{dt}\right)^2 = \left(r'\frac{d\phi}{dt}\vec{e}_2\right)^2 = r'^2\left(\frac{d\theta}{dt}\right)^2 \quad \text{by (4.3.5) as } \vec{e}_2^{\,2} = 1$$

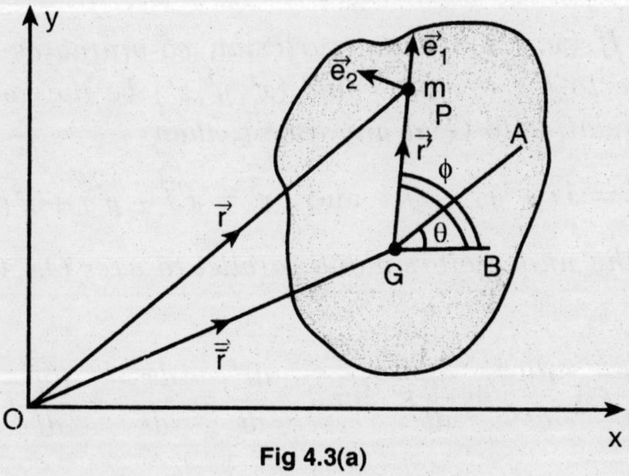

Fig 4.3(a)

Hence, from (4.3.4) we get

$$\begin{aligned}
\text{K.E.} &= \tfrac{1}{2}Mv^2 + \tfrac{1}{2}\sum m\,r'^2\left(\frac{d\theta}{dt}\right)^2 \\
&= \tfrac{1}{2}Mv^2 + \tfrac{1}{2}\left(\frac{d\theta}{dt}\right)^2\sum m\,r'^2 \\
&= \tfrac{1}{2}Mv^2 + \tfrac{1}{2}MK^2\left(\frac{d\theta}{dt}\right)^2 \qquad (4.3.6)
\end{aligned}$$

where K is the radius of gyration of the rigid body about the centre of inertia G.

Thus the kinetic energy of rigid body

= K.E. due to translation motion + K.E. due to rotation motion

Remark 4.3.1 *If $(\bar{x}, \bar{y})$ be the cartesian co-ordinates of the centre of inertia G relative to origin O and (x', y') be the co-ordinates of particle m at P relation to G at any time t, then*

$$\vec{\bar{r}} = \bar{x}\vec{i} + \bar{y}\vec{j} \quad and \quad \vec{r'} = x'\vec{i} + y'\vec{j}$$

where $\vec{i}, \vec{j}$ are the unit vectors along parallel to the axes Ox and Oy

$$\therefore \quad \frac{d\vec{\bar{r}}}{dt} = \frac{d\bar{x}}{dt}\vec{i} + \frac{d\bar{y}}{dt}\vec{j} \quad and \quad \frac{d\vec{r'}}{dt} = \frac{dx'}{dt}\vec{i} + \frac{dy'}{dt}\vec{j}$$

Thus,

$$\left(\frac{d\vec{\bar{r}}}{dt}\right)^2 = \left(\frac{d\bar{x}}{dt}\right)^2 + \left(\frac{d\bar{y}}{dt}\right)^2 \quad and \quad \left(\frac{d\vec{r'}}{dt}\right)^2 = \left(\frac{dx'}{dt}\right)^2 + \left(\frac{dy'}{dt}\right)^2$$

Then from (4.3.4) we get

$$Kinetic\ Energy = \tfrac{1}{2}M\left[\left(\frac{d\bar{x}}{dt}\right)^2 + \left(\frac{d\bar{y}}{dt}\right)^2\right] + \tfrac{1}{2}\sum m\left[\left(\frac{dx'}{dt}\right)^2 + \left(\frac{dy'}{dt}\right)^2\right]$$

Remark 4.3.2 *If $(\bar{x}, \bar{y}, \bar{z})$ be the cartesian co-ordinates of the centre of inertia G relative to the origin O and (x', y', z') be the co-ordinates of the particle m at P relative to G at any time t, then*

$$\vec{\bar{r}} = \bar{x}\vec{i} + \bar{y}\vec{j} + \bar{z}\vec{k} \quad \text{and} \quad \vec{r'} = x'\vec{i} + y'\vec{j} + z'\vec{k}$$

where $\vec{i}, \bar{j}, \bar{k}$ are the unit vectors along parallel to axes Ox, Oy and Oz.

Thus,

$$\frac{d\vec{\bar{r}}}{dt} = \frac{d\bar{x}}{dt}\vec{i} + \frac{d\bar{y}}{dt}\vec{j} + \frac{d\bar{z}}{dt}\vec{k} \quad and \quad \frac{d\vec{r'}}{dt} = \frac{dx'}{dt}\vec{i} + \frac{dy'}{dt}\vec{j} + \frac{dz'}{dt}\vec{k}$$

Then

$$\left(\frac{d\vec{\bar{r}}}{dt}\right)^2 = \left(\frac{d\bar{x}}{dt}\right)^2 + \left(\frac{d\bar{y}}{dt}\right)^2 + \left(\frac{d\bar{z}}{dt}\right)^2 \quad and \quad \left(\frac{d\vec{r'}}{dt}\right)^2 = \left(\frac{dx'}{dt}\right)^2 + \left(\frac{dy'}{dt}\right)^2 + \left(\frac{dz'}{dt}\right)^2$$

Therefore, from (4.3.4) we get

$$Kinetic\ Energy = \tfrac{1}{2}M\left[\left(\frac{d\bar{x}}{dt}\right)^2 + \left(\frac{d\bar{y}}{dt}\right)^2 + \left(\frac{d\bar{z}}{dt}\right)^2\right]$$
$$+\tfrac{1}{2}\sum m\left[\left(\frac{dx'}{dt}\right)^2 + \left(\frac{dy'}{dt}\right)^2 + \left(\frac{dz'}{dt}\right)^2\right]$$

= *K.E. of the mass placed at the centre of inertia G of the body + K.E. of the rigid body relative to the centre of inertia G.*

4.4.1 Vector Method (Moment of Momentum About the Origin at a Rigid Body Moving in Two Dimension)

Let, at any time t, $\vec{r}$ be the position vector of the particle of mass m at P and $\vec{\bar{r}}$ be the position vector of the centre of inertia of the rigid body of mass M relative to certain origin O. Let $\vec{r'}$ be the position vector of the particle at P relative to the centre of inertia G. Let $\vec{e}_1, \vec{e}_2$ be the unit vectors along and perpendicular to GP in the plane of motion and $\vec{v}$ be the unit vector perpendicular to the plane of the motion. Let the line GA fixed in the body makes an angle θ with the line GB fixed in space. Let the line PG makes an angle ϕ with the line GB.

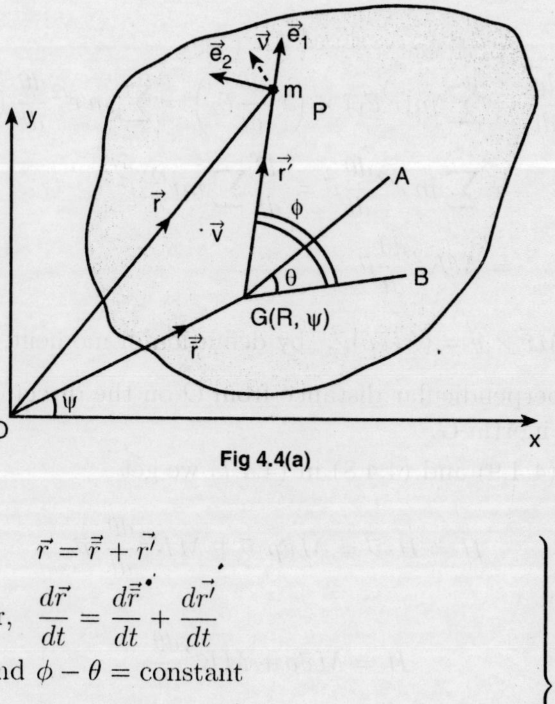

Fig 4.4(a)

Then

$$\left.\begin{aligned} &\vec{r} = \vec{\bar{r}} + \vec{r'} \\ \text{or,} \quad &\frac{d\vec{r}}{dt} = \frac{d\vec{\bar{r}}}{dt} + \frac{d\vec{r'}}{dt} \\ \text{and } &\phi - \theta = \text{constant} \\ \text{or,} \quad &\frac{d\phi}{dt} = \frac{d\theta}{dt} \\ \text{and } &\sum m\vec{r'} = 0, \ \sum m\frac{d\vec{r'}}{dt} = 0, \ \sum m = M \end{aligned}\right\} \qquad (4.4.6)$$

Thus, the angular momentum or moment of momentum of the rigid body about O, by definition, we get

$$\vec{H} = \sum\left[\vec{r} \times \left(m\frac{d\vec{r}}{dt}\right)\right] = \sum m\cdot\left(\vec{r} \times \frac{d\vec{r}}{dt}\right) = \sum m(\vec{\bar{r}} + \vec{r'}) \times \left(\frac{d\vec{\bar{r}}}{dt} + \frac{d\vec{r'}}{dt}\right)$$

$$= \sum m\vec{\bar{r}} \times \frac{d\vec{\bar{r}}}{dt} + \sum m\,\vec{\bar{r}} \times \frac{d\vec{r'}}{dt} + \sum m\,\vec{r'} \times \frac{d\vec{\bar{r}}}{dt} + \sum m\vec{r'} \times \frac{d\vec{r}'}{dt}$$

$$= \vec{\bar{r}} \times \frac{d\vec{\bar{r}}}{dt}\sum m + \vec{\bar{r}}\sum m\frac{d\vec{r'}}{dt} + \frac{d\vec{\bar{r}}}{dt}\sum m\,\vec{r'} + \sum m\vec{r'} \times \frac{d\vec{r'}}{dt}$$

$$= M\vec{\bar{r}} \times \frac{d\vec{\bar{r}}}{dt} + \sum m\,\vec{r'} \times \frac{d\vec{r'}}{dt} \quad \text{[by (4.4.6)]}$$

$$= M\,\vec{\bar{r}} \times \vec{v} + \sum \vec{r'} \times m\frac{d\vec{r'}}{dt} \tag{4.4.7}$$

where $\vec{v}$ is the velocity of centre of inertia G.

Again,

$$\vec{r'} = r' \cdot \vec{e}_1 \quad \text{and} \quad \frac{d\vec{r'}}{dt} = r'\frac{d\phi}{dt}\vec{e}_2 = r'\frac{d\theta}{dt}\vec{e}_2 \ \text{[by (4.4.6)]}$$

Now

$$\sum \vec{r'} \times \frac{d\vec{r'}}{dt} = \sum m(r'\vec{e}_1) \times \left(\vec{r'}\frac{d\theta}{dt}\vec{e}_2\right) = \sum m\,r'^2\frac{d\theta}{dt}(e_1 \times e_2)$$

$$= \sum m\,r'^2\frac{d\theta}{dt}\vec{\nu} = \frac{d\theta}{dt}\sum mr'^2 \cdot \vec{\nu}$$

$$= MK^2\frac{d\theta}{dt}\vec{\nu} \tag{4.4.8}$$

$$\text{and} \quad M\vec{\bar{r}} \times \vec{v} = (M\vec{v}p)\vec{\nu}, \ \text{by definition of moment,} \tag{4.4.9}$$

where p is the perpendicular distance from O on the direction of velocity $\vec{v}$ of the centre of inertia G.

Thus using (4.4.9) and (4.4.8) in (4.4.7) we get

$$\vec{H} = H \cdot \vec{\nu} = M\vec{v}p \cdot \vec{\nu} + MK^2\frac{d\theta}{dt} \cdot \vec{\nu}$$

Thus

$$H = M\vec{v}p + MK^2\frac{d\theta}{dt} \tag{4.4.10}$$

Again if $|\vec{\bar{r}}| = R$ and OG makes an angle ψ with Ox.

Thus the velocity of centre of inertia G perpendicular to $OG = R\dot{\psi}$.

Then the moment of momentum of the mass M placed at G about $O = M \cdot R.\ R\dot{\psi} = MR^2\dot{\psi}$.

Therefore from (4.4.10), we get, the moment of momentum or angular momentum of the rigid body of mass M

$$H = MR^2\dot{\psi} + MK^2\frac{d\theta}{dt}. \tag{4.4.11}$$

INDEX

INDEX

A

C

D

E

I

K

M

P

R

S

T

Calcutta University

2005

1. Find whether a given straight line is, at any point of its length, a principle axis of a material system and in case the line is a principle axis, find the other two principal axes.

   [HINTS : §1.16, pp. 50(a)]

2. A boat of mass m_B and length l is floating stationary on the still water of a lake. A man of mass m_M initially standing at one end of the boat walks to the other end of it and stops. Show that the distance through which the boat will move in the opposite direction is $\frac{m_M l}{m_M + m_B}$ assuming that water offers no resistance to the boat.

   [HINTS : Prob. 2.5.9, pp. 104]

3. Prove that the kinetic energy of a rigid body of mass M moving in two dimension is equal to the sum of the kinetic energy of a particle of mass M placed at the centre of inertia and moving with it and the kinetic energy of the body relative to the centre of inertia.

   [HINTS : §4.3, pp. 161]

5. A cylinder of mass M and radius R is held at rest in a plane inclined at an angle θ to the horizontal. When the cylinder is released it moves down the slope. Discuss the case of pure rolling motion of the cylinder. Show that when the cylinder rolls down a length l of the slope, the velocity of its centre of mass is $\frac{4}{3}gl \cdot \sin\theta$.

   [HINTS : Prob. 4.13.2, pp. 216]

6. A solid homogeneous cone, of height h and vertical angle 2α, oscillates about a horizontal axis through its vertex. Show that the length of the simple equivalent pendulum is $\frac{h}{5}(4 + \tan^2\alpha)$.

   [HINTS : Prob. 3.3.1, pp. 112]

7. Three equal uniform rods AB, BC and CD are hinged freely at their ends, B and C, so as to form three sides of a square and are laid on a smooth table. The end A is struck by a horizontal blow P at right angles to AB. Show that the initial velocity of A is 19 times that of D and that the impulsive actions at B and C are $\frac{5P}{12}$ and $\frac{P}{12}$ respectively.

   [HINTS : Prob. 5.8.12, pp. 315]

8. An elliptic lamina is rotating in its own plane about one of its foci with angular velocity ω. This focus is set free and the other focus is fixed at the same instant. Show that the resulting angular velocity is $\frac{\omega}{3}$, if the ecentricity of the ellipse is $\frac{\sqrt{2}}{3}$.

   [HINTS : Put $e = \frac{\sqrt{2}}{3}$, in Problem 5.8.13, pp. 317]

Calcutta University

2006

12. Explain what is meant by equimomental bodies.

Show that a uniform plane triangular lamina is equimomental with a system of three particles placed at the middle points of the sides, each equal to one third the mass of the triangle.

[HINTS : §1.77, pp. 52, Prob. 1.18.2, pp. 54]

13. (a) Establish the principle of independence of the motions of translation and rotation of a rigid body.

[HINTS : §2.4, pp. 91]

(b) A sphere of radius a, is suspended by a fine wire from a fixed point at a distance l from the centre. Show that the time period of a small oscillation is given by

$$2\pi\sqrt{\frac{5l^2+2a^2}{5lg}}\left(1+\frac{1}{4}\sin^2\frac{\alpha}{2}\right).$$

where α represents the amplitude of vibration.

[HINTS : Prob. 3.3.5, pp. 117]

14. (a) (i) Deduce the expression for the angular momentum of a rigid body relative to its centre of mass.

(ii) Deduce the principle of conservation of angular momentum for a rigid body.

[HINTS : §4.8, pp. 168]

(b) A rough uniform rod, of length $2a$ is placed on a rough table at right angles to its edge. If the centre of gravity of the rod be initially at a distance b beyond the edge, then show that the rod will begin to slide when it has turned through an angle

$$\tan\theta = \frac{\mu a^2}{a^2+9b^2}$$

where μ is the coefficient of friction.

[HINTS : Prob. 4.10.5, pp. 182]

Calcutta University

2007

12. (a) Find wheather a given straight line is, at any point if its length, a principal axis of a material system. If the line is a principal axis, find the other two principal axes at the point.

[HINTS : §1.16, pp. 50(a)]

(b) A uniform rod is held one end in contact with a rough horizontal table at an inclination of α to the horizon and then released. if μ be the coefficient of friction, show that it will commence to slide if $\mu < \frac{3\sin\alpha\cos\alpha}{1+3\sin 2\alpha}$

[HINTS : Prob. 4.10.6, pp. 185]

13. (a) Define compound pendulum and its equivalent simple pendulum. A solid homogeneous cone of height h and vertical angle 2α oscillates about a horizontal axis through its vertex. Show that the length of equivalent simple pendulum is $\frac{1}{5}h(3+\sec^2\alpha)$.

[HINTS : §3.2, pp. 107, Prob. 3.3.1, pp. 112]

(b) Define impulse of a force. Obtain equations of motion in two dimensions of a rigid body under inpulsive forces.

[HINTS : §5.1, pp. 290, §5.3, pp. 292]

14. (a) Show that the resultant kinetic energy of a rigid body moving in two dimensions under finite forces is equal to the sum of two kinetic energies, one due to translation and the other due to rotation.

[HINTS : §4.3, pp. 161]

(b) Two equal uniform rods AB and AC are freely jointed at A and they are placed on a table so as to be at right angles. The rod AC is struck by a blow at C in a direction perpendicular to AC. Show that the resulting velocities of the middle points of AB and AC are in the ratio 2 : 7.

[HINTS : Prob. 5.8.10, pp. 313]

Calcutta University

2008

12. (a) If A, B, C, D, E, F be the moments and products of inertia of rigid body about rectangular axes ox, oy, oz and if $l_1, m_1, n_1;\ l_2, m_2, n_2$ be the direction cosines of two lines OL, OK at right angles to each other, then find the moment of inertia of the body about OL and product of inertia about OL and OK.

[HINTS : §1.9(b), pp. 24]

(b) A rod of length $2a$, is suspended by a string of length l, attached to one end. If the string and rod revolve about the vertical with uniform angular velocity, and their inclinations to be the vertical be θ and ϕ respectively, then show that

$$\frac{3l}{a} = \frac{(4\tan\theta - 3\tan\phi)\sin\phi}{(\tan\phi - \tan\theta)\sin\theta}.$$

[HINTS : Prob. 2.5.4, pp. 98]

13. (a) Show that the rate of change of the angular momentum of a rigid body about the axis of rotation is equal to the sum of the moments about the same axis of all forces acting on the body.

[HINTS : Remark 2.4.1, pp. 94]

(b) A uniform rod, of length $2a$ and weight W, is turning about its one end O and starts from the position in which it was vertically above O. When it has turned through an angle θ, show that the horizontal and vertical reactions at O are respectively $\frac{3W}{4}\sin\theta(2-3\cos\theta)$ and $\frac{W}{4}(1-3\cos\theta)^2$.

[HINTS : Prob. 3.5.3, pp. 135]

14. (a) An imperfectly rough sphere moves from rest down a plane inclined at an angle α to the horizontal. Discuss the motion.

[HINTS : §4.12, pp. 210]

(b) AB, BC are two equal similar rods freely hinged at B and lie in a straight line on a smooth table. The end A is struck by a blow perpendicular to AB. Show that the resulting velocity of A is $\frac{7}{2}$ times velocity of B.

[HINTS : Prob. 5.8.3, pp. 299]

Calcutta University

2009

12. (a) Deduce the expression for the angular momentum of a rigid body relative to its centre of mass. What is the principle of coservation of angular momentum for a rigid body?

[HINTS : §4.8, pp. 168]

(b) A plank of mass M is initially at rest along a straight line of greatest slope of a smooth plane inclined at an angle α to the horizon, and a man. of mass M' starting from the upper end walks down the plank so that it does not move. Show that he gets to the other end in time $\sqrt{\frac{2M'a}{(M+M)g\sin\alpha}}$, where a is the length of the plank and g is the acceleration due to gravity.

[HINTS : Prob. 2.5.1, pp. 94]

13. (a) Two uniform rods AB and AC of masses m and m' respectively are freely jointed at A and laid on a smooth horizontal table in such a way that $\angle BAC$ is a right angle. The rod AB is suddenly struck by a blow P at B in a direction perpendicular to AB. Show that the initial velocity of A is $\frac{2P}{4m'+m}$.

[HINTS : Prob. 5.8.11, pp. 314]

(b) A sphere of radius a, is suspended by a fine wire from a fixed point at a distance l from the centre. Show that the time period of a small oscillation is given by $2\pi\sqrt{\frac{5l^2+2a^2}{5lg}}\left(1+\frac{1}{4}\sin^2\frac{\alpha}{2}\right)$, where α represents the amplitude of vibration.

[HINTS : Prob. 3.5.5, pp. 117]

14. (a) An elliptic lamina is rotating in its own plane about one of its foci with angular velocity ω. This focus is set free and the other focus is fixed at the same instant. Show that the resulting angular velocity is $\frac{\omega}{3}$, if the eccentricity of the ellipse by $\frac{\sqrt{2}}{3}$.

[HINTS : Put $e = \frac{\sqrt{2}}{3}$ in Prob. 5.8.13, pp. 317]

(b) Prove that the kinetic energy of a rigid body of mass M moving in two dimension is equal to the sum of the kinetic energy of a particle of mass M placed at the centre of inertia and moving with it and the kinetic energy of the body relative to the centre of inertia.

[HINTS : §4.3, pp. 161]

Calcutta University

2010

12. (a) What is meant by principal axes of a given material system at a point? Obtain the condition so that a given straight line may be a principal axis of the material system at any point of its length and if so find the other two principal axes at the point.

[HINTS : §1.16, pp. 50(a)]

(b) A homogeneous sphere of radius a is rotating with an angular velocity ω about a horizontal diameter. It is then gently placed on a table whose co-efficient of friction μ. Show that there will be slipping at the point of contact for time $\frac{2\omega a}{7\mu g}$ and then the sphere will roll with angular velocity $\frac{2\omega}{7}$.

[HINTS : §4.16.1, pp. 228]

13. (a) A rigid body of mass μ is moving about a fixed axes.
Find the (i) Equation of motion of the rigid body
(ii) Kinatic energy of the rigid body.

[HINTS : §3.1, pp. 106]

(b) A uniform rod OA of length $2a$, free to turn about its and O, revolves with uniform angular velocity ω, about the vertical oz through O and is inclined at a constant angle α to OZ. Find the inclination of the reaction at O to the horizontal.

[HINTS : §3.4, pp. 130]

14. (a) Define 'Impulse of a force'. Find the equation of motion of a rigid body moving under impulsive forces.

[HINTS : §5.1 and §5.2, pp. 290]

(b) A uniform circular plate is turning in its own plane about a point A on its circumference with an uniform angular velocity ω. Suddenly A is released and another point B of the circumference is fixed. Show that the angular velocity about B is $\frac{\omega}{3}(1 + 2\cos\alpha)$, where α is the angle that AB subtends at the centre.

Calcutta University

2011

12. (a) State D'Alembert's Principle. Prove that, for a rigid body its translational motion can be considered independently from its rotational motion.

[HINTS : §2.2, pp. 88, §2.4, pp. 91]

(b) Explain what is meant by equimomental bodies. Prove that a uniform triangular lamina of mass M is equimomental with three particles, each of mass $\frac{M}{12}$ placed at the angular points and a particle of mass $\frac{3M}{4}$ placed at the centre of mass of the triangle.

[HINTS : §1.17, pp. 52, Prob. 1.18.7, pp. 63]

13. (a) State and prove the principal of conservation of angular momentum. (For under impulsive forces and finite forces)

[HINTS : §4.8, pp. 168]

(b) A solid homogeneous cone of height h and vertical angle 2α oscillates about a horizontal axis through its vertex. Show that the length of the simple equivalent pendulum is $\frac{h}{5}(4+\tan^2\alpha)$.

[HINTS : Prob. 3.3.1, pp. 112]

14. (a) A sphere of radius a, is projected up an inclined plane with velocity V and angular velocity ω in the sense which would cause it to roll up, if $V > a\omega$ and the coefficient of friction μ exceeds $\frac{2}{7}\tan\alpha$. Show that the sphere will cease to ascend after time $\frac{5V+2a\omega}{5g\sin\alpha}$, where α is the inclination of the plane to the horizontal.

[HINTS : Prob. 4.16.6, pp. 240]

(b) Three equal uniform rods AB, BC, CD are freely jointed and placed in a straight line on a smooth table. The rod AB is struck at its ends by a blow which is perpendicular to its length. Find the resulting motion and show that the velocity of the centre of AB is 19 times that of CD and its angular velocity 11 times that of CD.

[HINTS : Prob. 5.8.9, pp. 312]

Calcutta University

2012

12. (a) Find the conditions that must be satisfied for a given line to be a principal axis of a material system at any point on its length. Find the other two principal axes.

[HINTS : §1.16, pp. 50(a)]

(b) A rod of length $2a$ is suspended by a string of length l attached to one of its ends. The string and the rod revolves about the partial with uniform angular velocity and their inclinations to the vertical are θ and ϕ respectively. Show that $\frac{3l}{a} = \frac{(4\tan\theta - 3\tan\phi)\sin\phi}{(\tan\phi - \tan\theta)\sin\theta}$.

[HINTS : Prob. 2.5.4, pp. 98]

13. (a) An elliptic lamina of eccentricty e is rotating in its own plane about one focus S' with angular velocity ω. S is set free and the other focus S' is fixed at the same time. Show that the lamina will now rotate with an angular velocity $\frac{2-5e^2}{2+3e^2}\omega$.

[HINTS : Prob. 5.8.13, pp. 317]

(b) A homogeneous sphere of radius a, rotating with angular velocity ω about horizontal diameter is gently placed on a table whose co-efficient of friction is μ. Show that there will be slipping at the point of contact for a time $\frac{2\omega a}{7\mu g}$ and then the sphere will roll with angular velocity $\frac{2\omega}{7}$.

[HINTS : Prob. 4.16.1, pp. 228]

14. (a) When a body rotates about a fixed axis show that the rate of change of the angular momentum of the body about the axis of rotation is equal to the sum of the moments about the same axis of all forces acting on the body.

[HINTS : §2.4.1, pp. 94]

(b) A bent lever, whose arms are of length a and b, the angle between them being α, makes small oscillations in its own plane about the fulcrum. Show that the length of the corresponding simple pendulum is $\frac{2(a^3+b^3)}{3\sqrt{a^4+2a^2b^2\cos\alpha+b^4}}$.

[HINTS : Prob. 3.3.2, pp. 113]

Calcutta University

2013

5. (a) Write down the general equations of motion of a rigid body in vector form. Establish the principle of independence of the motions of translation and rotation of a rigid body.

[HINTS : §4.2.1, pp. A-9]

(b) A circular homogeneous plate is projected up a rough inclined plane with velocity V without rotation, the plane of the plate being the plane of greatest slope of the inclined plane. Show that the plate stops slipping after a time $\frac{V}{g(3\mu\cos\alpha+\sin\alpha)}$, where μ is the coefficient of friction and α is the inclination of the inclined plane.

[HINTS : Prob. 4.16.1, pp. 237]

6. (a) Prove that the magnitude of the Angular Momentum of a rigid body moving in two dimentions about a fixed point can be expressed, with usual notations, in the form $Mvp + MK^2\frac{d\theta}{dt}$.

[HINTS : §4.4, pp. 163]

(b) Three uniform rods AB, BC, CD are hinged freely at their ends, B and C, so as to form three sides of a square and are laid on a smooth table. The end A is struck by a horizontal blow P at right angles to AB. Show that the angular velocity of A is nineteen times that of D and that the impulsive actions at B and C are respectively $\frac{5P}{12}$ and $\frac{P}{12}$.

[HINTS : Prob. 5.8.12, pp. 315]

7. (a) Show that for a compound pendulum the centre of suspension and centre of oscillation are interchangeable.

[HINTS : Remark 3.2.1, pp. 111]

(b) A circular plate rotates about an axis through its centre and perpendicular to its plane with an angular velocity ω. This axis is suddenly set free and a point on the circumference of the plate is fixed. Show that the resulting angular velocity will be $\frac{\omega}{3}$.

[HINTS : Prob. 5.8.15, pp. 318]

8. (a) Find the moment of inertia of a rigid body about a line having direction co-sines l, m, n.

[HINTS : §1.9, pp. 23]

(b) A solid homogeneous cone, of height h and the vertical angle 2α, oscillates about a horizontal axis through its vertex. Show that the length of the simple equivalent pendulum is $\frac{h(4+\tan^2\alpha)}{5}$.

[HINTS : Prob. 3.3.1, pp. 112]

Calcutta University

2014

5. (a) Define Principal Axes of a given material system. Obtain the conditions for a given straight line to be a Principal Axis of the system at some point of its length. If so find the other two Principal Axis.

[HINTS : §1.16, pp. 50(a)]

(b) A plank of mass M initially at rest along a line of greatest slope of a smooth plane inclined at an angle α to the horizon. A man of mass M', starting from the upper end walks down the plank so that it does not move. Show that the man gets the other end in time $\sqrt{\frac{2M'a}{(M+M')g\sin\alpha}}$, where a is the length of the plank.

[HINTS : Prob. 2.5.1, pp. 94]

6. (a) A rigid body a mass M is rotating about a fixed axis. Find the expression for :

(i) The kinetic energy of the rigid body;

(ii) the Angular Momentum about this axis of rotation of the rigid body

[HINTS : §4.3 and §4.4, pp. 161 and pp. 163]

(b) A rod of length $2a$, revolves with uniform angular velocity ω about a vertical axis through a smooth joint at one extremity of the rod, so that it describes a cone of semi-vertical angle α. Show that $\omega^2 = \frac{3g}{4a\cos\alpha}$. Prove also that the direction of reaction at the hinge makes with the vertical an angle $\tan^{-1}\left(\frac{3}{4}\tan\alpha\right)$.

[HINTS : Prob. 2.5.2, pp. 96]

7. (a) Establish the principle of Conservation of Angular Momentum under finite forces.

[HINTS : §4.8, pp. 168]

(b) An elliptic lamina can rotate about a horizontal axis passing through a focus and perpendicular to its plane. If the eccentricity of the ellipse be $\sqrt{\frac{2}{5}}$, prove that the centre of oscillation will be at the other focus.

[HINTS : Prob. 3.3.11, pp. 126]

Calcutta University

2015

5. (a) Prove that the kinetic energy of a body, moving in two dimensions, can be expressed in the form

$$\tfrac{1}{2}MV^2 + \tfrac{1}{2}MK^2\left(\frac{d\theta}{dt}\right)^2,$$

where the notations have their usual significance.

(b) A uniform circular disc is projected up a rough inclined plane with velocity V but no rotation, the plane of the disc being vertical and meeting the inclined plane along a line a greatest slope. Show that the disc ceases to slip after a time $\frac{V}{g(3\mu\cos\alpha+\sin\alpha)}$, where μ is the coefficienty of friction and α is the inclination of the inclined plane with the horizon.

6. (a) Prove that the time of a small oscillation of a compound pendulum is minimum when the axis of suspension is parallel to the maximum radius vector of the momental ellipsoid at the centre of inertia and the point of suspension is so taken that the centre of inertia bisects the join of the point of suspension and the centre of oscillation.

(b) AB, BC, are two equal similar rods freely hinged at B and lie in a straight line on a smooth table. The end A is struck by a blow perpendicular to AB. Show that the resulting velocity of A is $\frac{7}{2}$ times that of B.

7. (a) Show that the motion of the centre of inertia and the motion relative to the centre of inertia of a rigid body are independent of each other.

(b) An elliptic area of eccentricity e is rotating with angular velocity ω about one latus rectum. Suddenly this latus rectum is set free and the other is fixed. Show that the new angular velocity is $\frac{(1-4e^2)}{(1+4e^2)} \cdot \omega$.

8. (a) Derive the equation of momental ellipsoid at a point of a rigid body.

(b) A heavy uniform rigid straight rod OA of length $2a$ is free to turn about its end O. If the rod revolves with uniform angular velocity ω about the vertical through O making a constant angle $\frac{\pi}{4}$ with the vertical, show that $\omega^2 = \frac{3g}{2\sqrt{2}a}$.

Calcutta University

2016

5. (a) A rigid body is rotating about a fixed axis. Find the expressions for

(i) The K.E. of the rigid body

(ii) The angular momentum about the axis of rotation.

(b) ABC is triangular area and AD is perpendicular to BC and AE is a median. O is the middle point of DE. Show that BC is a principal axis of the triangle at O.

6. (a) Explain what you mean by the motion of a rigid body in two dimensions, defining properly the variables used which determine the motion. Find the moment of momentum of a rigid body moving in two dimensions about the centre of inertia.

(b) A boat of mass M and length ℓ is floating stationary on the still water of a Lake. A man of mass M' initially standing at one end of the boat walks to the other end of it, and stops. Show that the distance through which the boat will move in the opposite direction is $\frac{M'\ell}{M+M'}$, assuming the water offers no resistance to the boat.

7. (a) A thin circular disc of mass M and radius a, can turn freely about a thin axis OA, which is perpendicular to its plane and passes through a point O of its circumference. The axis OA is compelled to move in a horizontal plane with angular velocity ω about its end A. Show that the inclination θ to be the vertical of the radius of the disc through O is $\cos^{-1}\left(\frac{g}{a\omega^2}\right)$, uness $\omega^2 < \left(\frac{g}{a}\right)$ and then θ is zero. (Proper sketch of the diagram is necessary).

(b) Establish the principle of conversasion of angular momentum under impulsive forces.

8. (a) Find the Moment of inertia of a rigid body about a line having direction cosines ℓ m.n.

(b) Three uniform rods AB, BC, CD are hinged freely at their ends B and C, so as to form three sides of a square and are laid on a smooth table. The end A is struck by a horizontal blow P at right angles to AB. Show that the impulsive actions at B and C are respectively $\frac{5}{12}P$ and $\frac{1}{12}P$.

Calcutta University

2017

5. (a) State D'Alembert's Principle. Deduce the general equations of motion from D'Alembert's Principle.

(b) A solid homogeneous cone of height 'h' and vertical angle 2α oscillates about a horizontal axis through its vertex. Show that the length of the simple equivalent pendulum is $\frac{h}{5}(4 + \tan^2 \alpha)$.

6. (a) What is meant by principle axes of a given material system at a point? Find the condition so that a given straight line may be a principal axis of the material system at any point of its length and if so find the other two principal axes.

(b) A plank of mass M is initially at rest along a straight line of greatest slope of a smooth plane inclined at an angle 'α' to the horizon and a man of mass M', starting from the upper end walks down the plank so that it does not move. Show that he will reach the other end in time $\sqrt{\frac{2M'a}{(M+M')g \sin \alpha}}$.

7. (a) A rough uniform rod of length $2a$ is placed on a rough table at right angles to its edge; if its centre of gravity be initially at a distance b beyond the edge, show that the rod will begin to slide when it has turned through an angle $\tan^{-1} \frac{\mu a^2}{a^2+9b^2}$, where μ is the coefficient of friction.

(b) If a sphere be projected up an inclined plane, for which $\mu = \frac{1}{7} \tan \alpha$, with velocity V and an initial angular velocity ω (in the direction in which it would roll up), and if $V > a\omega$, show that the friction acts downwards at first, and upwards afterwards and prove that the whole time during which the sphere rises is $\frac{17V+4a\omega}{18g \sin \alpha}$.

8. (a) A uniform disc of radius a is rolling without slipping along a smooth horizontal table with velocity V, when the highest point becomes suddenly fixed. Prove that the disc will make a complete revolution round the point if $V^2 > 24ag$ where g is the acceleration due to gravity.

(b) Two equal rods AB and AC are freely jointed at A and are placed on a smooth table so as to be at right angles. The rod AC is struck by blow at C in a direction perpendicular to itself. Show that the resulting velocities of the middle points of AB and AC are in the ratio $2 : 7$.

Calcutta University

2018

5. (a) Explain what is meant by equimomental bodies. Show that a uniform plane triangular lamina is equimomental with a system of three particles placed at the middle points of the sides, each equal to one third the mass of the triangle.

(b) A non-homogeneous rod AB of length $2I$, whose density at any point is directly proportional to the distance from the point A, is roating with a uniform angular velocity ω about at vertical axis through A. If the rod is inclined at an angle α to the vertical, show that the value of α is either zero or $\cos^{-1}\left(\frac{2g}{3I\omega^2}\right)$. '$g$' being the acceleration due to gravity on the surface of the earth.

6. (a) Find the expression for moment of momentum about the origin of a rigid body moving in two dimensions.

(b) A uniform rod is held at an inclination α to the horizon, with one end in contact with a horizontal table whose coefficient of friction is μ. If it be then released, show that it will commence to slide if $\mu < \frac{3\sin\alpha\cos\alpha}{1+3\sin^2\alpha}$.

7. (a) Define 'Impulse of a force'. Obtain the equations of motion in two dimensions of a rigid body moving under impulsive forces.

(b) Two equal, uniform rods AB and BC, are freely joined at B and turned about a smooth joint at A. When the rods are in a straight line, ω is the angular velocity of AB and u is the linear velocity of the centre of gravity of BC. BC impinges on a fixed inelastic obstacle at a point D. Show that the rods are instantly brought to rest if $BD = 2a\cdot\frac{2u-a\omega}{3u+2a\omega}$, where $2a$ is the length of either rod.

8. (a) A lamina in the form of an ellipse is rotating in its own plane about one of its foci with angular velocity ω. This focus is set free and the other focus at the same instant is fixed. Show that the ellipse now rotates about it with angular velocity $\omega\left(\frac{2-5e^2}{2+3e^2}\right)$.

(b) Three equal uniform rods AB, BC and CD are freely jointed and placed in a straight line in a smooth table. The rod AB is struck at the end A by a blow which is perpendicular to its length. Show that the velocity of the centre of AB is 19 times that of CD, and its angular velocity is 11 times that of CD.

Calcutta University

2019

5. (a) Define Principal Axes of a given material system. Obtain the condition for a given straight line to be a Principal Axis of the system at some point of its length. If so, find the other two Principal Axes.

(b) Prove that the equation of the momental ellipsoid at a point on the circular edge of a solid cone is

$$(3a^2+2h^2)x^2+(23a^2+2h^2)y^2+26a^2z^2-10ahxz = \text{constant}$$

where h is the height and a is the radius of the base of the cone.

6. (a) A rigid body of mass M is rotating about a fixed axis. Find the expression for

(i) the Kinetic Energy of the rigid body.

(ii) the Angular Momentum about the axis of rotation of the rigid body.

(b) A sphere of mass m falls with velocity V on a perfectly rough inclined plane of mass M and angle α which rests on a smooth horizontal plane. Show that the vertical velocity of the centre of the sphere immediately after impact is $\frac{5(M+m)V\sin^2\alpha}{7M+2m+5m\sin^2\alpha}$, the bodies being all supposed perfectly inelastic.

7. (a) State D'Alembert's principle for a rigid body moving in any manner. Explain its difference with Newton's second law of motion for many particles.

(b) Define compound pendulum. Show that the centre of suspension and centre of oscillation are convertible for the pendulum.

8. (a) A rough uniform rod of length $2a$ is placed on a rough table at right angles to its edge, if its centre of gravity be initially at a distance b beyond the edge, show that the rod will begin to slide when it has turned through an angle $\tan^{-1}\frac{\mu a^2}{a^2+9b^2}$. where μ is the co-efficient of friction.

(b) A light rigid rod ABC has three particles each of mass m attached to it at A, B, C. The rod is struck by a blow P at right angle to it at a point whose distance from A is equal to BC. Prove that the kinetic energy set up is

$$\frac{1}{2}\cdot\frac{P^2}{m}\cdot\frac{a^2-ab+b^2}{a^2+ab+b^2}$$

where $AB = a$ and $BC = b$.

Kalyani University

2005

7. (a) Define momental ellipsoid. Show that the equation of the momental ellipsoid at the centre of an elliptic plate is $\frac{x^2}{a^2} + \frac{y^2}{b^2} + \left(\frac{1}{a^2} + \frac{1}{b^2}\right) z^2 =$ constant.

[HINTS : Prob. 1.15.1, pp. 43]

(b) If, from a rectangle lamina of sides a, b half the reactangle be recovered so as to have a right angle triangle of sides a, b, then prove that the value of for the principal axis at the right angle is given by

$$\theta = \frac{1}{2}\tan^{-1}\left(\frac{ab}{a^2 - b^2}\right)$$

[HINTS : Prob. 1.20.3, pp. 79]

8. (a) State D'Alebert's principle. Deduce the general equations of motion of a rigid body from D'Alember's principle.

[HINTS : §2.3, pp. 89]

(b) A thin heavy disc can turn freely about an axis in its own plane and this axis revolves horizontally with a uniform angular velocity ω about a fixed point on itself. Show that the inclination θ of the plane of the disc to the vertical is $\cos^{-1}\left(\frac{gh}{\omega^2 K^2}\right)$, where h is the distance of the centre of inertia of the disc from the axis and K is the radius of gyration of the disc about the axis.
if $\omega^2 < \frac{gh}{K^2}$, then show that the plane of the disc is vertical.

[HINTS : Prob. 2.5.6, pp. 101]

9. (a) What is a compound pendulum? Find the minimum time of oscillation a compound pendulum.

[HINTS : §3.2, pp. 109]

(b) A right cone of angle 2α can then freely about an axis passing through the centre of the base and perpendicular to the axis. If the cone starts from rest with its axis horizontal, than show that, when the axis is vertical, the ratio of the thrust on the fixed axis and the weight of the cone is $\{1 + (\cos^2\alpha)/2\} : \{1 - (\cos^2\alpha)/3\}$.

[HINTS : Prob. 3.5.4, pp. 139]

10. Deduce the dynamical two-dimensional equations of motion of a rigid body under finite forces. Also find the kinetic energy of the body and its moment of momentum about the origin.

[HINTS : §4.3 and §4.4, pp. 161, 163]

11. An imperfectly rough sphere moves from rest down a plane inclined at an angle α to the horizon. Discuss the motion.

[HINTS : §4.12, pp. 210]

Kalyani University

2006

7. (a) Define momental ellipsoid at a point O of a rigid body. Show that the momental elliposid at a point on the edge of the circular base of a thin hemispherical shell is $2x^2 + 5(y^2 + z^2) - 3xz =$ constant.

[HINTS : Prob. 1.15.6; pp. 50]

(b) Show that an extremity of the bounding diameter of a semi-circular lamina, the principal axis makes an angle $\frac{1}{2}\tan^{-1}\frac{8}{3\pi}$ to the diameter.

[HINTS : The polar equation of the semi-circle OAB of radius a reffered to the diameter OA is initial line and O as pole is $r = 2a\cos\theta$. Let ρ be the dencity of the semi-circular lamina and let $r d\theta\, dr$ be elimentary area $PQRS$ at $P(r, \theta)$.

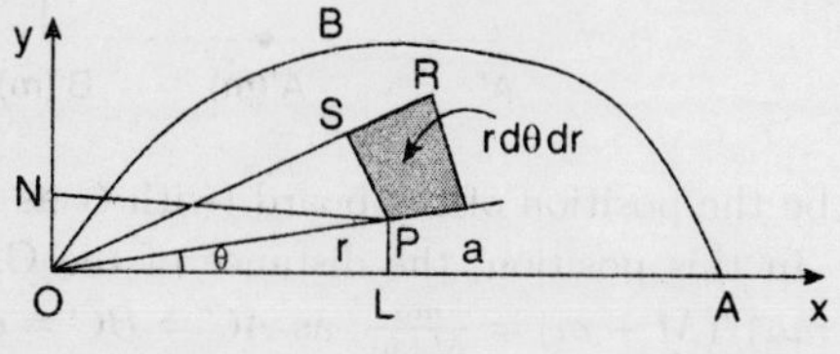

Let oy be perpendicular to OA. Let PL and PN be N perpendiculars from P on ox and oy respectively. Then $PL = r\sin\theta$ and $PN = r\cos\theta$. Now, we have

$$A = \text{M.I. of the semi-circular lamina about } ox$$

$$= \int_{\theta=0}^{\pi/2}\int_{r=0}^{2a\cos\theta} (r\sin\theta)^2 \rho\, r\, d\theta\, dr = \frac{1}{4}(2a)^4\rho \int_0^{\pi/2} \sin^2\theta\cos^4\theta\, d\theta$$

$$= 4\rho a^4 \cdot \frac{\Gamma(3/2)\Gamma(5/2)}{2\Gamma(4)} = \frac{\pi\rho a^4}{8}.$$

$$B = \text{M.I. of the semi-circular lamina abot } oy$$

$$= \int_{\theta=0}^{\pi/2}\int_{r=0}^{2a\cos\theta} (r\cos\theta)^2 \rho\, r\, dr\, d\theta = \frac{1}{4}(2a)^4\rho \int_0^{pi/2} \cos^6\theta a\theta$$

$$= 4\rho a^4 (5/6)\left(\frac{3}{4}\right)\cdot\left(\frac{1}{2}\right)\cdot\frac{\pi}{2} = \frac{5}{8}\pi\rho a^4$$

$$\text{and } H = \text{P.I. of the semi-circular lamina about } ox \text{ and } oy$$

$$= \int_{\theta=0}^{\pi/2}\int_{r=0}^{2a\cos\theta} (r\cos\theta)(r\sin\theta)\rho\cdot r\, d\theta\, dr$$

$$= \frac{1}{4}\rho\cdot(2a)^4 \int_0^{\pi/2} \cos^5\theta\sin\theta\, d\theta$$

$$= 4\rho a^4 \cdot \frac{1}{6} = \frac{2}{3}\rho a^4.$$

If α is the angle that a principal axis at O makes with ox, then

$$\tan 2\alpha = \frac{2H}{B-A} = \frac{8}{3\pi}, \quad \therefore \alpha = \frac{1}{2}\tan^{-1}\left(\frac{8}{3\pi}\right).$$

8. (a) Starting from the general equations of motion of a rigid body, show how its motion is composed of two parts-one, the motion of the centre of gravity and the other, the motion relative to the centre of gravity.

[HINTS : §2.4, pp. 91]

(b) A round uniform board of mass M and length $2a$, rests on a small horizontal plane and a man of mass m walls on it from one end to the other. Show that the distance through which the board moves $\frac{2Ma}{(M+m)}$.

[HINTS : Here the board and the man constitute a material system. Since there is no external horizontal force acting on the system, by D'Alembert's principle the centre of gravity of the system will remain at rest. In fact, as the man moves forwards, the board slips backwards, thus keeping the position of the centre of gravity unchanged.

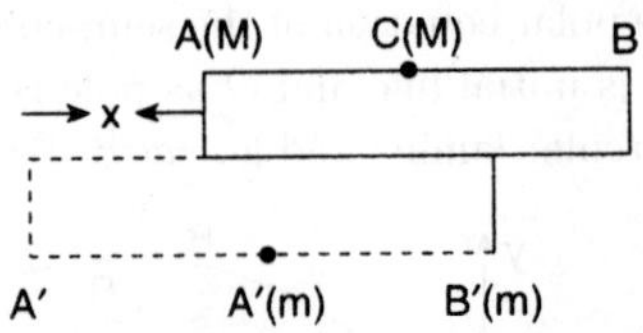

Let AB be the position of the board (with C as middle point) when the man is at A. In this position, the distance of the C.G. of the system from $A = (M.O + ma)/(M+m) = \frac{ma}{M+m}$, as $AC = BC = a$.

Let $B'B'$ be the position of the board when the man reaches the other end of the board. If the board slips through a distance x in this time (i.e., $AA = x$), then in this position, the distance of centre of gravity of the system from $A = \frac{M\cdot AB' + m\cdot AC'}{M+m} = \frac{M(2a-x)+m(a-x)}{M+m}$, C' being mid point of $A'B'$. But the distance of C.G. is the same from A in initial and final position, so we must have

$$\frac{ma}{M+m} = \frac{M(2a-x)+m(a-x)}{M+m} \quad \text{or } x = \frac{2Ma}{M+m}\ ]$$

9. (a) Obtain the two dismension equations of motion of a rigid body under finite forces.

[HINTS : §4.2, pp. 159]

(b) Show that then length of the simple equivalent pendulum of a solid hemisphere, where diameter of the base is $2a$, and with oscillates about a diameter which is horizontal is $\frac{16}{15}a$.

[HINTS : Prob. 3.3.13, pp. 119]

10. A imperfect rough sphere moves from rest down a plane inclined at an angle α to the horizontal. Discuss the motion.

[HINTS : §4.12, pp. 210]

Kalyani University

2007

7. (a) Show that for every body there exists at every point a set of three perpendicular axes such that the product of inertia of the body about them, taken two at a time, vanish.

[HINTS : Remark 1.13.1, pp. 39]

(b) Show that the momental ellipsoid at a point on the rim of the hemisphere can expressed in the form

$$2x^2 + 7(y^2 + z^2) - \frac{15}{4}zx = \text{constant}.$$

[HINTS : Prob. 1.15.4, pp. 45]

8. (a) Explain what is meant by the 'length of the simple equivalent pendulum'. Show that the centre of suspension and oscillation of a compound pendulum are interchangeable.

[HINTS : §3.2, pp. 109, Remark 3.2.1, pp. 111]

(b) A uniform thin sphereical shall is placed at rest on an inclined plane. If μ be the coefficient of friction between the shall and the plane, then show that the greatest inclination of the plane for which the shell will roll without slipping is $\tan^{-1}\left(\frac{5\mu}{2}\right)$.

[HINTS : Prob. 4.13.2, pp. 216]

9. Prove that, for a body of mass M moving in two dimensions, the expression for the kinetic energy is

$$\tfrac{1}{2}MV^2 + \tfrac{1}{2}MK^2\Omega^2$$

where V is the velocity of the centre of mass G, K is the radius of gyration of the body about a line through G perpendicular to the plane of motion and Ω is the angular velocity about this line.

[HINTS : §4.3, pp. 161]

Kalyani University

2008

7. (a) Show that the moment of inertia of a material system about any given straight line is equal to the sum of the moment of inertia of the system about the parallel straight line through the centre of mass and the moment of inertia about the given line, of the whole mass of the system supposed collected at the centre of mass.

[HINTS : §1.11(a), pp. 33]

(b) A wire is the form of a semi circle of radius a. Show that at the end of its diameter and the principal axes in its plane are inclined to the diameter at angles $\frac{1}{2}\tan^{-1}\frac{4}{\pi}$ and $\frac{\pi}{2}+\frac{1}{2}\tan^{-1}\frac{\pi}{4}$.

[HINTS : Prob. 1.20.5, pp. 81]

8. (a) State D'Alembert's principal and deduce the equations of motion of the centre of inertia of a rigid body and the equations of motion relative to the centre of inertia.

[HINTS : §2.4, pp. 91]

(b) A thin rod of length $2a$ revolves with uniform angular velocity ω about a vertical axis through a small joint at one extremity of the rod so that it describe a cone of semi-vertical angle α; show that $\omega^2 = \frac{3g}{4a\cos\alpha}$.

[HINTS : Prob. 2.5.2, pp. 96]

9. (a) A uniform rod of length $2a$ can therm freely in a vertical plane about an end O which is fixed. It is started from the position in which it hanges vertically with such an angular velocity as can just take it to the highest position.

Prove that the time of describing any angle $\theta < \pi$ is $\sqrt{\frac{4a}{3g}}\log\tan\left(\frac{\pi}{4}+\frac{\theta}{4}\right)$.

[HINTS : Prob. 3.3.4, pp. 116]

(b) A uniform sphere rolls down an inclined lane rough enough to present any sliding. Find the motion.

[HINTS : §4.12, pp. 210]

Kalyani University

2009

7. (a) Show that the principal axes at the corner of rectangular lamina of sides $2a, 2b$ makes with a side the angle θ and $\frac{\pi}{2}+\theta$, where

$$\tan 2\theta = \frac{3}{2}\cdot\frac{ab}{a^2-b^2}.$$

[HINTS : Prob. 1.20.2, pp 78]

(b) A rod of length $2a$ is suspended by string of length l attached to one end of the rod. If the string and the rod revolves about the vertical with uniform angular velocity and their inclinations to the downward drawn vertical be θ and ϕ respectively, then shown that $\frac{(4\tan\theta-3\tan\phi)\sin\phi}{(\tan\phi-\tan\theta)\sin\theta} = \frac{3l}{a}$.

[HINTS : Prob. 2.5.4, pp. 98]

8. (a) Deduce the equation $MK^2\ddot{\theta} = L$, for the motion of a rigid body rotating about a fixed axis where the symbols have their usual meaning. Show also that the kinetic energy of the body at the instant of time t is $\frac{1}{2}MK^2\dot{\theta}^2$.

[HINTS : §4.2, pp. 159. §4.3, pp. 161]

(b) A circular area of weight W can term freely about a horizontal axis which passes through a point O of its circumference and is perpendicular to its plane. If the motion commences when the diameter through O is vertically above 0, show that when the diameter has turned through an angle θ, the components of strain at O along and perpendicular to this diameter are respectively $\frac{W}{3}(7\cos\theta - 4)$ and $\frac{W}{3}\sin\theta$.

[HINTS : Prob. 3.5.1, pp. 132]

9. (a) A sphere of radius r, rotating with angular velocity ω about a horizontal diameter, is gently placed on a horizontal table with which the coefficient of friction is μ. Show that there will be slipping at the point of contact for a time $\frac{2\omega r}{7\mu g}$, and then the sphere will roll with angular velocity $\frac{2}{7}\omega$.

[HINTS : Prob. 4.16.1, pp. 228]

(b) A uniform rod of mass M and length $2a$ at rest is struck by a blow J at right angles to its length at a distance c from the centre. Find the point about which is will begin to turn when the ends are free.

[HINTS : Prob. 5.8.2, pp. 297]

Kalyani University

2010

7. (a) Show that the moment of inertia of a material system about any given straight line is equal to the sum of the moment of inertia of the system about the parallel straight line through the centre of mass and the moment of inertia about the given line of the whole mass of the system supposed collected at the centre of mass.

[HINTS : 1.11(a), pp. 33]

(b) A uniform rod is held at an inclination 'α' to the horizontal with one end in contact with the horizontal table whose co-efficient of friction is 'μ'. If it be then released then show that it will commence to size if $\mu < \frac{3\sin\alpha\cos\alpha}{1+3\sin^2\alpha}$.

[HINTS : Prob. 4.10.6, pp. 185]

9. (a) If A, B, C are the moment of inertia and D, E, F are the products of inertia for a right body with respect to the rectangular axes ox, oy, oz, prove that the moment of inertia about a line through O having direction cosines l, m, n is given by $Al^2 + Bn^2 + Cn^2 - 2Dmn - 2Enl - 2Flm$.

[HINTS : §1.9(a), pp. 23]

(b) State the principle of conservation of energy for a moving rigid body and prove it for two dimentional motion under finite forces. Is the principle applicable to a body rolling on a fixed rough surface under the force of gravity? Justify your answers.

[HINTS : §4.6, pp. 166]

Kalyani University

2011

7. (a) A bent lever, whose arms are of lengths a and b, angle between them being α moves small oscillation in its own plane about the fulcrum, show that the length of the corresponding simple pendulum is $\frac{2}{3}\frac{a^3+b^3}{\sqrt{a_2^4 a^2 b^2 \cos\alpha + b^4}}$.

[HINTS : Prob. 3.3.2, pp. 113]

(b) A plane, of mass M, is initially at rest along a line of greatest slope of the smooth plane inclined at angle α to the horizon, and a man, of mass M' starting from the upper end walks down the plank so that it does not move. Show that the gets to the other end in time $\sqrt{\frac{2M'a}{(M+M')g\sin\alpha}}$, where '$a$' is the length of the plank.

[HINTS : Prob. 2.5.1, pp. 94]

8. (a) If A and B be the moments of inertia of a uniform lamina about two perpendicular axes ox and oy, lying in its plane and F be the product of inertia of the lamina about these axes, show that the in principal moments of inertia at O are equal to $\frac{1}{2}\left[A+B\pm\sqrt{(A-B)^2+4F^2}\right]$.

[HINTS : Prob. 1.20.1, pp. 76]

(b) (i) Deduce the expression for the angular momentum of a rigid body relation to its centre of mass.

[HINTS : §4.4, pp. 163]

(ii) Deduce the principle of conservation of angular momentum for a rigid.

[HINTS : §4.8, pp. 168]

Kalyani University

2012

3. (a) Show that the equation of the momental ellipsoid at the corner of a cube of side $2a$ referred to its principal axes is $2x^2 + 11(y^2 + z^2) =$ constant.

[HINTS : Prob. 1.15.3, pp. 44]

(b) Prove that the kinetic energy of a rigid body moving in two dimensions is the sum of two kinetic energies, one due to translation and the other due to rotation.

[HINTS : §4.3, pp. 161]

(c) A uniform solid sphere rolls down an inclined plane, rough enough to prevents sliding. Discuss the motion if $\mu > \frac{2}{7}\tan\alpha$, where μ is coefficient of friction and α is the inclination of the plane to the horizon.

[HINTS : §4.15, Case III, pp. 227]

4. (b) State D'Alembert's principle and deduce the equation of motion of the centre of inertia of a rigid body and the equation of motion relative to the centre of inertia.

[HINTS : §2.4, pp. 91]

5. (a) A plank of mass M is initially at rest along a straight line of greatest slope of a smooth plane inclined at an angle α to the horizon and a man of mass m starting from the upper end walks down the plank so that it does not move. Show that the gets to the other end in the $\left\{\frac{2ma}{(M+m)g\sin\alpha}\right\}^{1/2}$.

[HINTS : Prob. 2.5.1, pp. 94]

7. (a) A sphere of radius a, is suspended by a fine wire from a fixed point at a instance l from the centre. Show that the time period of small oscillation is given by $2\pi\sqrt{\frac{5l^2+2a^2}{5lg}}\left(1+\frac{1}{4}\sin^2\frac{\alpha}{2}\right)$.

[HINTS : Prob. 3.3.5, pp. 117]

Kalyani University

2013

3. (d) A thin uniform rod has one end attached to a smooth hinge and is allowed to fall from a horizontal position. Show that the horizontal strain one of the hinge is greatest when the rod is inclined at angle $\frac{\pi}{4}$ to the vertical and that the vertical strain is then $\frac{11}{4}$ times of the weight of the rod.

[HINTS : Prob. 3.5.6, pp. 143]

(e) Two uniform rods AB, BC are freely jointed at A and laid on a smooth horizontal table so that the angle BAC is a right angle. The rod AB is struck by a blow P at B in a direction perpendicular to AB; show that the initial velocity of A is $\frac{2P}{4m'+m}$, where m, m' are the masses of AB, BC respectively.

[HINTS : Prob. 5.8.11, pp. 314]

7. (a) Find whether a given straight line is at any point of its length, a principal axis of a material system and in case the line is a principal axis, find the other two principal axes.

[HINTS : §1.16, pp 50(a)]

(b) A thin heavy disc can turn freely about an axis in its own plane and this axis revolves horizontally with a uniform angular velocity ω about a fixed point on itself. Show that the inclination θ of the plane of the disc to the vertical is $\cos^{-1}\left(\frac{gh}{\omega^2 k^2}\right)$, where h is the distance of the centre of inertia of the disc about the axis.

[HINTS : Prob. 2.5.6, pp. 101]

Kalyani University

2014

6. (b) A rod of length $2a$ is suspended by a string of length l, attached to one end. If the string and the rod revolves with uniform angular velocity about a vertical and their inclination with the vertical be θ and ϕ respectively, then prove that $\frac{3l}{a} = \frac{(4\tan\theta - 3\tan\phi)\sin\phi}{(\tan\phi - \tan\theta)\sin\theta}$.

[HINTS : Prob. 2.5.4, pp. 98]

7. (a) A square lamina of weight W can turn freely about a horizontal axis through the one corner O perpendicular to its plane. If the motion commences when the diagonal through O is vertically above O, then find the reaction at hinge at O when the diagonal has turned through an angle θ.

(b) State D'Alember's principle and deduce the equation of motion of the centre of inertia and the motion relative to the centre of inertia of a rigid body.

[HINTS : §2.4, pp. 91]

8. (a) A solid circular cylinder of radius 'a' rotating about its axis is placed gently with its axis horizontal on a rough plane, where inclination to the horizontal is α. Inertially the friction acts up the plane and the coefficeint of friction is μ. Show that the cylinder will move upwards, if $\mu > \tan\alpha$.

Also show that the time that elapses before rolling commences is $\frac{a\omega}{g(3\mu\cos\alpha - \sin\alpha)}$.

[HINTS : Prob. 4.16.3, pp. 236]

Burdwan University

2005

2. Define the centre of mass of a system of n articles moving under external forces.

(a) Using D'Alembert's principle obtain the equations of motion of the centre of mass and about the centre of mass of the system.

[HINTS : §2.4, pp. 91]

(b) Show that the kinetic energy of the system is equal to the K.E. of the total mass supported concentrated at the centre of mass and moving with it together with the K.E of the motion of the particles relation to the centre of mass.

[HINTS : §4.4, pp. 161]

3. (a) If A and B are the moment of inertia of a plane lamina about rectangular axes ox, oy in the plane of the lamina and F, the product of inertia with respect to these axis, find the moments of inertia A', B' and the product of inertia F' about any two perpendicular lines throgh O in the xy-plane. Also show that $A + B = A' + B'$, and $AB - F^2 = A'B' - F'^2$.

[HINTS : §1.9(c), pp. 27]

(b) Find, from the definition, the moments and product of inertia with respect to the coordinate axis of a uniform rectangular lane of mass M whose edges are given by $z = 0,\ x = 0;\ z = 0,\ x = a,\ y = 0$ and $z = 0, y = b$.
Write down the corresponding inertia matrix and obtain the principal moments about the origin.

[HINTS : §1.4(b), pp. 4]

4. (a) Obtain the equation of motion of a plane lamina rotating about a fixed axis perpendicular to the plane of the lamina in the form $MK^2\ddot{\theta} = L$, the symbols have to be explained by you.

[HINTS : §4.2, pp. 159]

(b) A uniform rod of mass m and length $2a$ can turn freely about a fixed end. Show that the last angular velocity with which it must be started from the lowest position so that it may just make a complete revolution is $\sqrt{\frac{3g}{a}}$, g being the acceleration due to gravity. With this angular velocity, show that the time to describe and angle θ is $2\sqrt{\frac{a}{3g}} \log \tan\left(\frac{\pi}{4} + \frac{\theta}{2}\right)$.

[HINTS : Prob. 3.3.4, pp. 116]

5. (a) A homogeneous sphere of radius 'a' rotating with angular velocity ω about a horizontal diameter, is gently placed on a table whose co-efficient of friction is μ. Show that there will be sliding at the point of contact for a time $\frac{2a\omega}{7\mu g}$ and that the sphere will roll with angular velocity $\frac{2\omega}{7}$.

[HINTS : Prob. 4.16.1, pp. 228]

Burdwan University

2006

2. A set of n particles are moving under central external forces and mutual action and reactions long the particles

(i) Write down the equation of motion of the ith particle and obtain the equation of motion of the centre of mass.

(ii) Define the angular momentum of the system about the origin O. Show that the time rate of change of angular momentum of the system about O is equal to the sum of the moments of the external forces about O obtain similar results about the centre of mass of the system.

[HINTS : §2.4, pp. 91]

3. (a) If the moments and products of inertia of a body (or, a system of particles) are given about a rectangular set of axes through any point O and OL be any line with direction co-sines l, m, n then find the moment of inertia about the line.

[HINTS : §1.9, pp. 23]

(b) Obtain the inertia matrix at a corner of a uniform solid cube taking edges as axes. Hence obtain the principal moments at that corner.

[HINTS : Prob. 1.12.4, pp. 37]

4. (a) Show that the kinetic energy of a rigid body moving in two dimensions is given by $\frac{1}{2}MV^2 + \frac{1}{2}MK^2\dot{\theta}^2$ is usual motions to be explained by you.

[HINTS : §4.3, pp. 161]

(b) A uniform rod OA of length $2a$ is free to turn about its end O. If the rod revolves uniformly with angular velocity ω about downward vertical oz making a constant angle $\frac{\pi}{4}$ with oz, show that $\omega^2 = \frac{3g}{2\sqrt{2}a}$. Also show that the reaction at O makes with the vertical an angle $\tan^{-1}\left(\frac{3}{4}\right)$.

[HINTS : Prob. 2.5.8, pp. 103]

5. (a) A solid circular disc of radius 'a' is rolling up a rough plane (along the line of greatest slope) inclined at an angle α to the horizontal. If V be the velocity of the centre of the disc at an instant show that the disc ascends a further distance $\frac{3V^2}{4g\sin\alpha}$ along the plane before coming to rest.

[HINTS : Prob. 4.13.5, pp. 219]

(b) A uniform rod is held at an inclination α to the horizontal with end in contact with a horizontal table whose co-efficient of friction μ. If it is then released show that it will commence to solid if $\mu < \frac{3\sin\alpha\cos\alpha}{1+3\sin^2\alpha}$.

[HINTS : Prob. 4.10.6, pp. 185]

Burdwan University

2007

3. (a) A system of n particles is moving under external forces and their mutual actions, shows that the kinetic energy of the system is equal to the K.E. of the total mass supposed concentrated at the c.m and moving with it together with the K.E. of motion of the particles relative to the centre of mass.

[HINTS : §4.5, pp. 165]

(b) A non-uniform rod OA of length l revolves with uniform angular velocity ω about a vertical axis through a smooth joint at O, so that its describes a cone of semi-vertical angle $\alpha(\neq 0)$. If its mass be per unit length at a distance x from O, show that $\omega^2 = \frac{3g}{4a\cos\alpha}$.

[HINTS : Consult, Prob. 2.5.2, pp. 96]

4. (a) Show that the moment of momentum about a fixed point O of a rigid body of mass M moving in two-dimensions is equal to $Mvp + MK^2\dot{\theta}$ explain moments involved. What does each term represent?

[HINTS : §4.4, pp. 163]

(b) A rod of length $2a$, is suspended by an elastic string of length attached to one end; if the string and the rod revolve about the vertical with uniform angular velocity, and their inclinations to the vertical be θ and ϕ respectively. Show that

$$\frac{3l}{a} = \frac{(4\tan\theta - 3\tan\phi)\sin\phi}{(\tan\phi - \tan\theta)\sin\theta}$$

[HINTS : Prob. 2.5.4, pp. 98]

5. (a) A uniform sphere, of radius a, is rotating about a horizontal diameter with angular velocity ω and is gently placed on a rough plane which is inclined at an angle α to the horizontal, the sence of rotation being such as to tend to cause the sphere to move up the plane along the line of greatest slope. Show that, if the coefficient of friction be $\tan\alpha$, the centre of the sphere will remain at rest for a time $\frac{2a\omega}{3g\sin\alpha}$ and will then moves downwards with acceleration $\frac{5g}{7}\sin\alpha$.

[HINTS : Prob. 4.16.10, pp. 255]

(b) A thin uniform rod has one end attached to a smooth hinge and is allowed to fall from a horizontal position; show that the horizontal strain on the hinge is greatest when the rod is inclined at an angle 45° to the vertical and the vertical strain is then $\frac{11}{8}$ times the weight of the rod.

[HINTS : Prob. 3.5.6, pp. 143]

Burdwan University

2008

2. (i) (a) Show that the time-rate of charge of the angular momentum of the system of n-particle referred to above abount in origin O is equal to sum of the moments of the external forces about O.

[HINTS : §2.4.1, pp. 94]

(ii) (a) A system of n particles of masses m_i $(i = 1, 2, 3, \cdots, n)$ moves under external forces and mutual actoms and reaction. Write down the equation motion of the ith mass and then deduce the equations of motion of the centre of mass.

[HINTS : §2.4, pp. 91]

(ii) (b) Show that the K.E. of a rigid body moving in two dimensions is given by $\frac{1}{2}MV^2 + \frac{1}{2}MK^2\dot{\theta}^2$ in usual motions.

[HINTS : §4.3, pp. 161]

(iii) (b) Calculate the moment of inertia of a circular disc about an axis perpendicular to the plane of the disc through its centre.

[HINTS : §104(d), pp. 6]

(iv) (a) A homogeneous sphere of radius a rotating with angular velocity ω about a horizontal diameter is gently placed on a table where coefficient of friction is μ. Show that there will be slipping at the point of contact for a time $\frac{2a\omega}{7\mu g}$ and that the sphere will roll with angular velocity $\frac{2\omega}{7}$.

[HINTS : Prob. 4.16.1, pp. 228]

(b) A circular disc rolls down an inclinded plane (perfectly rough) of inclination α to the horizontal. Find the acceleration of the disc down the plane.

[HINTS : Remark 4.12.2, pp. 213]

(v) (a) The lengths of the sides AB and CD of a rectangular $ABCD$ are respectively $2a$ and $2b$. Show that the inclination to AB of one of the principal axis at A is $\frac{1}{2}\tan^{-1}\left\{\frac{3ab}{2(a^2-b^2)}\right\}$

[HINTS : Prob. 1.20.2, pp. 78]

Burdwan University

2009

5. (i) A uniform solid sphere of radius 'a' rolls down an inclined plane rough enough to prevent sliding. Write down the equation of motion of the sphere. Find the acceleration of the centre down the plane and show that for pure rolling $\mu > \frac{2}{7}\tan\alpha$; α being the inclination of the plane to the horizontal and μ, the coefficient of friction.

[HINTS : §4.15, Cass III, pp. 227]

(iii) A uniform circular disc of radius 'a' is projected with its plane vertical along an imperfectly rough horizontal plane with a translation velocity V

and an angular velocity ω about its centre, where $V > a\Omega$. Write down the equation of motion of the disc. Find the time when rolling starts and obtain the velocity of the centre at this moment.

[HINTS : Prob. 4.16.2, pp. 231]

(iii) If A and B are the moments of inertia about the rectangular axis ox, oy, and F the product of inertia with respect to these axes, find the moments of inertia A', B' and the product of inertia F' about two mutually perpendicular lines through O. Show that $A + B = A' + B'$.

[HINTS : §1.20, pp. 76]

Burdwan University

2010

2. (i) (a) Show that the time-rate of change of angular momentum of a system of particles about a fixed point is equal to the sum of the moments of the external forces about the same fixed point.

[HINTS : Remark 2.4.1, pp. 94]

(b) A system of n-particles moves under external forces and their mutual actions and reactions. Prove that in internal forces form a system in equilibrium.

[HINTS : §2.3, pp.89]

(ii) (a) Show that the kinetic energy of rigid body moving in two dimensions is given by $\frac{1}{2}MV^2 + \frac{1}{2}MK^2\dot{\theta}^2$ where the symbols bear usual meaning.

[HINTS : §4.3, pp. 161]

(b) Calculate the moment of inertia of a circular disc of radius 'a' about a tangent at the extremity of a diameter.

[HINTS : Prob. 1.12.2, pp. 36]

(iii) (a) A non-uniform rod OA of length l revolves with uniform angular velocity ω about a vertical axis through a smooth joint at O so that it describe a cone of sine-vertical angle α ($\neq 0$). If its mass be λx per unit length at a distance x from O. Obtain the expression for Ω.

(b) A circular hoop of radius 'a rolls down a perfectly rough inclined plane of inclination α to the horizontal. Find the acceleration of the hoop down then plane.

[HINTS : Remank 4.12.3, pp. 214]

(iv) (a) A homogeneous sphere of radius 'a' rotating with angular velocity ω about a horizontal diameter is gently placed on a table whose coefficient of friction is μ. Show that there will be slipping at the point of contact for time $\frac{2a\omega}{7\mu g}$ and that then the sphere will roll with angular velocity $\frac{2\omega}{7}$.

[HINTS : Remark 4.1.6.1, pp. 228]

(b) Find the product of inertia of an ellipse lamina w.r.t. its major and minor axes.

[HINTS : §1.4(k), pp. 13]

Burdwan University

2011

2. (i) (a) Show that the kinetic energy of a system of n particles is equal to the kinetic energy of the total mass supposed to be collected at the centre of mass and moving with the velocity of the centre of mass plus the kinetic energy of the motion of the particles relative to the centre of mass.

[HINTS : §4.5, pp. 165]

(ii) (a) If the moments and products of inertia of a body about three perpendicular and concurrent axis are known, then find out the moment of inertia about any line passing through their meeting point. Then as a particular case, find the result for a plane lamina.

[HINTS : §1.9, pp. 23]

(b) Find the product of inertia of a semi-circular wire of uniform density about a diameter and tangent at its extremity.

[HINTS : §1.4(e); pp. 8]

(iii) A uniform solid cylinder is placed with its axis horizontal on a plane, whose inclination to the horizontal is α, show that the least coefficient of friction between it and the plane is $\frac{1}{3}\tan\alpha$ so that it may roll without sliding. If the cylinder be hollow and of small thickness, the least value is $\frac{1}{2}\tan\alpha$.

[HINTS : Prob. 4.13.1, pp. 214]

(iv) Discuss the motion of a uniform circular disc, projected with its plane vertical, along a rough horizontal plane with a velocity V of translation and an angular velocity ω about the centre.

[HINTS : Prob. 4.16.2, pp. 231]

(v) Show that $MK^2\ddot{\theta}$ = moment about centre of inertia of all extrernal forces acting on the system (symbol have their usual meanings).

[HINTS : §3.1, pp.106]

(vi) A uniform rod is placed with one end in contact with a horizontal table and is at inclination α to be horizontal and is then allowed to fall when it becomе horizontal, show that the angular velocity is $\sqrt{\frac{3g}{2a}}\sin\alpha$, where the plane be perfectly smooth or perfectly rough.

[HINTS : Prob. 4.10.4, pp. 178]

Burdwan University

2012

2. (i) (a) Show that the total angular momentum of a mechanical system about a point is the angular momentum of the system concentrated at the centre of mass, plus the angular momentum for motion about the centre.

[HINTS : §4.4, pp. 163]

(ii) Discuss the motion of a uniform heavy sphere (solid) along a perfectly rough inclined plane, If one considers a hallow sphere instead of solid sphere, then write down the differences observed.

[HINTS : §4.12 and Remark 4.12.1, pp. 210 and pp. 213]

(iii) (a) A uniform rod is held at an inclination λ to the horizon with one and in contact with a horizontal table whose co-efficient of friction is μ. If it then be released, show that it will commence to slide if $\mu < \frac{3 \sin \lambda \cos \lambda}{1+3 \sin^2 \lambda}$.

[HINTS : Prob. 4.10.6, pp. 185]

(b) Find the product of inertia of an elliptic lamina with respect to its major and minor axis.

[HINTS : §1.4(k), pp. 13]

(iv) A homogeneous sphere of radius r, rotating with angular velocity ω about a horizontal diameter, is gently placed on a table where coefficient of friction is μ. Show that there will be slipping at the point of contact for a time $\frac{2\omega r}{7\mu g}$ and that then the sphere will roll with angular velocity $\frac{2\omega}{7}$ (g is the acceleration due to gravity).

[HINTS : Prob. 4.16.1, pp. 228]

5. (i) Show that the moment of momentum about a fixed point O of a rigid body of mass M moving in two dimension is equal to $Mvp + MK^2\dot{\theta}$, explain the motions involved.

[HINTS : §4.4, pp. 163]

(ii) A rod revolving on a smooth horizontal plane about one end which is fixed, breaks in to two parts. What is the subsequent motion of the two parts.

[HINTS : Let the rod OA revolving around O break into two parts OB and AB. Let C be the middle point AB.

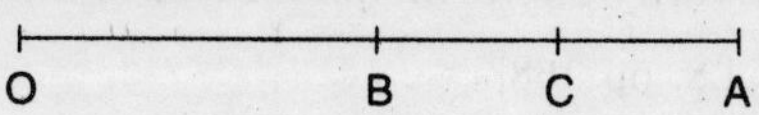

Now the part OB will continue to rotate about O with the same angular velocity.

The part AB whose C.G. is at C has a linear velocity. Therefore, this part will fly off in a tangent line at C to the circle with O as centre and OC as radius. Again, we know that the motion of a body about its centre of inertia is the same as if the centre of inertia were fixed and the same forces acted on it. Hence the part AB will continue to rotate about C with the same angular velocity.]

(iii) A rod of lengths $2a$, is suspended by a string of length l attached to other end; if the string and rod revolve about the vertical with uniform angular velocity, and thier inclination to the vertical be θ and ϕ respectively. Show that $\frac{3l}{a} = \frac{(4 \tan \theta - 3 \tan \phi) \sin \phi}{(\tan \phi - \tan \theta) \sin \theta}$.

[HINTS : Prob. 2.5.4, pp. 98]

Burdwan University

2013

2. (i) Find a condition so that a given straight line may be a principal axis of a rigid body at a point of its length.
Also draw that if the line passes through the centre of gravity of the body and be a principal axis at a point of its length, then it is a principal axis at all points of its length.

[HINTS : §1.14, pp. 39]

(ii) In two dimensional motion of a body, express the kinetic energy in terms of the motion of the centre of inertia and the motion relative to the centre of inertia.

[HINTS : §4.3, pp. 161]

(iii) A uniform solid cylinder is placed with its axis horizontal on a plane, whose inclination to the horizon is α. Show that the least co-efficient of friction between it and the plane is $\frac{1}{3}\tan\alpha$ so that it may roll without slipping. If the cylinder be hallow and of small thickness, the least value is $\frac{1}{2}\tan\alpha$.

[HINTS : Prob. 4.13.1, pp. 214]

(iv) Show that the angular momentum of a rigid body of mass M about a fixed point O, moving in two dimensions is equal to $Mvp + MK^2\theta$. Explain the different symbols used and interpret the different form.

[HINTS : §4.4, pp. 163]

5. (i) Explain the concept of momental ellipsoid and hence find out the momental ellipsoid at the centre of an elliptic plate.

[HINTS : §1.14, Remark 1.14.1; pp. 39 and 41]

(ii) State and verify the principle of conservation of moment of momentum.

[HINTS : §4.8, pp. 168]

(iii) A uniform rod of mass M and length $2a$, at rest, is struck by a blow J at right angles to its length at a distance C from its centre. Find the point above which it will begin to turn when the ends are free.

[HINTS : Prove. 5.8.2, pp. 297]

(iv) A uniform circular disc is projected up a rough inclined plane with velocity V, but no rotation, the plane of the disc being a plane of greatest slope. Show that the disc ceases to slip after a time $\frac{V}{g(3\mu\cos\alpha+\sin\alpha)}$: μ and α are respectively the co-efficient of friction and inclination of the plane.

[HINTS : Prob. 4.16.4, pp. 237]

(v) A rod of length $2a$ revolves with uniform angular velocity ω about a vertical axis through a smooth joint at one extremity of the rod so that it describes a cone of semi-vertical angle α, where $\omega^2 = \frac{3g}{4a\cos\alpha}$, g being the acceleration due to the gravity.

[HINTS : Prob. 2.5.2, pp. 96]

Burdwan University

2014

2. (i) (a) Show that the kinetic energy of a system of n particles is equal to the kinetic energy of the total mass supposed to be collected at the centre of mass and moving with the velocity of the centre of mass plus the kinetic energy of the motion particles relative to the centre of mass.

[HINTS : §4.3, pp. 161]

(b) Calculate the moment of inertia about OA of a wire AB of length $2a$ bent at right-angles at its middle point O.

[HINTS : §1.4, (a), pp. 3]

(ii) Discuss the motion of a uniform heavy sphere (solid) along a perfect rough inclined plane. If one considers a hallow sphere instead of solid sphere, then write down the difference observed.

[HINTS : §4.12, pp. 210, Remark 4.12.1, pp. 213]

(iii) (a) A uniform rod OA of length $2a$ is free to turn about its end O. If the rod revolves uniformly with angular velocity ω about the downward vertical oy making a constant angule $\frac{\pi}{4}$ with oy, show that $\omega^2 = \frac{3g}{2\sqrt{2}a}$.

[HINTS : Prob. 2.5.8, pp. 103]

(b) A uniform solid cylinder rolls, with generators horizontal, down an inclined plane of inclination α. Prove that the condition for pure rolling is that the co-efficient of friction must be $\geq \frac{1}{3}\tan\alpha$.

[HINTS : Prob. 4.13.1, pp. 214]

(iv) Show that the angular momentum of a rigid body of mass M about the point O, moving in two dimensions is equal to $Mvp + MK^2\dot{\theta}$. Explain the different symbols used and interpret the different terms.

[HINTS : §4.4, pp. 163]

5. (i) Prove that the moment of inertia of a right circular cone of height h and semi-vertical angle α about a line through the centre of inertia perpendicular to the axis is $3Mh^2(1 + 4\tan^2\alpha)$.

[HINTS : §1.3(g), pp. 10]

(ii) A uniform rod of length $2a$ is placed with end in contact with a horizontal table and is at an inclination α to the horizon and is allowed to fall. When it becomes horizontal, show that its angular velocity is $\sqrt{\frac{3g}{2a}}\sin\alpha$, whether the plane be perfectly smooth and perfectly rough.

[HINTS : Prob. 4.10.4, pp. 178]

(iii) A plank of mass M is initially at rest along a line of greatest slope of a smooth plane inclined at an angle α to the horizon, and a man of mass M' starting from rest from the upper end walks down the plank so that it does not more, show that he gets to the other end in time $\sqrt{\frac{2M'a}{(M+M')g\sin\alpha}}$, where a is the length of the plank.

[HINTS : Prob. 2.5.1, pp. 94]

(iv) Find the moment and products of inertia with respect to rectangular axis, parallel to the co-ordinate axes through the point $(a, 0)$ on the circumference for the uniform circular disc $x^2 + y^2 = a$ of mass M. Also write down the corresponding inertia matrix.

[HINTS : §1.4(d), pp. 6]

Burdwan University

2015

1. (a) The position of a moving particle at time t is given by $x = a\cos t$, $y = a\sin t$. Find the magnitude and direction of the velocity of the particle.

(b) Starting from rest, the acceleration of a particle, after t sec is $\frac{1}{2}e^{-2at}$. Find its maximum velocity at that time.

(c) What is the work done of a particle of mass 10 lbs during the 3rd second of its fall under gravity?

(d) A sphere impinges directly on an equal sphere at rest. If the coefficient of restitution be e, show that their velocities after impact are in the ratio $(1 - e) : (1 + e)$.

(e) A shot of mass 40 kg leaves a gun with a velocity of 1000 cm/sec. If the velocity of recoil is 10 cm/sec, find the mass of the gun.

(f) From what height a heavy elastic ball must be dropped on a floor so that after rebounding once it will reach a height of 1 meter?
(taking coefficient of restitution $e = 1/2$)

(g) A particle describes the parabola $p^2 = ar$ under a force which is always directed towards its focus. Find the law of force.

2. (a) (i) A particle moves on a straight line under an acceleration n^2x towards a point O on the line, where x is the distance from O.
If $x = a$ and $dx/dt = u$ when $t = 0$, show that the amplitude of the motion is $\sqrt{a^2 + \frac{u^2}{n^2}}$.

(ii) Two weights W and W^1 are connected by a light inextensible string passing over a smooth pully. If the pully moves vertically upwards with an acceleration equal to that of gravity, show that the tension of the string is $4WW^1/(W + W^1)$.

(iii) A bullet of mass m moving with velocity u, strikes a block of mass M, which is free to move in the direction of motion of the bullet and is embeded in it. Show that the loss of kinetic energy is $mMu^2/2(m + M)$

(b) (i) A particle is projected from a point O with a velocity V so as to just over the highest possible post at a horizontal distance d from O. Prove that greatest height above O attained by the particle in its flight is $\frac{V^6}{2g(V^4+g^2d^2)}$.

(ii) A spherical ball falls from a height h upon a fixed horizontal plane. If e be the coefficient of restitution, show that total distance described before the ball ceases to rebound is $\frac{1+e^2}{1-e^2}h$.

(iii) A particle describes a parabola under a force which is always directed perpendicularly towards its axis. Show that the force is inversely proportional to the cube of the ordinate.

(c) (i) Obtain the components of acceleration of a particle moving in a plane in polar co-ordinates along radial and cross-radial direction.

(ii) A particle describes a path $r^2 = a^2 \cos 2\theta$ under a force which is always directed towards the pole. Find the law of force.

(iii) Show that the height at which kinetic energy of a body falling from a height h is equal to its potential energy is $h/2$.

(d) (i) A gun of total mass M, free to recoil horizontally, fires a shot of mass m. If the gun be fired with the barrel inclined at an angle α to the horizontal, then prove that the shot is actually projected at an angle $\tan^{-1}\left\{\left(1+\frac{m}{M}\right)\tan\alpha\right\}$.

(ii) A particle describes the parabola $r = a\sec^2\frac{\theta}{2}$ such that the cross-radial component of the velocity is constant.
Show that $\frac{d^2r}{dt^2}$ is constant.

(iii) A particle projected vertically upwards with a velocity u is met with air resistance which varies as the square of the velocity.
Show that the greatest height attained by the particle is $\frac{V^2}{2g}\log\left(1+\frac{u^2}{V^2}\right)$ where V is the terminal velocity.

3. (a) With usual notation prove that, $$\frac{d^2u}{d\theta^2} + u = \frac{F}{h^2u^2}.$$

(b) If 'a' be the amplitude and T be the period of a particle executing simple harmonic motion in a straight line, show that the time taken by the particle to travel a distance x from the centre of force is $(T/2\pi)\sin^{-1}(x/a)$ and the velocity in that position is $(2\pi/T)\sqrt{a^2-x^2}$.

(c) An engine working at a constant rate H draws a load M against a resistance R. Show that the maximum speed is H/R and that the time taken to attain half this speed is $MH\left(\log 2 - \frac{1}{2}\right)/R^2$.

(d) Find the law of force to the pole when the path is the cardioide $r = a(1-\cos\theta)$ and prove that if F be the force at the apse and V, the velocity then $3V^2 = 4aF$.

(e) A particle moves in a straight line under a force which is always directed towards a point on it and proportional to its distance from the fixed point in a medium which offers a small resistance to its motion, proportional to its velocity. Discuss the motion.

Burdwan University

2016

1. (a) A particle moves along a straight line according to the law $S^2 = 6t^2 + 4t + 3$. Prove that the acceleration varies as $1/S^3$.
   (b) Show that the change of kinetic energy (K.E.) of a body is equal to the work done by the external forces acting on it.
   (c) Prove that the change of momentum in the direction of the force is equal to its impulse.
   (d) A particle is executing S.H.M. such that its period of oscillation is π seconds. If its maximum acceleration is $8\,\text{ft/sec}^2$, find its amplitude.
   (e) If the radial velocity is proportional to the transverse velocity, find the equation of the path of the particle in polar coordinates.
   (f) State Newton's Experimental law of impact.
   (g) Show that at an apse, the particle moves perpendicular to the radius vector.

2. (a) (i) Show that in a Simple harmonic motion of amplitude 'a' and period T, the velocity v at a distance x from the centre is given by the relation $v^2T^2 = 4\pi^2(a^2 - x^2)$.
   (ii) A particle of mass m slides down a rough inclined plane of inclination α to the horizon along the line of greatest slope. Find the acceleration of the particle down the plane.
   (iii) Show that for a particle of mass m moving down a smooth inclined plane under gravity, the sum of the kinetic energy and the potential energy of the particle at any point of the inclined plane is constant.
   (b) (i) A bullet of mass m is projected from a gun of mass M by an explosion which generates kinetic energy E.
   Prove that the initial velocity of the bullet is $\sqrt{\dfrac{2ME}{(M+m)m}}$
   (ii) A stone of mass m falls vertically downwards from rest from a height h and is then brought to rest by penetrating a distance d into the mud. Prove that the average resistance of mud is $m\left(1 + \frac{h}{d}\right) g$.
   (iii) Apply the principle of energy to show that the acceleration of two particles of masses m_1 and m_2 connected by a light string placed over a smooth pulley is $\frac{m_1 - m_2}{m_1 + m_2} g$.
   (c) (i) Prove that for a projectile the sum of kinetic energy and potential energy is constant throughout the motion.
   (ii) A insect crawls at a constant rate u along the spoke of a cartwheel of radius, a, the cart is moving with velocity v. Find its acceleration along and perpendicular to the spoke.
   (d) (i) If the orbit described by a particle under a central force to the origin be $r^n \cos n\theta = a^n$, find the law of force.
   (ii) A particle of mass m moves under attractive force mn^2x towards a fixed point O on the line, where x is the distance from O at any time t.

If it is projected with velocity V towards O from an initial distance 'a' form it, then prove that it reaches O after a time $\frac{1}{n}\tan^{-1}\left(\frac{na}{V}\right)$.

(iii) A particle moves under a force which is always directed towards a fixed point and is equal to $\mu \div (\text{distance})^2$ per unit mass. Show that its path is a conic.

3. (a) With usual notations, prove that $\frac{h^2}{p^3}\frac{dp}{dr} = F$.

(b) Define impulsive force. Give an example. Calculate the loss of kinetic energy in the direct impact of two perfectly elastic smooth spheres.

(c) A particle moves in a straight line with an acceleration K^2x towards a fixed point on it in a medium which offers a small resistance $\mu\frac{dx}{dt}$. Show that the particle execute a hormonic motion when $\mu^2 < 4K^2$. Also find its period.

(d) (i) What is the work done by gravity on a mass of m gm during t-th second of its fall.

(ii) Two perfectly inelastic boies of masses m_1 and m_2 moving with velocities u_1 and u_2 in the same direction impinge directly. Show that the loss of kinetic energy due to impact is $\frac{1}{2}\frac{m_1m_2}{m_1+m_2}(u_1 - u_2)^2$.

(e) Define Central force. If a particle describes an ellipse under a force to its centre, show that the force is directly proportional to the distance from the centre.

Burdwan University

2017

1. (a) If the position of a moving point is given by $x = a\cos t$ and $y = b\sin 2t$. Find its path.

(b) At what height would the kinetic energy of a falling particle be equal to half its potential energy?

(c) A wheel of radius 2 cm rolls 10 cm on a road in 5 seconds. Find the average angular velocity of the wheel.

(d) Two equal elastic spheres each of mass m move in the same direction with equal velocities u and v and collide directly. Show that they interchange their velocities.

(e) A particle describes a curve $r = ae^{\theta}$ with constant angular velocity. Show that its transverse acceleration varies as the distance from the pole.

2. (a) (i) A particle moving with S.H.M. has got velocities $u_1 \cdot u_2$ and corresponding acceleratoions f_1, f_2 in two of its positions. Show that the distance between these two positions is $(u_2^2 - u_1^2)/(f_1 + f_2)$.

(ii) A particle moves towards a centre of attraction starting from rest at a distance 'σ' from the centre. If its velocity when at any distance x from the centre varies as $\sqrt{\frac{a^2-x^2}{x^2}}$, find the law of force.

(iii) A bullet of mass m moving with a velocity u, strikes a block of mass M, which is free to move in the direction of motion of the bullet and is embeded in it. Show that the loss of kinetic energy is $mMu^2/2(m + M)$.

(b) (i) A particle moving in a straight line starts from rest and the acceleration at any time t is $(a - Kt^2)$ where 'a' and 'K' are positive constants. Show that the maximum velocity attained is $\frac{2}{3}\sqrt{\frac{a^2}{K}}$.

(ii) A particle executing S.H.M. in a straight line, has velocities v_1, v_2, v_3 respectively at distances x_1, x_2, x_3 from the mean position of the path. Prove that $x_1^2(v_2^2 - v_3^2) + x_2^2(v_3^2 - v_1^2) + x_3^2(v_1^2 - v_2^2) = 0$.

(iii) A particle describes the parabola $r = a\sec^2\frac{\theta}{2}$ such that the cross-radial component of the velocity is constant. Show that $\frac{d^2r}{dt^2}$ is constant.

(c) (i) Show that the path of a projectile in a vaccum is a parabola.

(ii) Obtain the components of acceleration of a particle moving in a plane in polar co-ordinates along radial and cross-radial direction.

(d) (i) A particle is moving in a straight line with an acceleration n^2x towards a fixed point on the line and is simultaneously acted on by a small resistance proportional to its velocity. Investigate the motion of the particle.

(ii) A spherical ball falls from a height h upon a fixed horizontal plane. If e is the co-efficient of restitution show that the total distance describe before the ball ceases to rebound is $\frac{1+e^2}{1-e^2}h$ and the total time taken is $\frac{1+e}{1-e}\sqrt{\frac{2h}{g}}$.

3. (a) With usual notation prove that $\frac{d^2u}{d\theta^2} + u = \frac{F}{h^2u^2}$.

(b) A gun of total mass M, free to recoil horizontally, fires a shot of mass m. If the gun be fired with the barrel inclined at an angle α to the horizontal, then prove that the shot is actually projected at an angle $\tan^{-1}\{(1 + \frac{m}{M})\tan\alpha\}$.

(c) A particle moves towards a centre of force, such that the acceleration at a distance x being given by $\mu(x + a^4/x^3)$. If it starts from rest at a distance 'a' from the centre, show that it will arrive at the centre in time $\frac{\pi}{4\sqrt{\mu}}$.

(d) If v_1 and v_2 are the linear velocities of a planet when it is respectively nearest and farthest from the Sun, prove that $(1 - e)v_1 = (1 + e)v_2$.

4. (a) State Lami's Theorem.

(b) What do you mean by 'torque',

(c) State conditions of equilibrium of a system of coplanar forces.

(d) Define astatic equilibrium of a system of coplanar forces.

5. (a) (i) If the position of the resultant of two like parallel forces P and Q is unaltered when the positions of P and Q are interchanged, then show that $P - Q$.

(ii) ABC is a triangle, D, E, F are the middle points of the sides BC, CA and AB respectively. Show that the forces acting on a particle and represented by the straight lines AD, BE, CF will maintain equilibrium.

(iii) Prove that if the algebraic sum of moments of a system of co-planar forces about any three non-collinear points on their plane vanish separately, then the system will be in equilibrium.

(b) (i) Define moment of a force about any point.

(ii) Prove that if each force of a co-planar system be rotated through an angle θ in its plane in the same sense that their resultant is also rotated through the same angle.

(iii) Three forces proportional to $AB, BC, 2CA$ act along the sides of a triangle taken in order. FInd the resultant and the point where it meets.

Burdwan University

2018

1. (a) A particle moves in a straight line according to the law $v^2 = 4(x \sin x + \cos x)$, where v is the velocity at a distance x. Find the acceleration.

(b) Show that the change of kinetic energy of a body is equal to the work done by the external forces acting on it.

(c) If the radial velocity is four times the transverse velocity, find the equation of the path of the particle in polar coordinates.

(d) State the principle of conservation of linear momentum.

(e) A sphere impinges directly on an equal sphere at test. If the coefficient of restitution be e, show that their velocities after impact are in the ratio $(1-e) : (1+e)$.

2. (a) (i) A particle of mass m is acted on by a force $m\mu\left(x + \frac{a^4}{x^3}\right)$ towards the origin. If it starts from rest at a distance a then show that it will arrive at the origin in time $\frac{\pi}{(4\sqrt{\mu})}$.

(ii) Show that in a simple harmonic motion of amplitude 'a' and period T, the velocity v at a distance x from the centre is given by the relation $v^2T^2 = (4\pi^2 a^2 - x^2)$.

(iii) A particle moves with S.H.M. its position of rest being at a distance 'a' from the centre. Find by the principle of energy, the velocity at the centre.

(b) (i) The weights W_1 and W_2 are connected by a light inextensible string passing over a smooth pulley. If the pulley moves vertically upwards with an acceleration equal to that of gravity, show that the tension of the string is $\frac{4W_1W_2}{W_1+W_2}$.

(ii) A stone of mass m falls vertically downwards from rest from a height h and is then brought to rest by peneterating a distance d into the mud. Prove that the average resistance of mud is $m\left(1 + \frac{h}{d}\right)g$.

(iii) A particle describes a path $r^2 = a^2 \cos 2\theta$ under a force which is always directed towards the pole. FInd the law of force.

(c) (i) Prove that for a projectile the sum of kinetic energy and potential energy is constant throughout the momentum.

(ii) A particle describes a circle of radius 'a' with uniform speed v. Show that at any instant its acceleration is directed towards the centre and is of magnitude $\frac{v}{a}$.

(iii) Show that the height at which kinetic energy of a body falling from a height h is equal to its potential energy is $\frac{h}{2}$.

(d) (i) A particle of mass m moves under attractive force mn^2x towards a fixed point O on the line, where x is the distance from O at any time t, if it is projected with velocity V towards O from an initial distance 'a' from it, then prove that it reaches O after a time $\frac{1}{x}\tan^{-1}\left(\frac{na}{V}\right)$.

(ii) A particle moves under a force which is always directed towards a fixed point and is equal to $\mu \div (\text{distance})^2$ per unit mass. Show that its path is a conic and discuss the cases that may arise.

3. (a) With usual notations, prove that $\frac{h^2}{p^3}\frac{dp}{dr} = F$.

(b) A bullet of mass m is projected from a gun of mass M by an explosion which generates kinetic energy E.

Prove that the initial velocity of the bullet is $\sqrt{\frac{2ME}{(M+m)m}}$

(c) Two perfectly inelastic bodies of masses m_1 and m_2 moving with velocities u_1 and u_2 in the same direction impinge directly. Show that the loss of kinetic energy due to impact is $\frac{1}{2}\frac{m_1m_2}{m_1+m_2}(u_1 - u_2)^2$.

(d) If ω be the angular velocity of a planet at the nearest end of the major axis, then prove that the period is $\frac{2\pi}{\omega}\sqrt{\frac{1+e}{(1-e)^3}}$, where e is the eccentricity of the elliptic orbit.

4. (a) Define moment of a force about an axis.

(b) State converse of Lami's theorem.

(c) Describe polygon of forces.

(d) State conditions of equilibrium of a system of coplanar forces.

5. (a) (i) The angle of inclination between two forces P and Q is θ. If P and Q be interchanged in position, show that the resultant will be turned through an angle ϕ, where $\tan\frac{\phi}{2} = \frac{P-Q}{P+Q}\tan\frac{\theta}{2}$.

(ii) The moments of a system of forces about the points $(0, 0)$ and $(a, 0)$ and $(0, a)$ are $a\omega$, $2a\omega$ and $3a\omega$ respectively. Find the components of their resultant parallel to the co-ordinate axes.

(iii) Define astatic equilibrium and astatic centre of a system of coplanar forces.

(b) (i) Define axis of a couple.

(ii) If two like parallel forces P and Q acting on a rigid body at A and B be interchanged in position, show that the point of application of the resultant will be displaced along AB through a distance d where $d = \frac{P-Q}{P+Q}AB(P > Q)$.

(iii) Prove that any system of coplanar forces acting on a rigid body can be reduced to a single force together with a single couple.